Surface Plasmon Resonance for Biosensing

Surface Plasmon Resonance for Biosensing

Editor

Alessandro Fantoni

MDPI • Basel • Beijing • Wuhan • Barcelona • Belgrade • Manchester • Tokyo • Cluj • Tianjin

Editor
Alessandro Fantoni
Centre of Technology and
Systems (UNINOVA-CTS)
FCT Campus
Portugal
Instituto Superior de
Engenharia de Lisboa (ISEL)
Portugal

Editorial Office
MDPI
St. Alban-Anlage 66
4052 Basel, Switzerland

This is a reprint of articles from the Special Issue published online in the open access journal *Biosensors* (ISSN 2079-6374) (available at: https://www.mdpi.com/journal/biosensors/special_issues/SPR_BIOSENS).

For citation purposes, cite each article independently as indicated on the article page online and as indicated below:

LastName, A.A.; LastName, B.B.; LastName, C.C. Article Title. *Journal Name* **Year**, *Volume Number*, Page Range.

ISBN 978-3-0365-4657-5 (Hbk)
ISBN 978-3-0365-4658-2 (PDF)

Cover image courtesy of Valentina Pibiri
"The slow path for the search of knowledge"

Contents

About the Editor

Alessandro Fantoni

Alessandro Fantoni was born in Rome, Italy, in 1966. He received a university degree in applied mathematics from the University of Camerino, Italy (1992), and a PhD in Material Engineering/Micro and Optoelectronics from the New University of Lisbon, Portugal (1999). Presently he is a Coordinator Professor at the Electronics, Telecommunications and Computer Department of the Engineering Institute of Lisbon (ISEL-ADEETC) and he is a member of the UNINOVA-CTS research centre (Centre of Technology and Systems), Caparica (Portugal). He currently teaches Electronics and Optoelectronics courses at both bachelor and master levels. His research interests are related to simulation and characterization of sensors and optoelectronic thin film devices.

Preface to "Surface Plasmon Resonance for Biosensing"

Point-of-care (POC) methods for medical screening and timely disease diagnosis allow for a continuous general health state assessment and are central for the future development of health systems. From this point of view, biosensors based on surface plasmon resonance (SPR) effects can play a major role because of their high sensitivity, reduced fabrication process complexity, and high level of integration. They offer the potential to move proteomic biology into the clinical setting as a routine diagnostic procedure and surpass the technical challenges of conventional methods. The light-generated SPR phenomena depend on the fine-tuning of the wavelength against the geometry of the resonant structures and the optical properties of the materials that are used. Targeting the optimization of this key point, a large panorama of different structure configurations can be proposed, based on different schemes for plasmon generation and sensor interrogation methods, ranging from local nanoparticle response (LSPR), optical waveguides, optical fibres, and interferometers, to advanced structures where the SPR is enhanced by the mutual influence of different nanostructures with 2D materials. The overall performance of the resulting biosensor depends on the ability to immobilize specific antibodies while maintaining their biological activity, as well as providing antibodies accessibility to the analyte. Surface Plasmon Resonance for Biosensing is a very interdisciplinary research topic and this book addresses the recent research results in this exciting area by proposing multiple perspectives related to this topic, ranging from material science to biochemistry, nanotechnology, and low-power electronic systems.

Alessandro Fantoni
Editor

Review

Gold Nanorods for LSPR Biosensing: Synthesis, Coating by Silica, and Bioanalytical Applications

Vincent Pellas [1,2,†], David Hu [1,†], Yacine Mazouzi [1], Yoan Mimoun [1], Juliette Blanchard [1], Clément Guibert [1], Michèle Salmain [2,*] and Souhir Boujday [1,*]

[1]   Laboratoire de Réactivité de Surface (LRS), Sorbonne Université, CNRS, UMR 7197, 4 Place Jussieu, F-75005 Paris, France; vincent.pellas@sorbonne-universite.fr (V.P.); david.hu@sorbonne-universite.fr (D.H.); yacine.mazouzi@sorbonne-universite.fr (Y.M.); yoan.mimoun@etu.sorbonne-universite.fr (Y.M.); juliette.blanchard@sorbonne-universite.fr (J.B.); clement.guibert@sorbonne-universite.fr (C.G.)

[2]   Institut Parisien de Chimie Moléculaire (IPCM), Sorbonne Université, CNRS, 4 Place Jussieu, F-75005 Paris, France

*   Correspondence: michele.salmain@sorbonne-universite.fr (M.S.); souhir.boujday@sorbonne-universite.fr (S.B.); Tel.: +33-1-4427-6732 (M.S.); +33-1-4427-6001 (S.B.)

†   These authors contributed equally to this work.

Received: 2 September 2020; Accepted: 13 October 2020; Published: 17 October 2020

Abstract: Nanoparticles made of coinage metals are well known to display unique optical properties stemming from the localized surface plasmon resonance (LSPR) phenomenon, allowing their use as transducers in various biosensing configurations. While most of the reports initially dealt with spherical gold nanoparticles owing to their ease of synthesis, the interest in gold nanorods (AuNR) as plasmonic biosensors is rising steadily. These anisotropic nanoparticles exhibit, on top of the LSPR band in the blue range common with spherical nanoparticles, a longitudinal LSPR band, in all respects superior, and in particular in terms of sensitivity to the surrounding media and LSPR-biosensing. However, AuNRs synthesis and their further functionalization are less straightforward and require thorough processing. In this paper, we intend to give an up-to-date overview of gold nanorods in LSPR biosensing, starting from a critical review of the recent findings on AuNR synthesis and the main challenges related to it. We further highlight the various strategies set up to coat AuNR with a silica shell of controlled thickness and porosity compatible with LSPR-biosensing. Then, we provide a survey of the methods employed to attach various bioreceptors to AuNR. Finally, the most representative examples of AuNR-based LSPR biosensors are reviewed with a focus put on their analytical performances.

Keywords: gold nanorods; silica coating; localized surface plasmon resonance (LSPR); surface functionalization

---

1. Introduction

The part taken by plasmonic nanoparticles in biotechnologies has been expanding steadily over the last few decades. The popularity of these nano-objects stems from the localized surface plasmon resonance (LSPR) phenomenon that leads to an intense absorbance band at certain resonance frequencies [1,2]. This outstanding property is at the origin of their implementation in multiple biomedical applications including biosensing [3,4], therapy, and theranosis [5,6]. Among the multiple features of plasmonic nanoparticles, an interesting property, namely the extreme sensitivity of the LSPR band to minor changes of the dielectric constant/refractive index (RI) of the local environment, gave rise to the LSPR-based biosensing [7]. This sensitivity allows for the detection of binding events through standard absorption spectroscopy measurements or even by visual detection [8–12].

Although the biomedical story originated with spherical gold nanoparticles, anisotropic particles, and in particular gold nanorods (AuNRs), are increasingly dominating the field. Indeed, while spherical gold nanoparticles exhibit a unique LSPR band around 520 nm, the two dimensions of AuNRs lead to two absorption bands, the first, also located around 520 nm, is due to the transverse localized surface plasmon resonance (t-LSPR), and the second, far more advantageous, is due to the longitudinal localized surface plasmon resonance (l-LSPR). The l-LSPR band is located at higher wavelengths, matching the first (650–950 nm) or the second (1000–1350 nm) biological window [13]. The position of the l-LSPR band can be tuned by modulating the aspect ratio (AR) of AuNRs rather than their length [14]. As a result, AuNRs have become the ideal candidates for a wide range of biomedical applications [13,15–18].

In the biosensing field, along with the other biomedical applications, the use of AuNRs is also expanding significantly and AuNRs are proving to be versatile and multipurpose actors allowing for the read-out of the transduction through multiple scenarios. For instance, they can be used to enhance the response of existing transducing techniques such as propagative SPR [19–21] or to quench and/or enhance the fluorescence signal [22,23]. They are also capable of achieving the transduction through an etching/growth balance [24]. Finally, they are extremely efficient candidates for LSPR-biosensing, the application on which we concentrate in this review and which, until quite recently, had been dominated by spherical nanoparticles. This boom is mainly the consequence of the fact that AuNRs' l-LSPR displays a higher sensitivity to variation of the local dielectric environmental compared to the LSPR band of spherical nanoparticles [25]. Nevertheless, despite their attractiveness, AuNRs have a significant number of disadvantages that limit their use on a larger scale as LSPR-biosensors. In particular, their synthesis seems to be subject to multiple approximations leading to difficulties in reproducing the ARs as well as a limitation of upscaling. In addition, the indispensable use of stabilizing agents strongly bound to the AuNRs' surface, such as CTAB, considerably complicates their surface chemistry and, by implication, the grafting of receptors necessary for the operation of the LSPR-biosensors. AuNRs coating with silica partially solves the second problem by conferring them a biocompatibility and a more practicable surface chemistry, however it also suffers from lack of reproducibility of their synthesis, which makes difficult the control of the thickness and, more generally, the quality of the silica shell.

In what follows, we review the recent progresses related to AuNRs for LSPR biosensing starting from their synthesis and coating by silica, then proceeding to their characterization and further surface functionalization, and ending with a selection of bioanalytical applications. In the past few decades, much progress has been made in the mastering of AuNRs synthesis and the underlying mechanism as summarized in these reviews dated in 2009 and 2013 [17,26]. In the first part of the section, we build on these reviews and discuss the more recent literature on the subject, with a special focus on the problems that may interfere with the subsequent use of these objects for biosensing. The second part of the synthesis section is devoted to silica coating although very little examples of LSPR biosensing using AuNR@SiO$_2$ can be found in the literature. This poor use of such promising objects is mainly driven by the difficulty to achieve thin layers of silica on top of the nanorods. When the silica shell is thick, the LSPR signal at its surface decays and, as a consequence, no change is recorded upon target recognition. However, herein, we intend to demonstrate that the silica coating of AuNRs has reached a level of maturity allowing for the precise tuning of the thickness of the silica layer down to the nanometer level and even below, which paves the way for their extensive use in LSPR biosensing. In the second section of this manuscript, we review the characterization methods allowing for the investigation of the structure and shape of gold nanorods and list, when possible, the means for the characterization of the outer silica shell. The third section is devoted to AuNRs surface functionalization, mandatory to attach the biological element responsible for target recognition prior to their use for LSPR-biosensing. We review in this part the strategies together with the relevant molecules allowing for AuNRs surface functionalization by physisorption, chemisorption, or conjugation. The last part of this manuscript is devoted to the applications of AuNRs in LSPR biosensing, i.e., the biosensors for which read-out is based on a shift in the position and/or intensity of the LSPR band. The main bioreceptors associated with AuNRs are antibodies and aptamers used either in solution or in solid phase. We recap the different strategies and discuss the analytical performances, i.e., the limit of detection (LoD) and the detection range (DR) put in perspective with other transduction techniques. We finally discuss the use of AuNRs as single molecules plasmonic biosensors.

2. Synthesis Methods

Most AuNR syntheses rely on the so-called seed-growth method as shown in Figure 1A. The seed-mediated growth process involves two separated steps: First, the synthesis of tiny spherical seed nanoparticles and second, their subsequent growth in a solution containing metal precursors, weak reducing agent such as ascorbic acid, and a capping agent, usually CTAB (hexadecyltrimethylammonium bromide), a cationic surfactant. CTAB adsorbs on AuNRs surface, and allows both their directional growth and electrostatic stabilization through the formation of a dense double layer [27]. The synthesis protocol is regulated by the balance of redox potentials (redox potentials of the main couples involved in AuNR synthesis are reported in Figure 1B). The TEM images shown in Figure 1C are extracted from the work of Jana et al., Sau et al., and Ye et al. [28–30], respectively, to highlight how the synthesis protocol has reached a level of maturity allowing for a quasi-perfect homogeneity of the rods. For detailed mechanistic investigation in AuNRs formation and parameters influencing their growth, we refer to the review published by El-Sayed et al. in 2009 [17] and Murphy et al. in 2013 [26,31]. In what follows, we will briefly introduce a historical background of AuNR synthesis, discuss the origins in the lack of reproducibility, difficulties encountered during the synthesis, storage issues, and finally, the complexity of scaling up the synthesis.

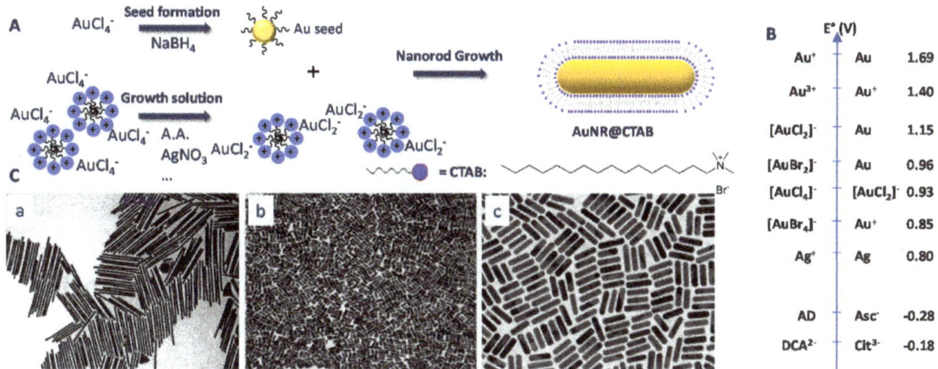

Figure 1. (**A**) Synthesis of gold nanorods (AuNRs) from crystal seed according to the seed-mediated growth method, and (**B**) standard potential in aqueous solution of different Red/Ox couples playing a role in the reaction of AuNR formation [32–34] and (**C**) TEM images of AuNRs synthesized following (**a**) Jana et al.'s method [28], (**b**) Sau et al.'s method [29], and (**c**) Ye et al.'s method [30].

2.1. Gold Nanorods

2.1.1. Historical Background

The very first reported method for the synthesis of AuNR, which dates back to 1994, was based on the electrochemical reduction of a gold precursor (HAuCl₄) within a porous membrane. However, this method had many limitations, among which was a very low yield [35]. The breakthrough in synthesis and applications of AuNRs really began at the dawn of the 21st century, when Jana et al. reported the synthesis of AuNRs by a seeding growth method [36]. In their initial protocol, a solution of 3.5 nm citrate-stabilized Au seeds is injected into a growth solution containing HAuCl₄, cetyltrimethylammonium bromide (CTAB), L-ascorbic acid (AA), and silver nitrate; the AR was varied by changing the amount of seed solution introduced in the growth solution. Shortly after this first publication, Jana et al. reported on the possibility to synthesize AuNRs with high AR (as high as 18, dimensions 400 × 25 nm) by performing successive (up to three) growth steps, in three identical growth solutions [28]. Later, in 2003, Nikoobakht et al. introduced the use of a seed solution containing smaller (ca. 1.5 nm), CTAB-stabilized, Au seeds and a growth solution containing a binary mixture of surfactants (benzyldodecylammonium chloride (BDAC) + CTAB), silver nitrate, and AA. This second synthesis protocol produces thinner AuNR (diameter ca. 12 nm) and with ARs in the range 1.5–4.5 [37]. The yield in NR (compared to other shapes of AuNPs) was relatively low in this initial report, and in the method developed by Jana et al., but Murray et al. significantly improved the yield toward AuNRs (~97%) of this second synthesis protocol for low AR (2–4) AuNRs [29]. Another protocol was developed shortly after that, which is often qualified as seedless, although it is rather a one-step seeded protocol. This third protocol, initially developed by Jana et al. [38], is actually a modification of the second one, where the seeds are grown in situ, in the growth solution, by the rapid addition of $NaBH_4$, a strong reducing agent. The use of this strong reducing agent allows for separating the nucleation step (fast reduction with $NaBH_4$) from the growth step (slower reduction with AA). Moreover, by using a growth solution containing a high concentration of gold, it allows the gram-scale production of AuNR. The AuNRs obtained with this protocol have significantly smaller dimensions (with diameter as low as 2.5 nm). This "seedless" protocol initially suffered from a large distribution in ARs and the presence of a high fraction of nanospheres (that were difficult to separate from these small AuNRs), but further developments and optimization allowed the synthesis of AuNRs with well controlled ARs and negligible contamination by nanospheres [39]. It is important to mention here that these small nanorods have lower extinction coefficients (for an l-LSPR band at 800 nm: ca. 26 times lower per AuNR and ca. 2.5 times lower per gold atom i.e., with normalization by the volume of the NRs) than the larger nanorods produced by the first and second methods, because they scatter light less strongly [39]. These lower extinction coefficients may not be favorable for their application as LSPR biosensors.

The synthesis of AuNRs has been the subject of two comprehensive reviews, first in 2009 by El-Sayed et al. [17] and then in 2013 by Murphy et al. [26]. At that time, the optimization of the protocols for the three synthetic routes was already well under way, including the identification of the main parameters allowing to control the AR of the AuNRs. We will focus this section dedicated to the synthesis of AuNRs on the new developments since 2013, based on the second synthesis protocol (CTAB-stabilized seeds/seeded growth) as (i) there has been very few new developments in the first synthesis protocol (citrate-stabilized seed/seeded growth) and (ii) the third protocol ("seedless" synthesis) is, as mentioned above, very similar to the second one except that the seeds are generated in situ, in the growth solution. We will concentrate on the practical aspects of the synthesis and highlight the main difficulties (mostly leaving aside the literature related to the mechanisms of formation of AuNRs discussed extensively in the above-mentioned reviews).

2.1.2. Origins of the Lack of Reproducibility in the Synthesis of AuNRs

Synthesis of AuNRs with different ARs (and, as a consequence different positions of their l-LSPR bands) does not require advanced synthetic skills or very complicated set-ups. For example, a shift of the position of the l-LSPR band over 300 nm can be obtained by changing the concentration of silver nitrate in the growth solution. Another simple way to alter the position of the l-LSPR band is to change the amount of added Au seeds. Nevertheless, one cannot deny that difficulties are often encountered when performing these syntheses. The main issue is certainly, as emphasized by Burrows et al., the lack of reproducibility [40]. While, for an (experienced) individual, it is possible to continuously adjust the position of the l-LSPR band from 600 to 1000 nm by gradually increasing the concentration of $AgNO_3$, for experiments performed by several individuals, variation of the l-LSPR band of about 150 nm can be observed for apparently identical synthesis conditions (Figure 2A). This is a major problem when the purpose is to produce AuNRs with a given l-LSPR position: The horizontal red line on Figure 2A shows that AuNRs with l-LSPR at a given position (785 nm, which corresponds to a typical laser wavelength used in biomedical applications) have been obtained for $AgNO_3$ concentration ranging between 60 and 220 µmol. This issue is actually to be expected considering the many components and steps that enter in the synthesis of AuNRs.

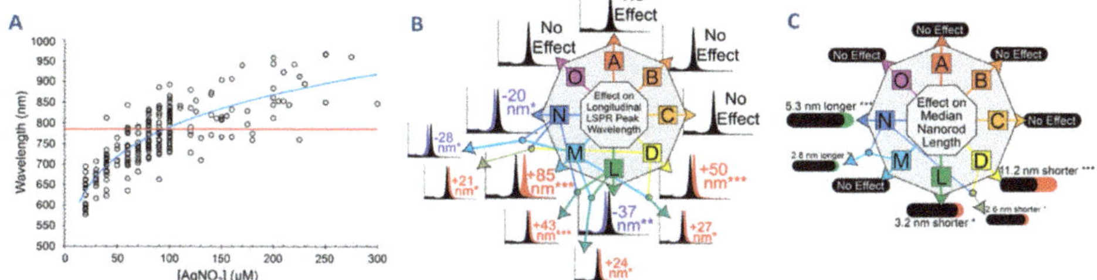

Figure 2. (**A**) Plot of the longitudinal surface plasmon resonance extinction peak wavelength as a function of silver nitrate concentration illustrating the synthesis variation by different individuals produced in the Murphy group over the last half-decade. (**B**) and (**C**) Graphical summary of the significant primary and secondary interaction effects on the l-localized surface plasmon resonance (LSPR) peak wavelength and of the median nanorod length: Amount of NaBH$_4$ (A), stirring rate of the seed solution (B), age of seed solution (C), amount of seed (D), temperature (L), amount of silver (M), amount of ascorbic acid (N), and age of reduced solution (O). P values (***) < 0.001 < (**) < 0.01 < (*) < 0.05 (from variance analysis). Adapted from [40].

As mentioned by Scarabelli et al. in their practical guide, it is of paramount importance to avoid sources of poor reproducibility such as a low purity of the CTAB source (a too high amount of iodide leads to a low shape selectivity toward AuNRs [41]) or of water, insufficient cleaning of glassware, inappropriate, or too long storage of the stock solutions [42]. These authors also emphasized that the addition of NaBH$_4$ to the AuCl$_4^-$/CTAB solution should be performed very fast in order to trigger a burst of nucleation and obtain small Au seeds with a good size homogeneity.

Regarding the possible sources of variations in the AR of AuNRs, Scarabelli et al. identified, based on an analysis of the literature that increasing the Ag$^+$ concentration or amount of Au seeds, or decreasing the pH or the AA concentration, leads to an increase in the AR of AuNRs [42]. Reza Hormozi-Nezhad et al. have identified, using experimental plans, that the position of the l-LSPR band can also be shifted by increasing CTAB concentration, but to a lower extent [43]. Also using experimental design, Burrows et al. [40] have identified the temperature of synthesis as another key parameter: The l-LSPR band of AuNRs blue-shifts upon increasing the temperature of reaction as shown Figure 2B (this last parameter is however limited by the fact that the properties of CTAB, and especially its solubility, strongly depend on the temperature). Burrows et al. also identified four synthesis parameters that had no influence on the formation of AuNRs: Amount of NaBH$_4$ used for the preparation of the seed solution; rate of stirring and age of seed solution; and aging of reduced growth solution prior to the addition of seeds. Moreover, they investigated the influence of these synthesis parameters on other aspects of the synthesis such as the yield in gold, the fraction of AuNRs (vs. particles of other shape) and the width and length of the AuNRs. Some of these results are, however, questionable: For example, the fact that the concentration of silver strongly affects the position of the l-LSPR band (expected result) while it does not modify the length and width of AuNRs (and hence, the AR, Figure 2B,C), a conclusion that is really unexpected based on previous literature; the absence of effect of seed aging is also at variance with previous works, especially the work of Watt et al., which clearly established, based on SAXS experiments, an aging of the seeds through Oswald Ripening within a few hours and its detrimental impact on the formation of AuNRs [44]; the absence of detrimental impact of an increase of the concentration of AA on the selectivity toward the formation of AuNRs is also at variance with previous works [29,45], but this is due to the fact that the authors deliberately choose to work with an sub-stoichiometric amount of AA (with respect to Au).

It is important to mention here that Hormozi-Nezhad et al. [43] and Burrows et al. [40] both concluded that the identified parameters did not simply act additively but also, for some of them, synergistically. The existence of secondary interactions between some of the synthesis parameters indicates that these parameters are mechanistically connected.

Another important conclusion of the work of Burrows et al. is that the variations in the position of l-LSPR band that are observed during their series of experiments cannot fully explain the large discrepancy (ca. 150 nm), for apparently identical experimental conditions in the position of the l-LSPR band for samples prepared by different experimenters, and that further work is needed to explore other, less obvious, parameters.

2.1.3. Other Difficulties Often Encountered in the Preparation of AuNRs

In addition to the lack of reproducibility in the position of the l-LSPR band, the other aspects of the synthesis of AuNRs that hamper their industrial development are:

- Low yield in reduced gold

It is important to mention here that many authors use the term yield to describe the fraction of AuNRs (vs. AuNPs of other shapes). Hence, in many publications, "high yield" means high selectivity toward AuNRs. This can be all the more misleading as yield in gold is often not reported [29,38]. Using AA as reducer, Au is only partially reduced, leading usually to yield in reduced gold of about 15–20% [40,46]. Indeed, during the growth phase, it is important to avoid the nucleation of new Au seeds and the growth of existing Au seeds into isotropic Au NPs. Hence, the rate of reduction should be kept slow, and with AA, this can only be achieved by keeping the concentration of AA close to stoichiometry ($[AA]/[AuCl_4^-] \approx 1.1$), while higher concentrations of AA leads to large amounts of spherical particles [45]. Several authors have proposed to replace AA by other, milder reductants such as hydroquinone (used by Vigderman et al. in 10× excess with regards to $HAuCl_4$, yield in gold close to 100% based on ICP [45]), H_2O_2-NaOH (used by Xu et al. with an up to 300× excess with respect to $HAuCl_4$, yield in reduced gold is not reported [47]), hydroxylamine (used by Leng et al. in 20× excess, yield in gold is not reported, [48] or 3-amino-phenol (used by Wu et al. in 10x stoichiometric excess, yield in gold close to 100%).

Other approaches have been proposed to increase the yield in reduced gold (while keeping a high selectivity toward AuNRs): for example Canonico-May et al. have proposed to recycle the growth solution, that is, after the separation of the AuNR from the supernatant by centrifugation, AA is added to the supernatant and the obtained solution is seeded again with Au seed and used to grow a second series of AuNR. This can be repeated up to 5 times allowing to use about 75% of the $HAuCl_4$ (compared to about 17% during the first growth step based on ICP-MS measurements), while producing AuNRs with similar ARs and dimensions [49]. Similarly, Kozek et al. also increased the fraction of reduced gold by slowly adding (with a syringe pump), to the first growing solution and after a first conventional growth step, an AA aqueous solution. The yield in gold is not calculated but it is assumed to reach 100%, based on the fact that addition of more AA does not lead to a further growth of the AuNRs [50]. Chang and Murphy were able to significantly increase the fraction of reduced gold (up to 89%), while keeping the fraction of AuNR high (>95%) using AA as reducer in a 1.6 molar excess by decreasing the pH (which has the consequence to decrease the reducing power of AA). This lower pH however also leads to the formation AuNRs with smaller dimensions (diameter below 9 nm) and with relatively small AR (below 4) [51].

- Presence of Au particles with other shapes and/or too large distribution of AR.

The most obvious solution to this difficulty is to use protocols with a high selectivity toward nanorods and with a narrow distribution in their dimensions. Among the many attempts at optimising the synthesis conditions in order to produce AuNRs with high shape selectivity and a narrow distribution in AR, one can cite the recent and very promising work of González-Rubio et al.: one origin of polydispersity in AuNR dimensions is the fact that symmetry breaking (from cuboctahedral NPs to nanorods via the emergence of (110) and (250) facets) occurs asynchronously during the growth step. González-Rubio et al. have developed a protocol including two growth steps: the Au seeds are first fully transformed to very small AuNRs (first growth step) and these small AuNR are further grown to larger AuNR (second growth step) [52,53]. By doing so, they dissociate the symmetry breaking step (which occurs during the first growth step) from the anisotropic growth step (second growth step). The key step in this protocol is the formation of small AuNR thanks to the addition of n-decanol, a cosurfactant. This protocol leads to AuNRs with a narrow l-LSPR band whose position can be tuned over a wide spectral range (from 600 to 1270 nm). The gain in absorbance, at similar wavelength between the colloidal suspensions of AuNRs prepared with this method compared to the conventional two-step method, is close to a factor 2 (Figure 3A).

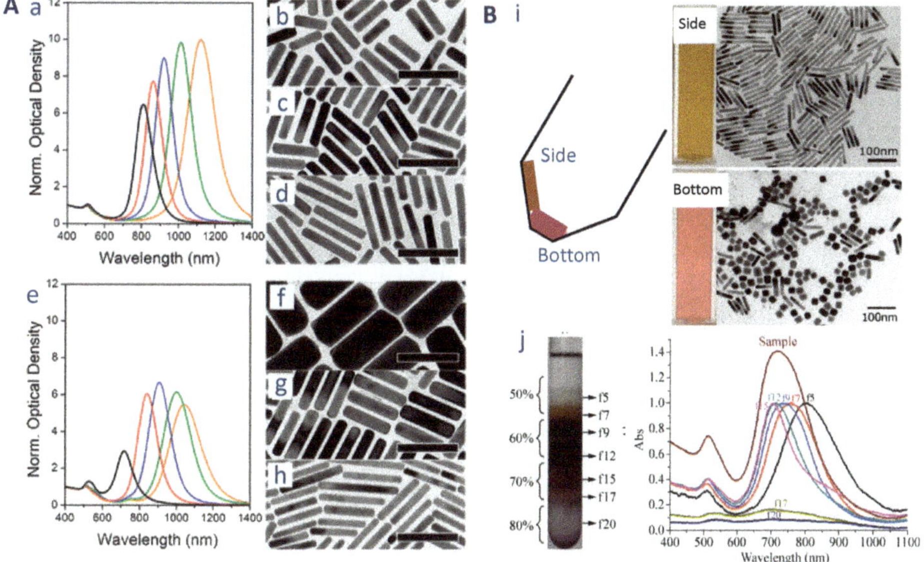

Figure 3. (**A**) Increasing AuNRs monodispersity: (**a–d**) Comparison between the use of small AuNRs seeds in the two step growth method and (**e–h**) standard Au nanocrystal seeds of 1–2 nm in conventional method in the growth of AuNRs at increasing [HCl]/[HAuCl$_4$] ratios and their respective TEM images and normalized absorption spectra. Adapted from [53]. Scale bars: 100 nm. Purification methods: (**B**) (**i**) Separation of AuNRs from Au nanospheres after conventional centrifugation; TEM pictures of the particles deposited (top) on the side wall and (bottom) at the bottom of the centrifugation tube. Adapted from [54] and (**j**) picture of AuNR suspension after gradient centrifugation in aqueous cetyltrimethylammonium bromide (CTAB)-ethylene glycol solution and UV-vis spectra of the colloidal suspensions recovered at different positions in the centrifugation tube. Adapted from [55].

Post-synthesis removal of the small fraction of AuNRs of other shapes, or separation of AuNRs as a function of their size can also be used to increase sensitivity. Purification of AuNR can be performed by several methods. The very small AuNPs (<5 nm) will usually remain in the supernatant. The easiest method to separate "large" spherical particles from AuNRs has been proposed by Sharma et al. They observed that, after centrifugation, two spots of different colors could be observed in the centrifugation tube. The analysis by TEM and UV-vis reveals that the part of the sample located in the walls is highly enriched in NRs, while the part of the sample located at the bottom of the centrifugation tube contains mostly isotropic AuNPs [54] (Figure 3B). This protocol of separation by conventional centrifugation was further refined by Boksebeld et al. who introduced a succession of three short (1 min) centrifugation steps at 6700 g, in order to selectively precipitate gold nanospheres while leaving gold nanorod in the supernatant. This protocol was successfully applied to rather "short" (AR 2.4, 3.7 and 5.3) AuNRs [56]. More complex separation can be performed using depletion-induced aggregation [57]. This method is very efficient but requires the addition of large amounts of CTAB and therefore increases further the total cost of the synthesis (see next section). A fine purification of AuNRs can also be achieved using density gradient centrifugation: In this protocol, a centrifugation tube is filled with layers of solutions of different densities prepared by mixing ethylene glycol with aqueous CTAB solution (50–80% of EG by volume), and the raw suspension is deposited on the top of these layers. After centrifugation, the AuNPs will remain in different layers depending on their dimensions [55] (Figure 3C).

- High cost of synthesis

When preparing gold nanoparticles, one would certainly expect the main source of cost to be the gold precursor. However, as emphasized by Xu et al., due to the high concentration of surfactant required for AuNRs synthesis and to the necessity of using high purity surfactant, the estimated cost of surfactant is 85% of the total cost of production of AuNRs in the initial synthesis process (pure CTAB as surfactant, Nikoobakht and El-Sayed protocol [37]) and decreases to 50% when a mixture of CTAB and NaOL is used, following the protocol of Ye et al. [30]). In their estimation of the contribution of the surfactant to the total cost of the synthesis, Uson et al. calculated a slightly lower contribution for surfactant (ca. 58%) for a synthesis protocol using only CTAB [58]. These two calculations indicate that, in order to achieve industrial scale production of AuNRs, the contribution of surfactant to the total cost must be reduced. To this purpose, Xu et al. have successfully replaced CTAB by a gemini surfactant (maleic acid diethyl bis(hexadecyl dimethylammonium bromide (P16-8-16)) of industrial grade and with a much lower critical micellar concentration (CMC). With this surfactant, the gold precursor becomes by far the major source of cost (99.3%). Moreover, this surfactant participates in the reduction of the gold precursor thanks to its C = C double bond [59]. Reaching a high yield in reduced gold is, of course, also one of the keys to decrease the cost of synthesis, but this aspect has already been discussed in this review. It is also important to remind here that the major cost may come from the post-modification of AuNRs: For example, Uson et al. have estimated that a PEGylation step can account for 2/3 of the cost [58].

2.1.4. Short Shelf-Life of AuNRs/Poor Stability of AuNRs in Oxidizing Conditions

Storage conditions play an important role in the shelf-life of AuNRs. Kaur and Chudasama have confirmed that AuNRs should not be stored in their growth solution, because they continue to age and this leads to a progressive blue-shift of their l-LSPR band; storage in a CTAB-rich solution is not desirable either as it also leads to a progressive blue-shift (although slower and less pronounced) of the l-LSPR band; on the other side, dispersion of AuNRs in water (leading, in fact, to a CTAB-poor solution) allows preserving the properties of AuNRs over a period of 30 days [60].

One potential limitation of the application of AuNRs as sensors is their instability in an oxidative environment. This instability results in a progressive blue-shift of the l-LSPR band associated with a broadening of this band and a decrease of intensity. Ultimately, complete dissolution of AuNR to $AuBr_4^-$ occurs. Vassilini et al. have observed that the oxidative dissolution of AuNR by H_2O_2 is strongly slowed down for AuNR with larger diameter. They also observed that reducing the amount of Br^- is key to produce oxidant resistant AuNR and explain the negative impact of Br^- on the stability of AuNR by a lowering of the oxidation potential of gold in presence of this anion. They demonstrate that, thanks to this improve in stability, AuNRs can be used as SERS substrate for monitoring the oxidation of crystal violet (CV) and that the AuNRs can be recycled (up to 10×) without modification in the intensity of the SERS signal of adsorbed CV [61].

2.1.5. Difficulties in Scaling up the Synthesis

The amount of AuNRs produced during a conventional synthesis is fairly small: ca. 7 mg from 100 mL of aqueous solution [62]. This is due to the fact that this synthesis is performed in very dilute solutions and that the yield in reduced gold is often relatively low. Hence, syntheses that are described as "large-scale" often produce ca. 100–200 mg of AuNRs for 1 L of aqueous solution [50]. The very first report of gram-scale synthesis of AuNRs was by Jana et al. in 2005 (using a "seedless" synthesis protocol) but this gram-scale production was obtained at the expense of a loss of control over product properties (higher polydispersity in the dimension of the nanorods) [38]. Scaling-up of the synthesis of AuNRs could, in principle, be achieved by two different approaches: Either by a volumetric scale up (that is, by keeping all concentrations identical to the standard protocol) or by increasing the concentrations of reactants in the solution.

An obvious limitation of the first of these two approaches (that is, increasing the volume of the solution) is that the subsequent post-processing steps (centrifugation and washing) will require handling large volumes of solution. Moreover, an attempt to increase the volume of reaction is hampered by the fact that AuNRs produced from larger volumes usually have a higher polydispersity in their dimensions, probably because higher volumes alter rates of thermal transport and reagent diffusion [62]. Successful scaling up of the synthesis to 1 L solution has been reported by Kozek et al. [50] and Chang and Murphy [51] for batch synthesis, while Lohse et al. reported a production rate of ca. 0.5 mg/min using a millifluidic set-up [62]. Synthesis of AuNRs using microfluidic reactors (either using a chip reactor [44,63] or micrometric tubings [58]) have also been reported. However, as mentioned by Lohse et al. reaching gram-scale with a microfluidic reactor would require numbering up a large number of reactors [62]. Beside scaling-up, using micro- or milli-fluidic reactors has two added advantages. Flow synthesis usually leads to a lower polydispersity in the dimensions of the nanorods [62] and it is possible to integrate all the steps of the synthesis in the same flow device (Figure 4A), that is the preparation of the seeds and growth solution [63] or post-synthesis modifications such as PEGylation [58], hence decreasing the number of synthesis steps. In the example of postsynthesis PEGylation, the cost of the synthesis is reduced by a factor of 100 in proportion of the amount of PEG-SH used. Integration of the preparation of growth and seeds solutions in the flow device probably participates in the reduced polydispersity and higher reproducibility of the AuNRs because of the continuous use of freshly prepared solutions [44].

Increasing the concentration of reactants is even more challenging as a simple increase of all concentrations results in AuNRs with high polydispersity in their dimensions and to the formation of spherical AuNPs. Park et al. have examined this approach by analyzing the influences of increasing the concentration of gold precursor in the growth solution and the amount of seeds on the quality (extent of polydispersity) and purity (extent of formation of spherical nanoparticles) for the batch synthesis of AuNRs. Their analysis confirmed that the concentration of gold in the growth solution could only be marginally increased (ca. by a factor of 3) and that the concentration of Au in the growth solution and the amount of seeds should be both increased to keep a reasonably high purity (see Figure 4B). Park et al. assigned the observed decrease in quality and purity of AuNRs upon increasing the concentration of gold precursor in the growth solution to the disruption of the balance between reactant reduction and micelle adsorption on very small Au nanoparticles that is at the origin of symmetry breaking (from spherical- to nanorod-shaped NPs). Hence, they propose a two-step growth protocol (quite similar to the protocol proposed by González-Rubio et al. and described above [53]) that allows separating the symmetry breaking step (growth of isotropic AuNP into AuNRs seeds) to the growth of these anisotropic seeds (see Figure 4B). The scaling-up of the synthesis was possible up to a 200× increase in the amount of seeds and concentration of gold precursor (corresponding to a 100× higher production of AuNRs) with only minor changes in the dimensions of the nanorods [64].

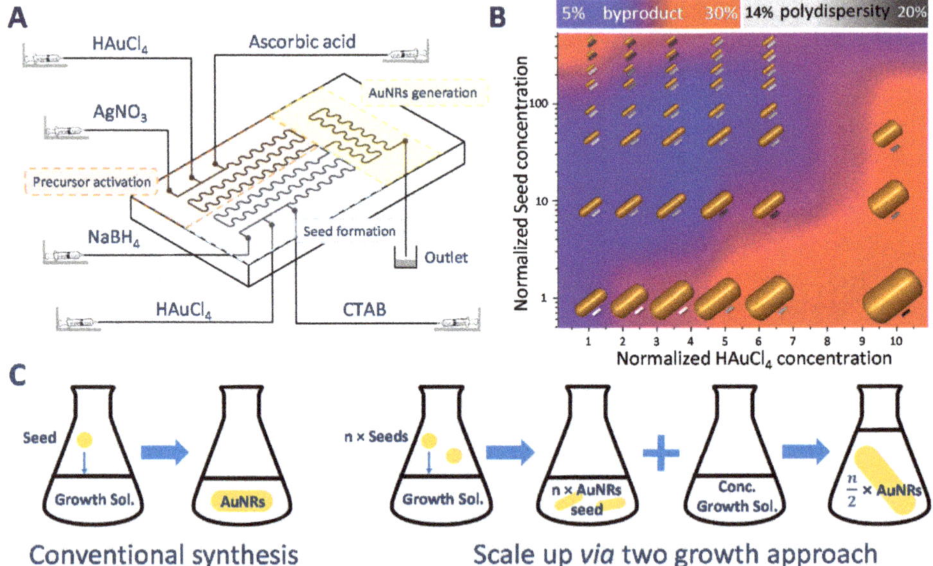

Figure 4. (**A**) Microfluidic flow device allowing better control of reaction parameters and reduction of AuNRs polydispersity. Adapted from [63]. (**B**) Impact of increasing seed and reactant concentration on the structural characteristics of the obtained AuNRs. AuNRs obtained by the conventional seed-mediated protocol is represented as 1/1 ratio in seed/HAuCl$_4$ normalized concentration (polydispersity of AR is reflected by the color of the bar next to the AuNR and product purity by the shade of background color) Adapted from [64]. (**C**) Conventional one-step growth approach (left) compared to the two-step growth approach developed by Park et al. (right) with increasing seeds and second growth solution concentration Adapted from [64].

One can also mention here the work of Khanal and Zubarev who reported the successful gram-scale synthesis of AuNRs using the El-Sayed protocol by (i) increasing the concentration of gold by a factor of 2; (ii) increasing the volume of aqueous solution by a factor of 600 (from 10 mL to 6 L), and (iii) increasing the yield in reduced gold to ca. 100 by slow addition of AA after the first growth step [65].

2.2. Silica Coating of Gold Nanorods (AuNR@SiO$_2$)

Silica is widely used as a coating material for AuNRs. The benefits of silica coating are multiple, among them: An improved colloidal and thermal stability, an increase of the rods surface area while preserving the optical properties of the gold core, and the adjunction of a controllable porosity. Moreover, silica improves AuNRs biocompatibility, while its reactive surface silanols enable drug loading and surface conjugation by functional ligands or biomolecules for an efficient and selective probing of targeted molecules. In addition, SiO$_2$ shell determines the refractive index of AuNRs surrounding environment, thus, thickness control is a crucial parameter affecting the LSPR response and sensitivity.

In the following subsection, we discuss the variety of synthetic methods that have been developed to encapsulate AuNRs with silica. Three possible strategies depicted in Figure 5 are utilized for AuNRs capping by silica: CTAB exchange by a functional primer, coating through a primer on top of CTAB bilayer, and direct coating. We also list other protocols such as microemulsion and biphasic growth at the end of this section.

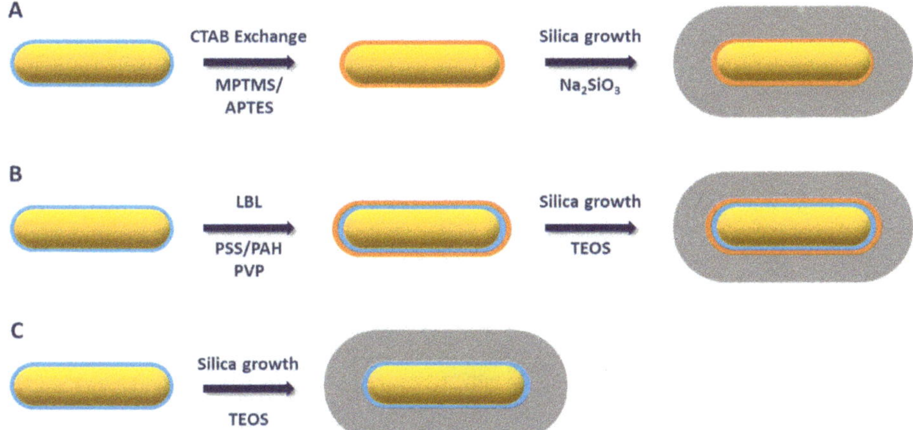

Figure 5. Strategies commonly used for AuNRs capping by silica. (**A**) CTAB exchange by a functional primer followed by silica growth, (**B**) coating through a primer on top of CTAB bilayer then silica growth, and (**C**) direct coating of silica on CTAB-stabilized AuNRs.

2.2.1. CTAB Exchange by a Functional Primer

Due to the strong interaction between CTAB and AuNRs surface and also to the stabilizing role that CTAB plays on colloidal AuNRs suspensions, its replacement by other molecules through ligand exchange is a complex reaction. The first reports on the coating of AuNRs with silica relied on priming the metallic surface with silane coupling agents or polymers to increase the affinity between gold surface and silica. These protocols are derived from the one developed by Liz-Marzàn et al. for citrate-stabilized gold nanospheres, in which the gold surface is modified, in water, by exchanging citrate ligands by (3-aminopropyl)trimethoxysilane (APTMS), a silane coupling agent.

APTMS creates a silanol-rich surface on the AuNR, that favors the deposition of a silica layer by condensation of sodium silicate at pH = 10–11 [66]. In 2001, Murphy and co-workers reported the encapsulation of high AR AuNRs with a thin (5–10 nm) silica shell using (3-mercaptopropyl)trimethoxysilane (MPTMS) instead of APTMS (Figure 6A,B). This primer was chosen because of the stronger affinity of sulfur toward gold enabling the displacement of strongly adsorbed CTAB molecules [67]. Pérez Juste et al. applied a similar procedure to AuNRs with short AR (Figure 6C). They observed the formation of a thin but irregular silica shell (thickness 5–7 nm) and highlighted the role of CTAB micelles in the nucleation of free silica and as a consequence the importance of particle washing prior to silica encapsulation [68]. Later on, Li et al. reported on the formation of ultra-thin silica shell (UTSC) on MPTMS-capped AuNRs. The thickness of the silica shell was finely tuned in the 0.8–2.1 nm range by changing the amount of Na_2SiO_3 [69]. These authors also investigated the rate of silica-shell growth using LSPR. They observed an increase in the constant rate of formation of the silica shell upon increasing the Na_2SiO_3 concentration that was attributed to a rise of pH (from ca. 7 to ca. 9). A consequence of this high condensation rate of silica was the formation of core-free silica nanoparticles at high Na_2SiO_3 concentration. To avoid the formation of these undesired silica aggregates, a silica-shell precursor was added in a stepwise matter for the highest Na_2SiO_3 concentrations. Using this stepwise protocol, a thicker (up to 3.5 nm) silica shell was obtained. The texture of the silica shell grown using thiol-silane as primer, to the best of our knowledge, has never been characterized using N_2-sorption. Nevertheless, it is usually considered as non-porous. Indeed, on the basis of Raman spectroscopy analysis of AuNR@SiO2 exposed to pyridine that strongly adsorbs on gold and not on silica, Li et al. have concluded to the absence of pinholes (exposing gold surface to reactants) in the shell of Au@SiO2 composites with a shell thickness above 4 nm [70].

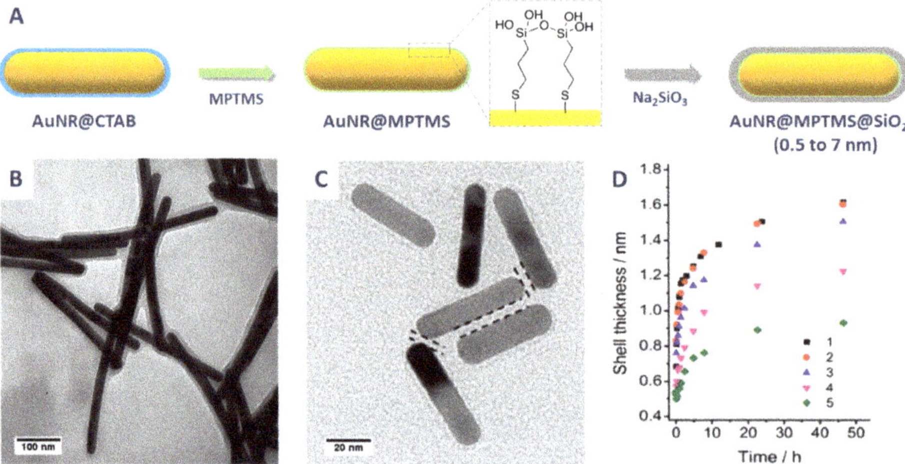

Figure 6. (**A**) CTAB replacement by (3-mercaptopropyl)trimethoxysilane (MPTMS) for the growth of thin layers of silica. (**B**) and (**C**) TEM images obtained for different aspect ratio (AR) from ref [67] (**B**) and ref [69] (**C**). (**D**) Kinetics of silica shell thickness for decreasing concentrations of Na$_2$SiO$_3$ from ref [69].

Table 1 summarizes the reaction conditions and the silica shell aspects obtained when using a silane primer for silica growth. The source of silica is exclusively sodium silicate and the thicknesses are very low, ranging from 0.5 to 8 nm. Silane strategy is an effective way to grow, in aqueous solution, a thin silica layer shell on AuNRs surface. However, such protocol requires an extended time duration from one to six days for the silica shell to form and stabilize the particles (see Table 1), unlike the next primers we will be discussing.

Table 1. Summary of the reaction conditions for silica coating on AuNRs using a surface primer exchanged with CTAB and respective silica shell aspects.

Method	AR	Solvent	[CTAB]	Primer	Silica Source	Reaction Time	Shell Thickness (nm)	Shell Feature
Obare et al. [67]	13.0	Water	Minimized	MPTMS	Na_2SiO_3	24 h	8	Thin
Perez-Juste et al. [68]	1.9–3.8	Water	Minimized	MPTMS	Na_2SiO_3	2 d	5–7	Thin
Li et al. [70]	3.0	Water	Minimized	APTMS	Na_2SiO_3	1–2 d	4	Thin
Li et al. [69]	NC	Water	Minimized	MPTMS	Na_2SiO_3	2–6 d	0.5–3.5	Ultrathin
Fernández-López et al. [71]	3.5–4.8	EtOH/Water	Without	PEG-SH	TEOS	2 h	4–31	Thin/Thick and dense
Wang et al. [72]	3.2	EtOH	Without	PEG-SH	TEOS	3 h	nc	Thick and mesoporous

Fernandez-Lopez et al. used a ligand exchanged-based protocol but with a thiol-modified poly(ethylene glycol) ((O-[2-(3-mercaptopropionylamino)ethyl] O'-methyl-poly(ethylene glycol, mPEG-SH) [71]. The action of this primer is similar to the one described above for the thiol-silane: The thiol function interacts strongly with the gold surface while the oxygen-rich poly(ethylene)glycol favors the nucleation of the silica shell. Ligand exchange is performed in water with an excess of mPEG-SH (150 molecules/nm^2) and a minimum of CTAB (~1 mM). After this step, particles are washed to remove free CTAB micelles and excess of reactant and then transferred in a NH_3/water/EtOH mixture for silica coating by the Stöber method. The resulting particles exhibit a non-porous shell of very well controlled thickness, which can be tuned between 4 and 31 nm by adjusting the amount of added tetraethyl orthosilicate (TEOS) (Figure 7). Surface composition of mPEG-SH functionalized AuNRs in classical conditions is expected to contain both mPEG-SH and CTAB molecules. In order to fully remove cytotoxic CTAB, Kinnear et al. proposed a two-step procedure that enhances PEGylation: The first step is performed by adding 10 mPEG-SH/nm^2 to the AuNRs solution in water with ~1 mM CTAB and complete functionalization of the AuNR surface is secondly performed in ethanol/water solution (90% *v/v* EtOH in water) containing another 10 mPEG-SH/nm^2. The use of EtOH is reported to increase the critical micellar concentration (CMC) of CTAB, hence destabilizing the bilayer and facilitating its desorption [73].

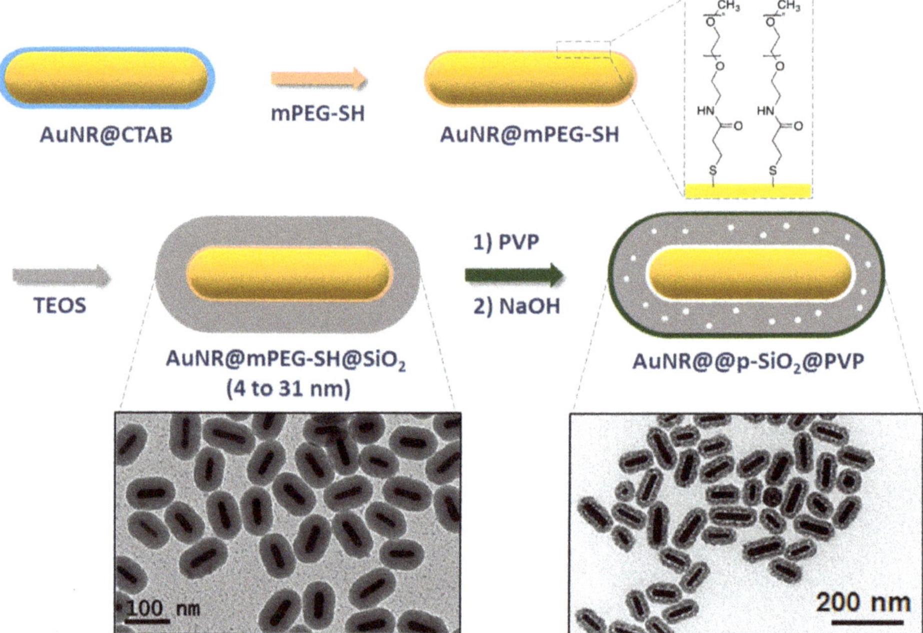

Figure 7. CTAB replacement by (O-[2-(3-mercaptopropionylamino)ethyl] O'-methylpolyethylene glycol (mPEG-SH) for silica growth and further induced porosity with the corresponding TEM images from ref [71,72].

The use of PEG-SH/Stöber process requires shorter reaction times as reported in Table 1. In addition, while silica shell grown using PEGylation/Stöber process is essentially nonporous, Wang et al. recently reported the use of NaOH as etching agent to induce porosity in the silica shell. The silica outer layer is first wrapped with PVP to prevent complete dissolution of the shell, then by adjusting NaOH concentration and etching time, different structures were obtained (Figure 7). Surface area, pore volume, and average diameter were determined by N_2 physisorption and estimated to be 183 m^2.g^{-1}, 0.32 cm^3.g^{-1}, and 3.2 nm, respectively [72].

2.2.2. Coating through a Primer on Top of CTAB Bilayer

Graf et al. proposed in 2003 a very general protocol based on the intermediate adsorption of a layer of polyvinyl pyrrolidone (PVP) for the coating of colloidal nanoparticles with silica [74]. The role of this intermediate layer is to allow the transfer of water stable colloidal nanoparticles in the ethanol/H$_2$O/NH$_3$ solution used for classical Stöber process. However, according to Pastoriza-Santos et al., this protocol cannot be directly applied to CTAB-capped AuNRs. These authors also mentioned the difficulty to apply the protocol proposed by Murphy and coworkers to AuNRs with small AR (because of tip-to-tip aggregation of NR after ligand exchange of CTAB with the thiol silane). Therefore, they developed an alternative protocol whose purpose is to screen the CTAB molecules [75,76]. This screening is obtained by the layer-by-layer (LBL) adsorption of charged polyelectrolytes (Figure 8). Typically, AuNRs were successively wrapped in aqueous media with a negatively charged polystyrene sulfonate (PSS) and positively charged polyallylamine chloride (PAH). Finally, a last layer of PVP was added in order to reduce the surface charge (zeta potential below 20 mV). This reduction of the surface charge is required to preserve the colloidal stability of the AuNRs during the last step, which is the growth of a silica shell by base-catalyzed hydrolysis and condensation of TEOS in a 2-propanol/water mixture (Stöber process). Between each functionalization step, the particles are washed to remove excess reagents as well as free CTAB micelles leading to the formation of a non-porous shell. The control of the shell thickness from 15 to 40 nm (Table 2) was achieved by modulating the amount of TEOS added. Although efficient, this procedure may be laborious and time consuming because of the multistep process. Moreover, reaction conditions such as polyelectrolyte concentrations and molecular weight as well as the ionic strength of the solution should be carefully controlled because of potential risk of particle aggregation or heterogeneous silica shell.

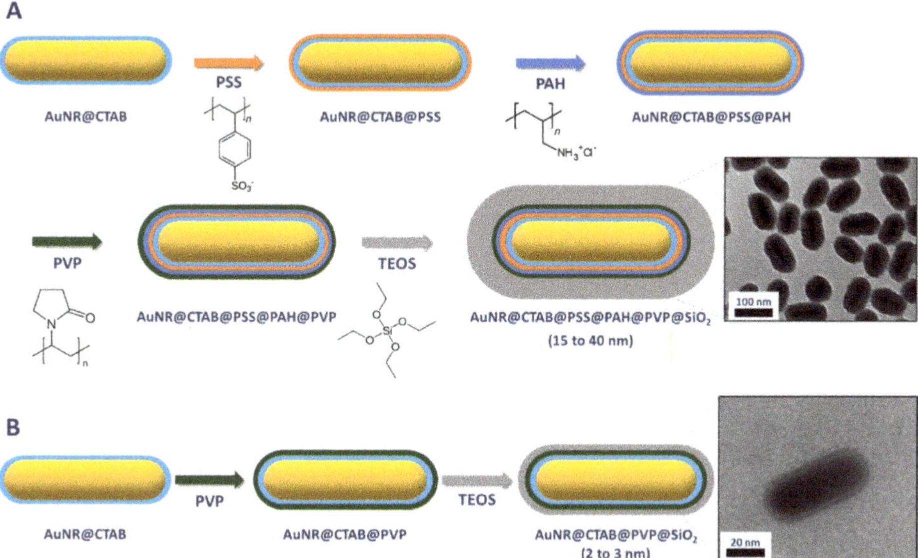

Figure 8. Silica coating on top of CTAB using an layer-by-layer (LBL) approach adapted from ref [75] (**A**) and ref [77] (**B**).

Table 2. Summary of the reaction conditions for silica coating on AuNRs using a surface primer with respective silica shell aspect. In both cases, the silica source is tetraethyl orthosilicate (TEOS) with no CTAB.

Method	AR	Solvent	Primer	Reaction Time	Shell Thickness (nm)	Shell Feature
Pastoriza-Santos et al. [75]	4.0	2-propanol/Water	PSS, PADH, PVP	2 h	15–40	Thick and dense
Nallathamby et al. [77]	3.6	EtOH	PVP	24 h	2–3	Thin

More recently, Nallathamby et al. achieved PVP-mediated silica coating of CTAB-capped AuNRs, without the need of intermediate polyelectrolyte layers. Using this protocol, they obtained AuNR coated with a thin and homogeneous silica shell whose thickness was modulated by changing the PVP molecular weight [77].

2.2.3. Direct Coating of AuNR with a Mesoporous Silica Shell

Direct coating of AuNR with silica was achieved by several groups as summarized in Table 3. Table 3 also reports the reaction conditions and the respective silica shell size and aspect.

Gorelikov et al. [78] were the first to report the direct coating of AuNR with silica. A detailed description of the synthesis protocol can be found in ref. [76]. The protocol is quite simple. Briefly, a solution of the CTAB-capped nanorods redispersed in water is basified to pH = 10–11 and a solution of TEOS in an alcohol (methanol, ethanol, etc.) is added to this solution under stirring. The suspension is then aged during several hours to days. As mentioned above, the initial attempts to directly coat the AuNR with silica failed and led to the formation of free silica nanoparticles, while leaving the AuNR uncoated. Gorelikov managed to solve this issue by simply adding a centrifugation step between the synthesis and coating steps. The purpose of centrifugation is to drastically decrease the concentration of CTAB in the AuNR solution before adding the silica precursor, TEOS. A total removal of CTAB molecules must, however, be avoided because they stabilize the AuNR suspension by providing an electrostatic repulsion thanks to their positively charged headgroups. According to the protocol of Gorelikov et al., the CTAB concentration should be about 1.5 mM (slightly above the CMC in water). The negative impact of extensive washing of the AuNR was emphasized by Cong et al. who observed that repeating twice the washing step leads to aggregation of AuNR together with core-free silica particles [79]. The residual CTAB in the solution, by interaction with the negatively charged silicate species formed by basic hydrolysis of TEOS, leads to the formation of a porous shell made of disordered pores (with an estimated pore size, based on TEM images of about 4 nm). These good textural properties have been later confirmed and refined by Liu et al. based on N_2 physisorption measurements (pore diameter: 2.6 nm, surface area: about 510 $m^2.g^{-1}$) [80].

The detailed mechanism leading to the capping of the AuNR by this mesoporous silica shell was later explored by several authors and the composition of the synthesis solution was altered in order to control the thickness of the silica shell. Parameters such as CTAB, alcohol, AuNR and TEOS concentrations, pH, and aging times have been analyzed and most of them have a strong influence, not only on the thickness, but also on the shape (continuous coating vs. lollipop and dumbbell coatings) of the silica coating.

Table 3. Summary of the reaction conditions for direct silica coating on AuNRs with respective silica shell size and aspect.

Method	AR	Solvent	[CTAB] (mM)	Precursor	Reaction Time	Shell Thickness (nm)	Shell Feature
Gorelikov et al. [78]	3.5	MeOH/Water	Minimized	TEOS	2 d	15–60	Thick and mesoporous
Wu et al. [81]	3.6	EtOH/Water	0.1	TEOS	20 h	3–20	Thin/Thick and Mesoporous
Cong et al. [79]	3.0	Isopropanol/Water	0.2	TEOS	20 h	60–150	Thick and dense
Liu et al. [80]	4.3	EtOH/Water	Minimized	TEOS	20 h	10–40	Thick and mesoporous
Abadeer et al. [82]	1.1	MeOH/Water	0.4–1.2	TEOS	20 h	11–26	Thick and mesoporous
Yoon et al. [83]	NC	MeOH/Water	0.4–50	TEOS	24 h	8–21	Thick and mesoporous
Rowe et al. [84]	4.0	MeOH/Water	1.7	TEOS	20 h	NC	Uniform to Dumbbell
Wang et al. [85]	3.9	EtOH/Water	1–9	TEOS	12 h	13–20	Uniform to Dumbbell

An expected increase in the silica shell thickness has been observed by several authors upon increasing the amount of added TEOS (Tian et al.: From 13 to 30 nm [86]; Wu et al.: From 3 to 17 nm [87] (Figure 9); Zhang et al.: From 3–38 nm [88]; Mohanta et al.: From 8–16 nm [89]). Wu et al. have also concluded that the thickness of the silica shell is, over a reasonable range of added TEOS, consistent with the full consumption of added TEOS (Figure 9). However, adding more TEOS either results in no modification in the thickness of the silica coating or, when a large excess of TEOS is added, to the formation of an irregular shell of silica, which only partially coats the gold nanorods.

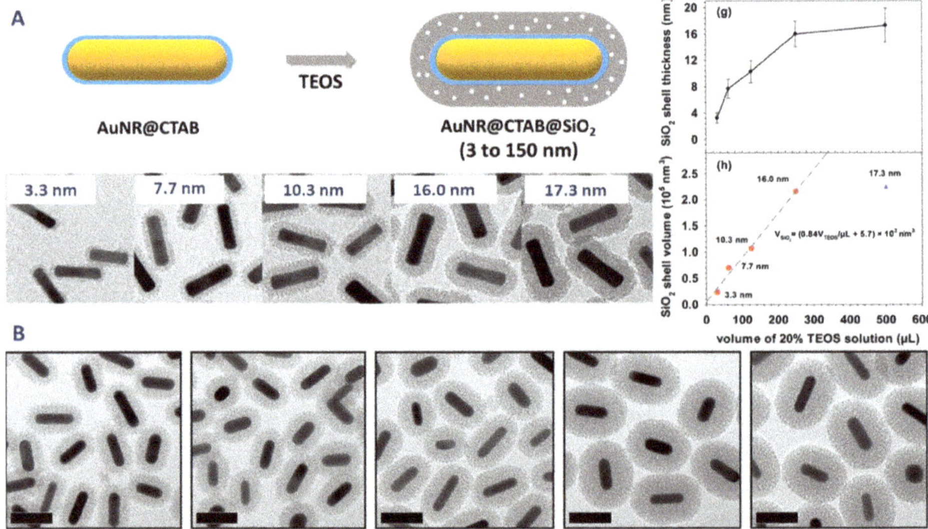

Figure 9. (**A**) Variation of silica coating thickness with amount of added TEOS: Left: TEM images (numerical values reported on each TEM are amount of TEOS and average coating thickness); right: Average thickness (top) and average volume (bottom) of silica coating vs. volume of added TEOS illustrating the complete conversion of TEOS to SiO_2 shell up to 250 µL of 20% TEOS (red dots) and incomplete conversion above (blue triangle) (from [81]). (**B**) Variation of silica coating thickness with CTAB concentration: From left to right 1.2, 1.0, 0.9, 0.7, 0.4 (from [82]), the bar scale corresponds to 50 nm.

Another less obvious parameter that is commonly used to modulate the thickness of the silica coating is the concentration of CTAB. Indeed, several authors observed that increasing the concentration of CTAB in solution results, for the same amount of added TEOS in thinner silica shell (Abadeer et al. [82]: From 26 to 11 nm (see Figure 9B); Yoon et al. [83]: From 30 to 13 nm).

Fine tuning of the thickness of the silica layer can also be, to a certain extent, achieved by modulating the duration of the synthesis. However, the addition of a terminating agent (PEG silane) before completion of the silica shell growth resulted in the production of a thin (as thin as 2 nm), smooth, and homogeneous shell [81], while quenching the reaction before completion leads to an inhomogeneous coating [90]. Other parameters have been less investigated such as the pH of the solution, but they also seem to play a role. For example, Huang et al. observed that decreasing the pH from 11 to 10.3 leads to dumbbell-shaped particles [90]. Another parameter whose influence has long been underrated is the amount of alcohol used to dilute TEOS. This parameter has been recently examined by Rowe et al. [84] They observed that, when the TEOS fraction in the TEOS/methanol solution was in the range 3–7 v%, a dumbbell-shaped silica coating was obtained (see below).

2.2.4. Other Strategies

Other strategies relying either on a biphasic process or microemulsions were also utilized for silica capping of AuNRs. Xu et al. have proposed a modification of the thiol-silane/Stöber protocol, that allows growing a highly porous silica shell. In this biphasic process, inspired by the work of Shen et al. [91], MPTMS is added first to a AuNRs/TEA/CTAC (TEA: Triethylamine; CTAC: Cetyltrimethylammonium chloride) aqueous solution (25 wt% CTAC) and a solution of TEOS in cyclohexane is added on top of this aqueous solution [92] (Figure 10A).

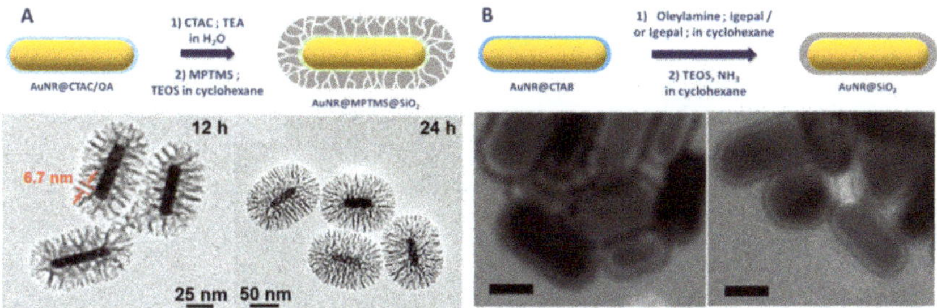

Figure 10. (**A**) AuNR@SiO$_2$ with a highly porous silica shell grown in a biphasic solution [92]. (**B**) Silica coating of gold nanorods by reverse microemulsion from ref [77].

A second strategy relying on microemulsion was also used for thin layer growth. Microemulsions are transparent and thermodynamically stable systems formed through the mixture of surfactant, water, and organic solvent (oil). The use of these homogeneous dispersions, especially the water-in-oil (W/O) or reverse microemulsion, is appealing to synthesize monodisperse nanoparticles or core-shell nanocomposites [93,94]. Reverse micelles dispersed in the continuous oil phase consist of water droplets stabilized by surfactant layer act as "nanoreactors" in which chemical reactions occur. In the case of silica coating, the cores are enclosed within the reverse micelles during the coating step thus preventing aggregation through steric stabilization, while the silica precursor (e.g., TEOS) is dissolved in the oil phase and hydrolyses at the water/oil interface. The major advantage of using reverse microemulsion is to restrict silica condensation to the limited water domain around the cores. This leads to the formation of dense and uniform silica coatings (shell thickness from 2 nm to tens of nm) on a variety of hydrophobic cores [95–98].

Reverse microemulsion is usually used to coat hydrophobic cores as they are readily dispersible and stable in the oil phase. As CTAB-capped gold nanorods are hydrophilic, they tend to aggregate in organic medium; so reverse microemulsion is not the favored method to coat gold nanorods with silica. Recently, Nallathamby et al. reported the silica coating of gold nanorods (AR of 3.6) by reverse microemulsions composed of IGEPAL/water/cyclohexane (IGEPAL is a trademark for polyoxyethylene (n) nonylphenyl ether). Their strategy and obtained TEM images are shown in Figure 10B. They reported that either hydrophilic CTAB-capped gold nanorods and hydrophobic oleylamine-capped gold nanorods (obtained after ligand exchange) can be, respectively, coated by a 3.8 nm and 6.1 nm thin silica shell [77].

2.2.5. Commonly Encountered Difficulties and Their Remedies

- Core-free silica particles

Quite often TEM micrographs indicate the presence of undesired core-free silica particles. Yoon et al. proposed to proceed by successive injections of small amounts of TEOS (up to 16) in order to keep the concentration of silicate anions in solution below the nucleation threshold to avoid the formation of these undesired silica particles [83], while, for other authors, the formation of core-free silica nanoparticles may not be a problem because these particles are small (and light) and mostly remain in the supernatant during the centrifugation step [82]. Moreover, according to Wu et al., the formation of these core-free particles mostly occurs because the control of the thickness of the silica shell is obtained by increasing the CTAB concentration in the solution: Using conditions with low CTAB concentration and controlling the thickness of the silica shell by adjusting the TEOS amount allows converting all TEOS in shell coating of the AuNR [81].

- Dumbbell-shaped coatings

Under certain experimental conditions, the silica shell grows exclusively/mostly on the two tips of the nanorods, leading to dumbbell/peanut-shaped coating (that can be either desired or undesired), see Figure 11. The origins of this shape are still not completely understood. According to Wang et al. dumbbell-shaped particles are formed because of the energy barrier that TEOS needs to cross to access the hydrophobic space located in the middle of the CTAB double layer that surrounds the AuNR. This energy barrier is stronger on the side of the AuNRs than on the tips (leading to a favorable nucleation of the silica coating around AuNR tips) and varies with the composition (concentration of TEOS, CTAB, and ethanol) of the reaction medium [85].

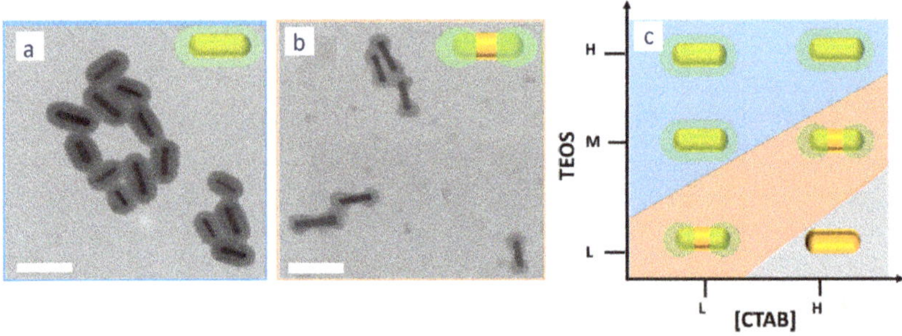

Figure 11. TEM images of AuNR@SiO$_2$ obtained at constant TEOS concentration and with (**a**) 1 mM CTAB and (**b**) 9 mM CTAB. (**c**) Shape of the silica coating under various conditions according to Wang et al. [85].

Rowe et al. also investigated the influence of the amount of methanol on the shape of the silica coating (Figure 12). More precisely, they observed dumbbell-shell coating for TEOS-methanol compositions ranging from 3 to 7 vol% of TEOS (above and below these values, the silica shell forms a continuous and homogeneous coating). They also concluded, based on the shape of the coating after a short reaction time, that SiO$_2$ is probably initially deposited all over the surface of the nanorod and that a reshaping of the coating from continuous to dumbbell occurs with time.

- Removal of CTAB

AuNR@SiO$_2$ prepared using this type of protocol can still contain a fairly large amount of CTAB, as CTAB molecules are located not only at the AuNR surface, but also in the mesopores of the silica shell. The best-suited protocol to fully remove CTAB from the pores of the silica shell, while avoiding shape-transformation of AuNRs is, according to Feng et al., an extraction of CTAB using a NH$_4$NO$_3$/methanol solution [99].

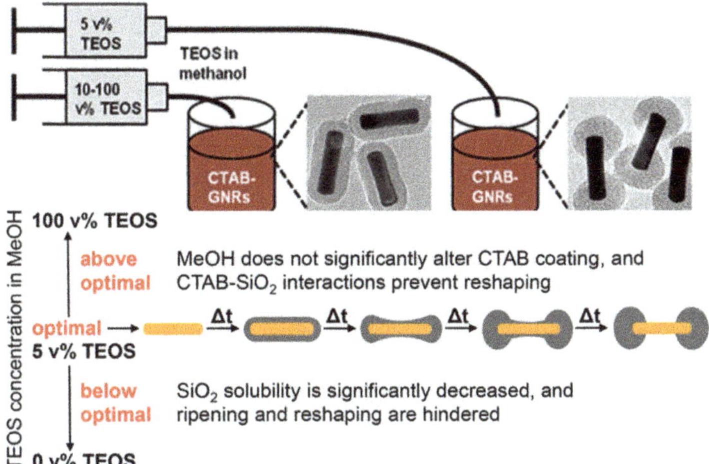

Figure 12. Top: TEM images showing continuous coating and dumbbell-shaped coating of Au NRs. Bottom: Cartoon showing the mechanism proposed by Rowe et al. to explain the influence of methanol on the formation of dumbbell shaped silica coating (from [84]).

2.2.6. Concluding Remarks

This brief overview of the literature on the coating of AuNR with porous silica reveals that it is possible, with this method, to prepare silica-coated AuNR free of core-free silica and with a fine control of the thickness of the silica coating. It also shows that a complete mastering of the coating does require some expertise: (i) Tuning the thickness of the silica coating through an adjustment of the concentration of CTAB may lead to the formation of undesired core-free silica particles, and, although most of them can be eliminated by an appropriate centrifugation step, it is not unusual that a small fraction remains in the final sample; (ii) prediction of the thickness of the silica coating based on the amount of added TEOS is made difficult by the fact that the concentration of gold nanorods is not always precisely known. Moreover, parameters leading to a thin coating of AuNR with silica are close to those leading to a dumbbell-shaped coating.

3. Methods of Characterization

Several common techniques are used to characterize the structure and shape of gold nanorods. In the following, the main ones will be described and, for each of them, the possibility of characterizing an outer silica shell will be discussed.

3.1. UV-Visible Spectroscopy

Gold nanoparticles are well known and used for their absorption in the visible range that is characteristic of LSPR. The associated resonance frequencies are related to the characteristic dimensions of the particle. In the case of gold nanorods, their two dimensions lead to two absorption bands in their UV-visible spectrum [100]: One due to the transverse localized surface plasmon resonance (t-LSPR) and one to the longitudinal localized surface plasmon resonance (l-LSPR) (see Figure 13A). If both are characteristic of the gold nanorods, l-LSPR is the one that is used and studied for nanorods. For example, because the l-LSPR band is subjected to shifts (usually a red-shift followed by a blue shift, see section Figure 13A) when the nanorods grow, it allows a simple monitoring of the growth of the rods [101]. However, this technique cannot be used as an absolute characterization of gold nanorods, as the position of the l-LSPR band is in fact correlated to the AR of the rods and not directly to their length (Figure 13B, [14]). As emphasized by Scarabelli et al. [42], other qualitative and quantitative information can be extracted from the UV-Vis spectrum:

1. The absorbance at 400 nm can be used for the quantitation of reduced gold (see details in the section of this chapter dedicated to quantitation);
2. The intensity ratio between l- and t-LSPR bands is a good qualitative indication of the polydispersity (a high ratio indicates a low polydispersity in the dimensions of the nanorods);
3. A shoulder close to the t-LSPR peak indicates the presence of AuNPs of other shapes (e.g., spherical AuNPs);
4. The width and the symmetry of the l-LSPR band are related to the polydispersity of the sample (a larger width and/or an asymmetric shape indicates a higher polydispersity).

These pieces of information are, however, only qualitative because the broadening of the l-LSPR band has also been shown to be related to the dimensions of the rods [102].

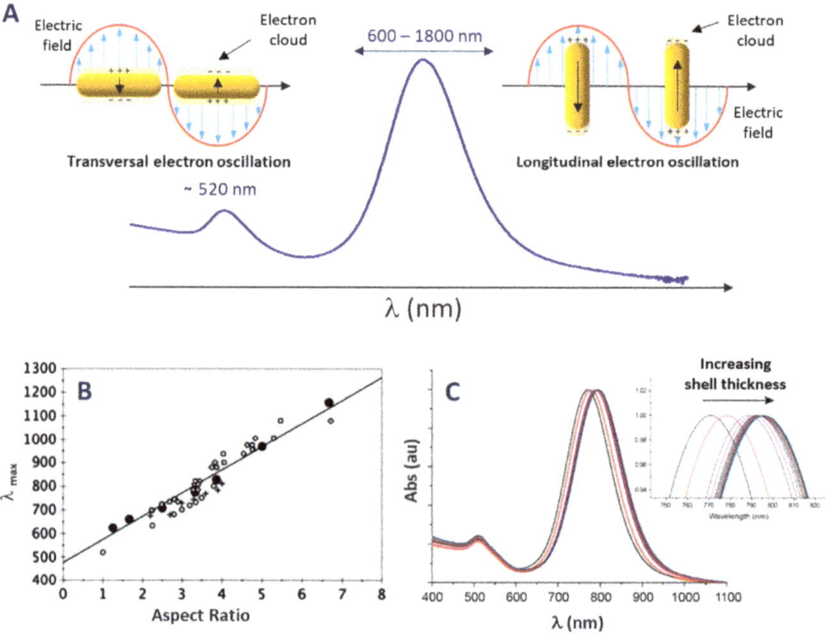

Figure 13. (**A**) Typical extinction spectrum of AuNR and the schematic representation of the electronic oscillations adapted from ref [103]. (**B**) Position of l-LSPR band maximum λ_{max} as a function of AuNRs AR, Simulation results using the DDA (Discrete Dipole Approximation) method (black circles) [14] and experimental data from the works of Al-Sayed et al. (open circles) [104], Pérez-Juste et al. (diamonds and squares) [105] and Brioude et al. (crosses) [14]. (**C**) l-LSPR band shift with increasing silica shell thickness adapted from ref [83].

In addition, the position of the l-LSPR peak is extremely sensitive to the adsorption of molecules [106]. This feature is at the origin of LSPR biosensing as it will be discussed in Section 5. As a consequence, the position of the l-LSPR band also allows silica shell growth monitoring. Indeed, a red shift in its position has been observed in several studies and correlated to the thickness of silica shell up to a certain value [83]. Figure 13C illustrates the l-LSPR shift upon successive additions of TEOS resulting in silica shell thicknesses ranging from 8 to 20 nm.

3.2. Electronic Microscopy

As illustrated throughout this manuscript, electron microscopy is one of the most useful techniques to characterize gold nanorods. Indeed, it provides direct images of the nanorods. With transmission electron microscopy (TEM), the difference between the contrast of the gold core and the silica shell allows to characterize them separately. The quantitative analysis of micrographs displaying enough particles provides the distribution of the nanorod dimensions. Yet, unbiased and reproducible analysis of the micrographs is not an easy task as Grulke et al. showed in their work where they propose a standardized protocol to analyze such images [107].

The main limitations of electron microscopy characterizations are:

1. The handling of dried samples to analyze them in vacuum in the microscope chamber can also introduce multiple artefacts;
2. The selection of a necessarily limited number of micrographs of the samples can introduce a biased overview of the sample or, at least, a measurement uncertainty. General statistics like Sturges' rule [108] or specific studies dedicated to nanoparticle distributions [109] can help to evaluate the accuracy of this limited sampling. These problems are amplified by the difficulty of the manual or automatic analysis of the dimensions of such objects on micrographs;
3. It is usually advised to analyze 200–300 nanorods to obtain a reasonable evaluation of the average dimensions and polydispersity in size of the nanorods, which can be time consuming if performed manually with a software such as ImageJ and automatic analysis is often not efficient (especially when particles of different shapes are present). New developments in this field are oriented toward the automatic analysis of TEM images by the development of appropriate algorithms [110].

To overcome these obstacles, another type of measurement can be performed, using some properties of the waves scattered by nanorods.

3.3. Scattering Techniques

Two main scattering techniques are commonly used to characterize gold nanorods: Dynamic light scattering (DLS) and small angle X-ray scattering (SAXS) (Figure 14). Both techniques operate in liquid and allow to analyze a large number of particles simultaneously, which provides, in general, a faithful description of the samples.

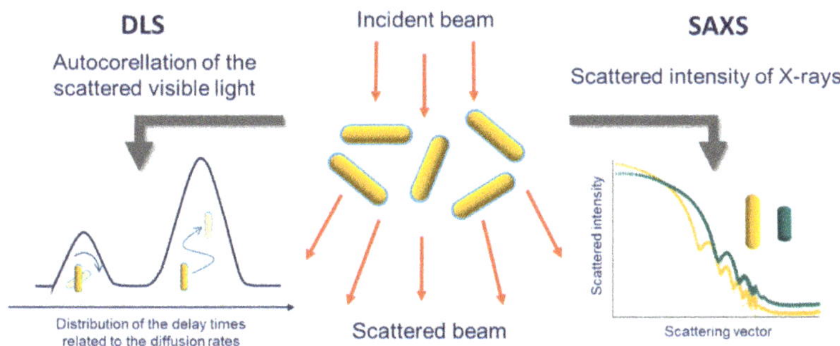

Figure 14. Scattering techniques commonly used to characterize AuNRs.

3.3.1. DLS

DLS measurement is relatively easy to perform and this technique is commonly found in laboratories that deal with soft matter. Indeed, nanorod dispersions can most of the time be directly analyzed, although sometimes after dilution, and a quick and automatic measurement gives a result within a few minutes, provided that the laser used does not display a similar wavelength as one of the LSPRs of the nanorods. However, interpreting the results of this technique is not straightforward for anisotropic objects.

Indeed, DLS is based on the analysis of the dynamics of the loss of scattered light autocorrelation due to the Brownian motion of the studied particles. In other words, it consists of measuring the characteristic time taken by the particles to rearrange in a completely different random pattern that yields a different scattered intensity. In the case of isotropic particles, this time can be linked to their size through their diffusion constant and the Stoke–Einstein equation. When it comes to anisotropic particles, two characteristic times arise that correspond to two motions: Rotational and translational (Figure 12) [111]. These two times give rise to the observation of two signals in the particle size distribution, one of low intensity, at relatively low dimensions (a few nm) and the other at larger dimensions. The low intensity peak should not be mistakenly interpreted as indicating the presence of small particles, as it is actually due to the rotational diffusion of nanorods, while the larger peak at higher dimension is associated with translational diffusion [112].

Experimentally, using a system of polarizer/analyzer, it is possible through depolarized DLS (DDLS) to distinguish both contributions [113,114]. Several models and theories have been proposed to reproduce the experimental observations, but they are still debated [114]. Therefore, if these measurements are indubitably characteristic of the nanorod structure, they can mainly be used as a signature of the colloidal stability of the nanorod dispersions but not to directly infer their dimensions. DLS can also be used to highlight the stages of NR modification as well as the formation of aggregates: In particular, an increase in the peak corresponding to translational diffusion, if it is associated with the disappearance of the rotational scattering peak, indicates the formation of isotropic aggregates, whereas if it is associated with an increase in the intensity of the rotational scattering peak, it indicates a tip-to-tip aggregation of NRs [112].

3.3.2. SAXS

Another relevant scattering technique to study nano-objects is small angle X-ray scattering (Figure 14). It is less commonly used as the corresponding setups are rarely present in laboratories and the synchrotron beamlines displaying this technique can only be accessed upon acceptance of a proposal. The use of SAXS seems perfectly relevant to characterize nanorods though, as the high density of gold gives a strong contrast with the dispersing phase. Hence, especially with synchrotron light sources, measurements can be performed quickly enough, to follow the growth of nanorods in real time [101,115]. The main limitation of this technique to characterize the length of rods is the smallest scattering vector q_{min} (corresponding to the smallest probed angle) that can be measured with the equipment that is used. Indeed, q_{min} is related to the biggest size that can be characterized: $l_{max} \approx 2\pi/q_{min}$. Hence, two rods displaying the same radius but different lengths, both superior to l_{max}, would not be distinguished.

Once obtained, the scattering curves need to be fitted with suitable model form-factors to determine the dimensions of the rods [116]. To precisely model nanorods, some authors had to use a combination of form-factors [117]. To take into account interactions between nanorods, a structure-factor can also be combined to the form-factor [118]. Unfortunately, very thin silica coating cannot be efficiently characterized through this technique, as its scattering contribution is not significant enough, but thicker layers should be quantifiable if modelled with the appropriate form factor, as well as more complicated structures (e.g., core/satellite superstructures in [119]).

3.4. Other Characterizations

3.4.1. Zeta Potential Measurement

Zeta potential measurements (usually based on electrophoretic mobility measurements) can be a precious help to monitor the changes that take place at the surface of the nanorods. Indeed, this technique gives indirect information about the surface charge that is strongly correlated with the type of material at the surface (gold, silica) and the molecules adsorbed at the interface (surfactants, polyelectrolytes, functionalization agents). This technique can be helpful to monitor the formation of a silica shell onto gold nanorods and also to follow their functionalization. For example, Gorelikov and Matsuura used it to monitor the effect of washing cycles on the gradual removal of CTAB surfactant from the surface of silica-coated gold nanorods (Figure 15) [78].

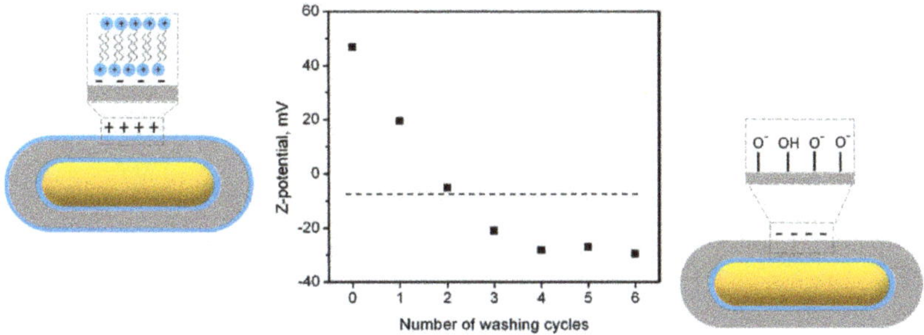

Figure 15. Monitoring CTAB removal by zeta potential measurements, adapted from [78].

3.4.2. Quantitation of Reduced Gold in AuNRs

Spherical gold nanoparticles are obtained under strong reduction conditions, leading to a complete reduction of Au^{3+} to Au^0 and, as a consequence, a 100% reduction yield can be postulated. However, this is not the case for the preparation of AuNRs. Indeed, their synthesis requires the use of weak reducing agents, thus the hypothesis of a 100% reduction of the gold salts no longer applies. Therefore, the determination of the reduction yield (i.e., efficiency of gold incorporation into the AuNR) is crucial information for the optimization of the synthesis protocol, the investigation of AuNRs formation mechanism, the surface modification procedure, and some applications. As mentioned above, the reduction yields are still scarcely reported in the literature. The main techniques utilized for these characterizations are inductively coupled plasma coupled to optical emission spectroscopy or mass spectrometry (ICP-OES/MS), UV-Vis spectroscopy, and X-ray absorption near edge spectroscopy (XANES) measurements. ICP-OES/MS are highly sensitive but destructive elemental analysis techniques (AuNRs samples are digested in strong acids before analysis) that have been used to determine Au^0 concentration in AuNR dispersions [46,120–122]. The extinction coefficient of l-LSPR band could also, in principle, be used for the quantitation of AuNRs concentration, provided that the value of the extinction coefficient at each wavelength can be calculated. Orendorff et al. [46] found that extinction coefficients of AuNRs depend on their AR (calculated from TEM images) and, as a consequence, on the energy of their l-LSPR band. Park et al. [122] later came to a different conclusion, which is the fact that the extinction coefficient of AuNRs depends mostly on the effective radius of the AuNR ($R_{eff} = (3V/4\pi)^{1/3}$) and only marginally on its AR. However, the extinction coefficient is not only linked to the dimensions of the AuNRs, as dispersions containing particles with similar average dimensions but different polydispersities display significantly different absorptions [42]. To determine the concentration of reduced gold based on UV-Vis spectroscopy, several authors have proposed to monitor the absorbance at 400 nm, assuming that interband transitions of Au^0 at this wavelength are the only contribution to the absorption and are independent of the particle size, shape, and surface chemistry [42,123–125]. Edgar et al. found a reliable correlation (within 20% discrepancy) in Au^0 concentration using ICP-MS or optical measurement by analyzing the data published by Orendorff et al. [126]. Scarabelli et al. [42] found a linear correlation between absorbance at 400 nm and the concentration of reduced gold in solution and established that a value of 1.2 for the absorbance at 400 nm corresponds to a Au^0 concentration of 0.5 mM. Hence, UV-vis spectroscopy can be used for quantitation of reduced gold with a relatively good precision with the added advantage that it can be used in situ. Finally, XANES studies were also conducted to monitor the Au^{3+}, Au^+, and Au^0 concentrations and their incorporation during AuNR growth as this technique allows to discriminate the different redox states of gold [101,115]. However, the limited accessibility to a synchrotron X-ray source necessary for this kind of measurement remains a major drawback for routine analysis.

3.4.3. Silica-Shell Porosity Assessment

The porosity of a 20 nm-thick mesoporous silica shell coated on AuNR was directly measured by N_2 physisorption analysis and was found similar to that of mesoporous silica nanoparticles templated with the same CTAB surfactant. Porosity was also indirectly probed by SERS with aromatic thiols of increasing bulkiness. Owing to the strong dependence of the SERS signals intensity on the distance to the gold surface, only the molecules smaller than the pores and able to penetrate within the silica core can chemisorb to the gold surface and will then provide a Raman signal. The absence of a Raman signal for the AuNR@SiO$_2$ exposed to the largest thiol of the series indicated that the molecule was unable to reach the bottom of the mesopores but rather adsorbed to their walls [127].

4. Functionalization Methods of Gold Nanorods

Prior to their use for biosensing, AuNRs require a functionalization step during which the biological element responsible for target recognition becomes attached to the particles. The literature survey shows that three general strategies are employed for this purpose, namely physisorption, chemisorption, or conjugation.

4.1. Physisorption

Physisorption is by far the easiest method to immobilize a bioreceptor on AuNR. Conversely to citrate-coated gold nanospheres, the organic capping of AuNR, namely the CTAB double layer, is strongly attached to the gold surface, thereby not easily displaced by competing molecules. The highly positive surface charge conferred by the CTAB layer enabled antibodies to be physisorbed at the surface of AuNR probably by electrostatic interaction [128,129]. Let us note that the long-term stability of the resulting nanoprobes has not been investigated.

Alternatively, the LbL technique, previously discussed in Figure 8, was applied to electrostatically physisorb antibodies to CTAB-capped AuNR in solution [130–132] or immobilized on a glass slide [133]. The principle of the LbL technique relies in the successive coating of negatively and positively charged polyelectrolytes to alternatively confer positive or negative surface charge to the nanorods. This procedure conveniently masks the cytotoxic CTAB layer and enables to cover the entire surface of the nanorods by the bioreceptor and not only the tips [130]. When coated by silica, the surface charge of nanoparticles becomes negative at physiological pH due to the low PZC of silica, allowing for an opposite electrostatic adsorption [134]. Antibodies have been physisorbed to negatively charged silica-coated AuNR [135]. Here, again, no study on the long-term stability of the resulting antibody-AuNR bioconjugates was performed.

4.2. Chemisorption

Another relatively straightforward strategy to conjugate bioreceptors to AuNR is to take advantage of the strong affinity of sulfur for gold, particularly in the form of thiols and disulfides. This approach has been employed to conjugate various thiol-containing bioreceptors including various thiol-terminated DNA oligomers [136–143], a cysteine-containing peptide [144] and a thiolated lactose derivative [145] to CTAB-capped AuNR in solution or immobilized on glass slides (Figure 16A). The anionic surfactant SDS can be added during chemisorption of aptamer to prevent NP aggregation [142].

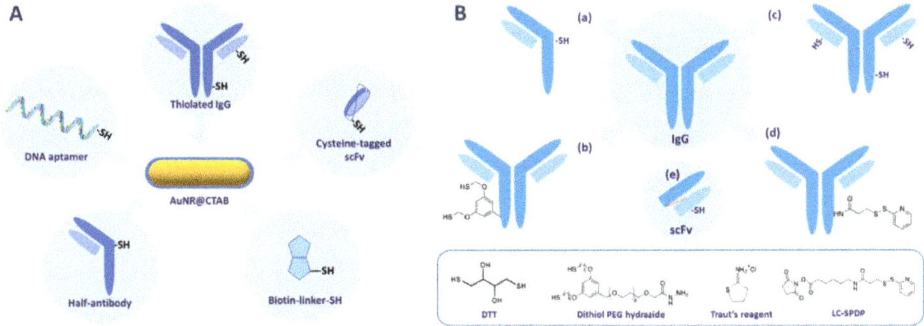

Figure 16. (**A**) Chemisorption of various thiol-containing biomolecules to CTAB-capped AuNR. (**B**) Introduction of thiol groups on IgG-type antibodies, (**a–d**), starting from an IgG using: (**a**) DTT (dithiotreitol); (**b**) NaIO$_4$ then dithiol PEG hydrazide; (**c**) Traut's reagent; (**d**) N-Succinimidyl-6-(3'-(2-PyridylDithio)-Propionamido)-hexanoate (LC-SPDP); and (**e**) scFv.

Since IgG-type antibodies do not naturally contain any free cysteine in their sequence, several pathways have been used to create thiol groups (Figure 16B) prior to chemisorption to AuNR.

Initially, the heterobifunctional cross-linker N-Succinimidyl-6-(3'-(2-PyridylDithio)-Propionamido)-hexanoate (LC-SPDP) enabled the introduction of reactive disulfide bonds on an IgG type antibody [121]. Later on, Traut's reagent (2-iminothiolane), a popular protein thiolation reagent was employed to introduce sulfhydryl groups by reaction of some of the amino groups of antibodies [146–148]. Occasionally, CTAB molecules capping AuNR were temporarily replaced by PVP and SDS [147]. Other strategies have been implemented to introduce sulfhydryl groups on antibodies in a site-selective fashion, away from the antigen binding site. For instance, selective reduction of disulfide bridges located at the hinge region of IgG using a mild reducing agent like DTT (dithiotreitol) afforded half-antibodies with 2 or 3 sulfhydryl groups [146,149]. Alternatively, the oligosaccharide residues located on the Fc domain of most IgG were mildly oxidized with NaIO$_4$ to generate aldehydes followed by reaction with a dithiol PEG hydrazide heterobifunctional cross-linker [146,150]. Finally, a single chain Fv (scFv) fragment engineered to include a single cysteine tag was genetically produced and readily chemisorbed onto AuNR immobilized on glass slides [151]. Generally, the conjugation step is followed by a blocking step, typically with mPEG-SH, 11-mercaptoundecanol, or 6-mercaptohexanol, to prevent further nonspecific binding.

Owing to the known strong binding of CTAB to the AuNR, the question of complete or incomplete displacement of CTAB molecules by competing thiolated bioreceptors in solution arises. Regarding the chemisorption of thiolated lactose, it was clearly shown to be incomplete since the zeta potential of the nanorods remained positive after reaction [145]. The same conclusion was drawn in the case of an iminothiolane-treated IgG [148] and a thiol-terminated DNA aptamer [152].

To make sure all physisorbed CTAB molecules are properly displaced during ligand exchange and/or to prevent particle aggregation during the functionalization process, or even to make sure that the whole nanorod surface is covered with bioreceptor (and not only the tips, see below), more sophisticated procedures were put in place to chemisorb thiol-terminated DNA oligomers onto CTAB-capped AuNR.

A "round-trip" phase transfer method was introduced to prepare a DNA-AuNR bioconjugate (Figure 17) [153]. It relies on the exchange of CTAB molecules covering AuNR with mercaptoalkylcarboxylic acids followed by a second exchange with thiol-terminated DNA oligomer. The addition of dodecanethiol (DDT) to a suspension of CTAB-capped AuNR followed by the addition of acetone resulted in extraction of the NP in the organic phase while the CTAB molecules remained in the aqueous phase. After removal of excess DTT by addition of toluene and methanol, the NP were taken up in toluene and the solution was heated to 70 or 95 °C in the presence of mercaptoalkylcarboxylic acid until aggregation occurred. After washing and deprotonation with isopropanol, the NP were taken back into an aqueous phase where they were fully soluble. The mercaptohexanoic acid-coated AuNR were used to prepare a DNA-AuNR conjugate by exchange of the MCA ligand with 5'-thiolated DNA oligomer. Progressive addition of salt was necessary to prevent particle aggregation during DNA adsorption for charge screening. The final bioconjugate displayed very good stability over time and a fluorescence assay gave an average of 28 DNA strands per NP.

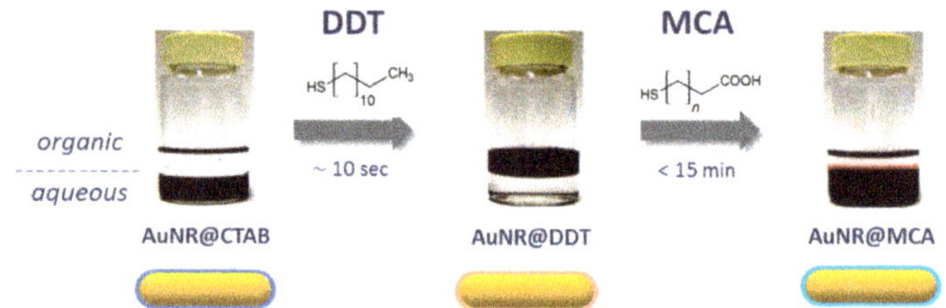

Figure 17. Round-trip" phase transfer method to prepare a DNA-AuNR bioconjugate [153].

Gates and coll. reported an alternative two-step procedure where the initial CTAB layer is temporarily replaced by loosely bound PVP and SDS, which are in turn replaced by thiolated-ssDNA applying salt screening as above. The number of DNA strands per NP depended on the initial DNA/AuNR ratio and reached saturation at a ratio of ≥25,600:1 with a maximum loading of 870 ± 60 DNA strands per NP [154].

Another strategy involved the chemisorption of mPEG-SH in the presence of the nonionic detergent Tween 20 followed by addition of thiolated DNA and citrate to screen the charge repulsion between AuNR and DNA. Tween 20 helped to displace CTAB molecules and further stabilized the nanoparticles against aggregation. It was found that the DNA/AuNR ratio was inversely related to the initial amounts of mPEG-SH and Tween 20 used at the intermediate step. An average of 200 DNA strands per particle was determined by fluorescence measurement [155].

4.3. Conjugation

Unlike simple chemisorption, and although it includes chemical bond formation, we refer to bioconjugation when the nanoparticles surface is modified and somehow tethered to attach the biomolecule. Two-step functionalization methods were also applied to attach various bioreceptors to AuNR and AuNR@SiO$_2$ as illustrated in Figure 18.

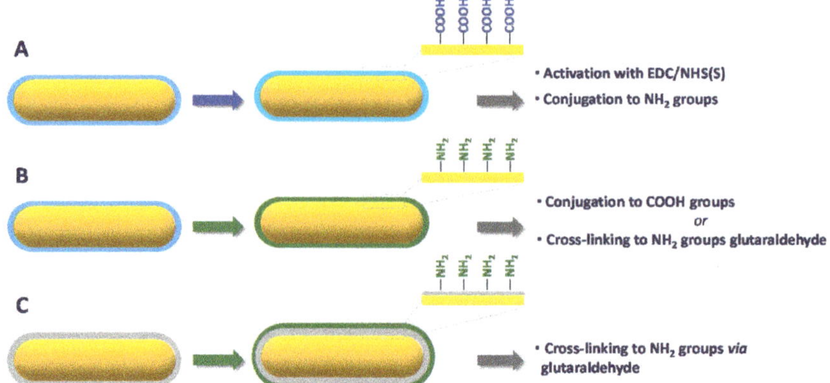

Figure 18. Covalent conjugation of bioreceptors to AuNR. (**A**) Grafting of carboxylic acids via ligand exchange; (**B**) grafting of primary amines via ligand exchange; (**C**) grafting of primary amines to silica-coated AuNR with (3-aminopropyl)trimethoxysilane (APTMS) or polyethylenimine (PEI).

The most popular method involves intermediate grafting of carboxyl functions at the surface of the nanoparticles, followed by coupling with reactive amine groups carried by the bioreceptor in the presence of a mixture of N-(3-Dimethylaminopropyl)-N′-ethylcarbodiimide (EDC) and N-Hydroxysuccinimide (NHS) (Figure 18A). Figure 19 depicts the structure of compounds used to introduce carboxyl groups on AuNR. A characteristic that have in common is to include a terminal thiol function to create strong Au-S bonds and a carboxylic acid function at the other end, typically MUA [20,156–159], MHDA [160,161], or mercaptoPEG acid derivatives [162–167]. This general strategy has been applied to decorate AuNR with bioreceptors both in solution or once deposited on solid substrates (essentially glass slides) using the mercaptocarboxylic acids alone or in mixture with other thiols like mPEG-SH or 11-mercaptoundecanol in various proportions. Occasionally, grafting of MUA was achieved by the "round-trip" phase transfer ligand exchange method [132,168,169] described in part 4.2. The disulfide derivative DTNB was also used to attach a peptide aptamer to AuNR via EDC/NHS coupling while providing a convenient Raman reporter for SERS biosensing [170].

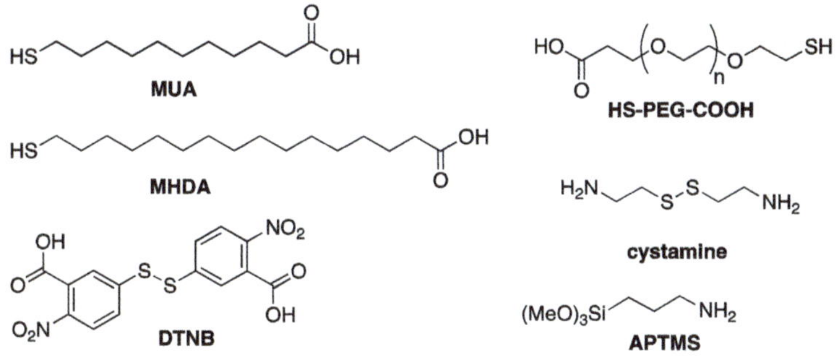

Figure 19. Bifunctional reagents for conjugation of bioreceptors to CTAB-capped AuNR and SiO$_2$@AuNR.

Another route involved the reaction of cystamine (Figure 18B) with CTAB-capped AuNR to introduce reactive amine groups at the surface of gold. This strategy was used to graft biotin [171] and folic acid [172] via EDC/NHS activation or antibodies via glutaraldehyde cross-linking [173,174]. A similar strategy was employed to attach antibodies to silica-coated AuNR using APTMS to graft amino groups to the surface instead of cystamine (Figure 18C) [175].

4.4. Selective Grafting at the Ends or the Sides of AuNR

Selective grafting strategies take advantage of the shape and anisotropic features of gold nanorods. On the one hand, they allow to generate high regular superstructures of nanoparticles thanks to ligand-bioreceptor recognition (see below [176]). Alternatively, selective grafting of the ends of nanorods is motivated by the heterogeneous distribution of the electric field enhancement at the gold nanorods, with enhancement being more pronounced at the nanoparticle tips as compared to the side thus allowing highly sensitive optical detection down to the single molecule level [177].

It had been noticed that the molecules of CTAB located at the ends of gold nanorods were relatively labile and therefore more easily exchangeable by thiols or disulfides [178]. This feature made it possible to selectively attach bioreceptors to the tips of AuNR. The site-selective chemisorption of thiolated biotin molecules was achieved at the tips of AuNR in solution [179]. Similarly, biotin molecules were also chemisorbed at the tips of AuNR immobilized on glass slides silanized by MPTMS. This was performed by exposure of UV/ozone cleaned, AuNR-functionalized slides to a solution of CTAB to form a dense bilayer at the surface of the nanoparticles, followed by treatment with a solution containing both thiolated biotin and CTAB [177,180]. The latter was necessary to ensure site-selective grafting at the tips [181]. Following this procedure, it was estimated that the amount of biotin molecules at the tips of the AuNR was seven times larger than on the side [180]. Along the same line, 5′- and 3′-thiolated DNA oligonucleotides were site-selectively chemisorbed to the tips of AuNR by displacement of CTAB [182,183].

Yu and Irudayaraj used the same property to selectively attach IgG F_{ab} to the ends of AuNR through preliminary chemisorption of MUA followed by coupling via EDC/NHS. By choosing an initial F_{ab}-to-AuNR ratio of 2, an average of ca. 1 F_{ab} per AuNR was grafted regardless of the AR [184].

Kotov and coll. reported two original conjugation strategies to graft proteins at the tips or on the side of the gold nanorods [176]. Grafting at the tips was ensured by ligand exchange followed by covalent attachment of target proteins. In practice, carboxyl groups were predominantly introduced at the tips of AuNR by reaction of thioctic acid followed by MC-LR-OVA antigen or anti-MC-LR Ab conjugation by EDC/NHS coupling (Figure 20A). This procedure afforded a nanoprobe with an Ab/AuNR ratio of ca. 10. Such a site-selective functionalization had previously been reported by Chang et al. to selectively graft anti-mouse IgG antibody to the tips of CTAB-capped AuNR [185]. The authors also pointed out that preferential binding of thioctic acid was due to its cyclic thus rigid structure. A similar procedure was employed by Xu and coll. to site-selectively anchor MC-LR-OVA antigen or anti-MC-LR Ab to the tips of AuNR except that thioctic acid was conjugated to both proteins prior to chemisorption (Figure 20B) [186].

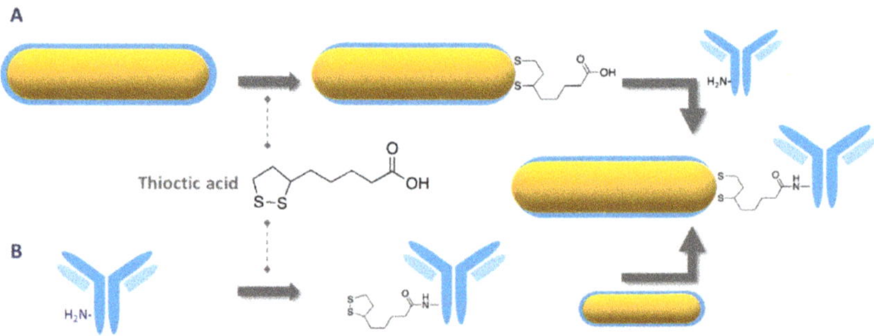

Figure 20. Tip-selective covalent conjugation bioreceptor to CTAB-capped AuNR. (**A**) Grafting of thioctic acid to AuNR followed by conjugation via N-(3-Dimethylaminopropyl)-N′-ethylcarbodiimide (EDC)/N-Hydroxysuccinimide (NHS); (**B**) coupling of thioctic acid to antibody via EDC/NHS followed by grafting to AuNR.

On the other hand, selective grafting on the side of the rods was achieved by electrostatic binding of proteins, taking advantage of the larger area of contact affording stronger electrostatic interactions [176]. In this case, the Ab/AuNR ratio was estimated to be 31. When the two nanoprobes resulting from electrostatic binding of MC-LR-OVA antigen or anti-MC-LR Ab were mixed together, the rods assembled to predominantly form ladders (side-by-side assemblies; figure). Conversely, when the two nanoprobes resulting from covalent binding of MC-LR-OVA antigen or anti-MC-LR Ab were mixed together, the rods assembled to form strings (end-to-end assemblies; Figure 21). Interestingly both these assemblies were stable in the long term. The same strategy was used by Wang and coworkers to prepare two nanoprobes whose side was coated by gentamicin-OVA antigen and anti-gentamicin antibody [187].

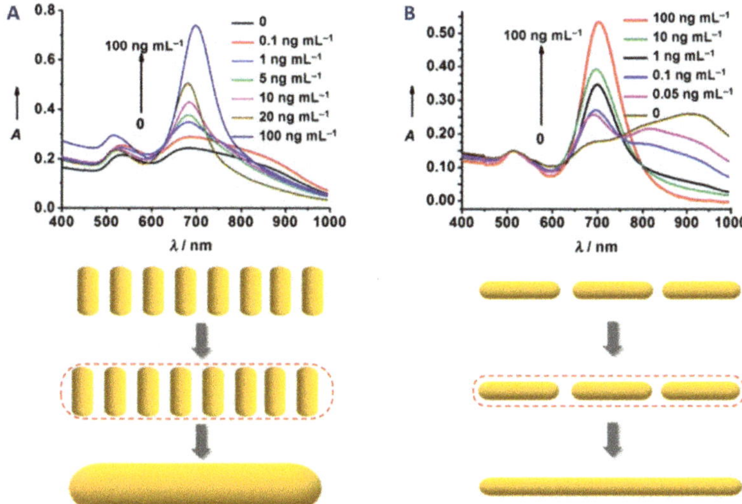

Figure 21. Evolution of surface plasmon resonance spectra of the nanorods upon increasing concentrations of microcystin-LR (indicated in the graph) and graphical representation of a plasmon system and corresponding nanowire approximation for side-to-side (**A**) and end-to-end (**B**) assemblies, from ref. [176].

Finally, a recombinant protein resulting from the fusion of staphylococcal protein A and gold binding polypeptide (GBP-SpA) was predominantly chemisorbed at the tips of CTAB-capped AuNR. Most importantly, activated charcoal was added to the reaction mixture in order to absorb released CTAB molecules and prevent particle aggregation. FT-IR spectroscopy analysis of the bioconjugate revealed the presence of remaining CTAB molecules on the nanoparticles [188].

5. Applications of AuNRs in LSPR Biosensing

An important asset of LSPR biosensors is the simplicity of the readout as the measurements can be done with a benchtop UV-visible spectrophotometer [189], an equipped smartphone [190], and sometimes even simply by naked-eye readout [12,191]. Gold nanorods plasmonic biosensors operate following four main mechanisms summarized in Figure 22.

In what follows, we recap the optical properties of AuNR in relation to LSPR biosensing, then survey the applications classified according to the biorecognition element. Most of the papers we review in this section deal with gold nanorods without silica coating. Indeed, up to now, only two examples of biosensors based on AuNR@SiO$_2$ have been reported in the literature [135,175]. Both of them involve antibodies as bioreceptors and in one of the cases, it was demonstrated that the presence of a thin SiO$_2$ shell not only provided a protective layer and increased the shelf life of the particles, but also improved the overall efficiency of the biosensor [135].

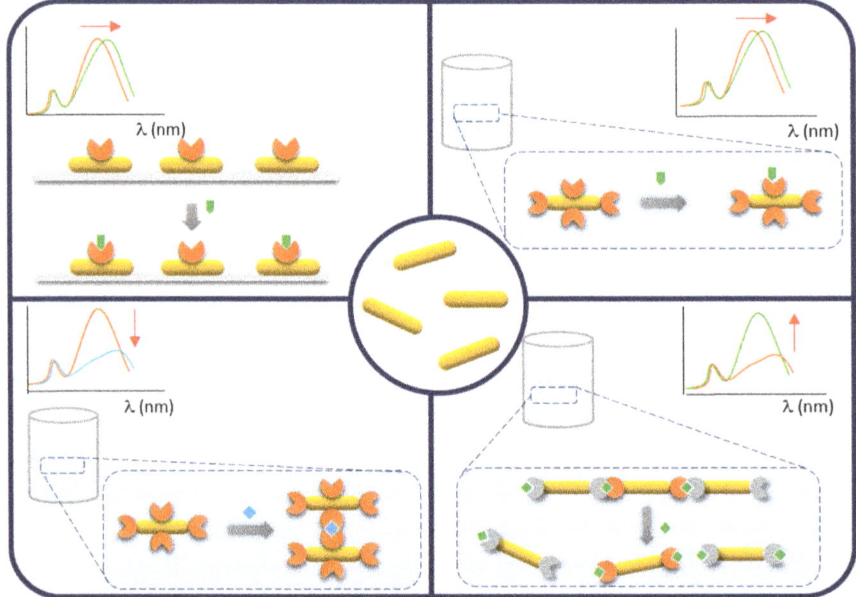

Figure 22. Summary of the operating configurations for Gold Nanorods nanoplasmonic biosensors.

5.1. Optical Properties of AuNR in Relation to LSPR Biosensor Development

Gold nanorods possess unique optical properties owing to the localized surface plasmon resonance phenomenon [103]. They exhibit anisotropic plasmonic responses translated into longitudinal and transverse plasmon modes with extremely high extinction coefficient giving rise to bright colours from purple to brown owing to light absorption and scattering processes (Figure 13A). The position of the longitudinal plasmon band can easily be tuned from the visible to the near IR spectral range just by changing the AR (Figure 13B, [14]). Another extremely useful property of AuNR regarding LSPR biosensor development is certainly the sensitivity of the longitudinal plasmon wavelength to the refractive index of the surrounding medium.

The plasmon band shift $\Delta\lambda$ is governed by Equation (1),

$$\Delta\lambda = m(n_{adsorbate} - n_{medium}) \times \left\{ 1 - e^{\left(-\frac{2d}{l_d}\right)} \right\} \tag{1}$$

where n is the refractive index, l_d is the decay length of the electric field, d is the thickness of the adsorbate layer, and m is the intrinsic RI sensitivity factor.

The parameter m can be experimentally determined by measuring the extinction spectrum of AuNR in aqueous solutions containing increasing concentrations of sucrose or glycerol. It also depends on the size and shape of the gold nanoparticles [192].

Another important property is the exponential decrease in electrical field enhancement making the plasmon band shift highly distance dependent (parameter l_d in Equation (1)). This feature enables separation-free assays since only the events occurring in the neighborhood of the particles will contribute to the plasmon band shift. The electromagnetic decay length l_d, defined as the distance from the nanorod surface at which the electric field enhancement is reduced by a factor of e [193], has been experimentally determined for AuNR of various sizes using a LbL approach [194]. The strategy adopted in this study consists in treating AuNR-coated glass slides with polycations and polyanions alternatively to create an LbL structure of increasing thickness while monitoring the position of the LSPR band (Figure 23). This enlightening study revealed that l_d is linearly correlated to both diameter and length of the particles, but the l_d/diameter slope is 5 times larger than the l_d/length slope. For instance, an AuNR of 24×50 nm had a decay length l_d of 30 nm. This value should be taken into account when designing an LSPR biosensor for a given target. On the whole, increasing AR results in red shifts in λ_{max}, higher m, and longer l_d [195].

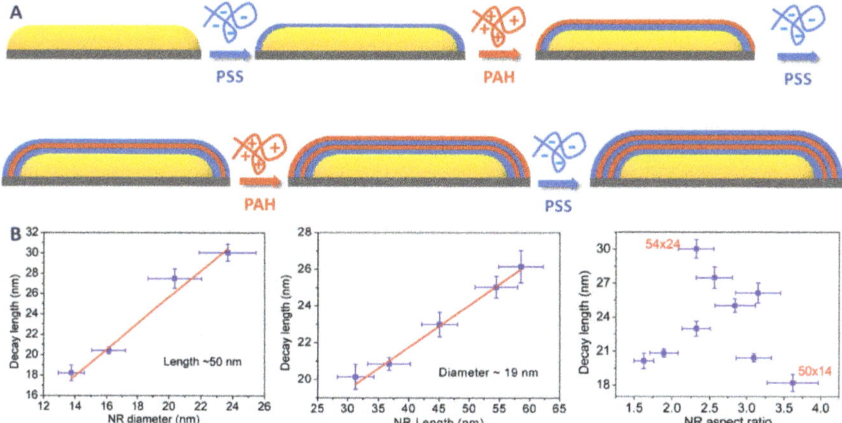

Figure 23. (**A**) Schematic representation of AuNR on a glass substrate covered with polyelectrolyte multilayers by the LBL approach. (**B**) Electromagnetic decay lengths of AuNR with different diameters, lengths and AR of AuNR, adapted from ref [194].

Characterization of the sensing capabilities of metal nanostructures is best described by the figure of merit (*FoM*), which is calculated according to Equation (2) where FWHM is the full width at half maximum [192]:

$$FoM = \frac{m}{FWHM} \tag{2}$$

The survey of the literature shows that LSPR biosensors including gold nanorods can be roughly classified into two categories, either solution phase- or solid phase-based configurations. They can also be classified according to the type of bioreceptor responsible for analyte recognition and capture. We will successively survey all these configurations.

5.2. Immunosensors

Immunosensors are the class of biosensors using an antibody as bioreceptor. Antibodies are essentially immunoglobulins G produced in various animal species to bind antigens with high affinity and specificity. IgGs are glycoproteins with a molecular weight of 150 kDa, a Y-shaped 3D structure, and an approximate size of $12 \times 14 \times 7$ nm, comprising two antigen binding sites located at the two extremities of the F_{ab} domains. Their association to spherical nanoparticles has been widely studied [196]. AuNR-antibody bioconjugates operate either in homogeneous solution-phase or in solid-phase when deposited on planar substrates as discussed in what follows.

5.2.1. Solution-Phase Based Immunosensors

In solution, the optical response of an AuNR nanoimmunoprobe to a given analyte depends on (1) the type of antibody and (2) the size of the analyte. For high molecular weight analytes such as proteins having more than one epitope, the use of a polyclonal antibody as bioreceptor results in nanoparticles aggregation, which translates into a decrease of the LSPR band intensity (Figure 24A), whereas a monoclonal antibody will yield a red shift of the LSPR band as a result of change of the local refractive index due to analyte binding (Figure 24B). LSPR biosensors for low molecular weight analytes having only one epitope are based on the assembly of antibody and antigen nanoprobes forming aggregates. Addition of analyte leads to a progressive disruption of the nanoparticle network, which translates into an increase of the LSPR band (Figure 24C).

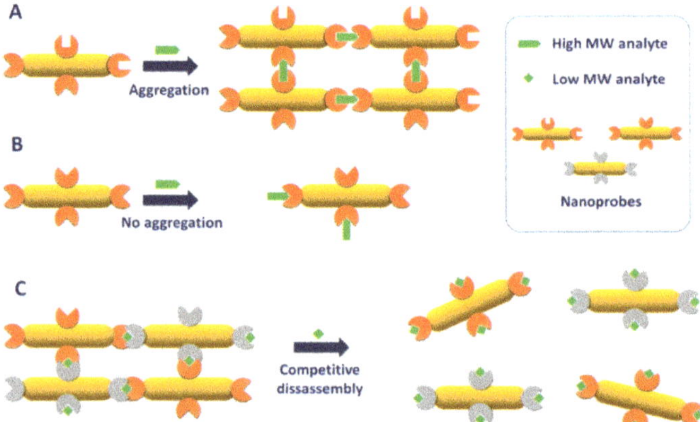

Figure 24. Immunosensor configurations. (**A**) Polyclonal antibody and high molecular weight analyte; (**B**) monoclonal antibody and high molecular weight analyte; (**C**) antibody and low molecular weight analyte.

Solution phase immunosensors have been applied to various analytes (Table 4) giving rise to biologically meaningful limits of detection. Signal enhancement was achieved by optimizing the distance between the analyte and the surface of the nanorods. For instance, Tang et al. showed that covalent binding of antibody to AuNR via conjugation to a self-assembled monolayer (SAM) of MUA significantly improved the sensitivity as compared to physically bound Ab via coating to PSS layer on CTAB [132]. This was rationalized by the shorter distance between the analyte and the rod surface that places it within the sensing volume. The same group also used magnetic nanoparticle (MNP) immunoprobe in conjunction to AuNR immunoprobe to both selectively extract cTnI and enhance the LSPR signal owing to the high RI of MNP [168]. Another way to enhance sensor response was to coat the nanorods with a thin layer of silica before grafting the antibody [135]. This resulted in an increase of RI sensitivity (parameter m in Equation (1)) and in turn to an extremely low LoD in the detection of the food pathogen *E. coli O157:H7*.

Table 4. Strategies, analytes, and analytical performances for solution-phase based immunosensors.

Analyte	Format	Analytical Performances	Ref.
hIgG	Direct; aggregation	LoD = 60 ng/mL (0.4 nM)	[130]
Goat anti-hIgG	direct	LoD = 0.4 nM DR = 0.4–100 nM	[156]
HBsAg	Direct	LoD = 0.01 IU/mL DR = 0.01–1 IU/mL	[128]
cTnI	Direct; aggregation	LoD = 10 ng/mL DR = 1–200 ng/mL	[131]
	Sandwich with MNP-Ab	LoD = 1 ng/mL DR = 1–20 ng/mL	[168]
	Direct	LoD = 1 ng/mL DR = 1–20 ng/mL	[132]
CRP	Direct	LoD = 6.2 nM DR = 10–100 nM	[188]
E. coli O157:H7	Direct	LoD = 10 CFU Linear response to 5×10^4 CFU	[135]
MC-LR	Competitive disassembly	LoD = 0.03 or 0.6 ng/mL DR = 0.05–1 ng/mL or 1–100 ng/mL	[176]
Aflatoxin B1	Competitive disassembly	LoD = 0.16 ng/mL DR = 0.5–20 ng/mL	[157]
Gentamicin	Competitive disassembly	LoD = 0.05 ng/mL DR = 0.1–20 ng/mL	[187]
E. coli O157:H7 *S. typhimurium*	Direct; aggregation; multiplex	DR = $10–10^8$ CFU/mL	[173]
Mb cTnI	Direct; multiplex	DR (Mb) = 25–250 ng/mL DR (cTnI) = 1–10 ng/mL	[169]

LoD = limit of detection; DR = detection range.

It has been shown that the LSPR associated electric field is anisotropically distributed, being largest at the ends of the rods with respect to the sides [13,180]. This feature was exploited to build up an immunosensor of the environmental toxin MC-LR [176]. For this toxin, having only one epitope, the competitive disassembly format was chosen and two pairs of nanoprobes were synthesized for which the binding partners (antigen and antibody) were grafted at the sides or on the ends of AuNR. In the absence of competitive analyte, the pairs of nanoprobes were assembled into ladder (sides) or chain (tips) patterns (see Figure 21), which were progressively disrupted upon addition of MC-LR. Remarkably, the sensor configuration based on end-to-end nanoparticle assembly displayed much better analytical performances.

As mentioned above, one of the most remarkable properties of AuNR is the ability to tune the position of the l-LSPR wavelength across a large range of the spectrum, i.e., from ca. 600 to 1000 nm by changing the AR of the rods. This property was exploited to set up multiplex assays, where each analyte is associated with a gold nanorod of a given AR so that overlap between the longitudinal band of each rod is minimal. In this way, a multiplex biosensor was reported for the simultaneous detection of two pathogenic bacteria [173] and two myocardial infarction biomarkers [169].

5.2.2. Solid-Phase-Based Immunosensors

Solid phase AuNR-based immunosensors present several benefits with respect to solution-based ones. First they generally give higher RI sensitivity and FoM [159]; second, they enable real-time monitoring of binding events and therefore determination of kinetic and equilibrium constants [161]; third, they are easier to handle. Moreover, since the nanorods are firmly attached to their substrate, the formation of antibody–analyte pairs always result in a red shift of the LSPR band whose magnitude is dependent on analyte concentration until saturation occurs at full occupation of the binding sites. In the following, we will focus on bottom-up processes to prepare AuNR-based chips. However, it is important to note that several assays were performed with top-down generated LSPR-substrates [197], but as these platforms do not include colloidal particles in their engineering, they are out of the topic of this review and will not be included in the discussion below.

Construction of solid phase AuNR-based biosensors requires the immobilization of AuNR onto transparent substrates generally made of silica (glass slide). Adhesion of CTAB-capped AuNR to glass was performed in different ways either by (1) physisorption via electrostatic interaction to negatively charged piranha- (+ O_2 plasma) or NaOH-treated substrates or (2) covalent bonding by silanization with APTMS or MPTMS to introduce primary amines or thiols, respectively, resulting in the formation of Au-N or Au-S bonds. Alternatively, thiol groups have been introduced by sequential reaction of APTMS, succinic anhydride, EDC/NHS, and cysteamine [151]. Immobilization of AuNR@SiO$_2$ on quartz slides was achieved by coating of the substrates with PVP [175]. Grafting of the antibody to the AuNR is then performed by one of the methods described in part 4. Occasionally, the reverse process by which the antibody is first grafted onto the AuNR in solution then the conjugate is immobilized onto glass slides treated with APTES and glutaraldehyde was also performed [164].

In addition to the model analyte human IgG (Table 5), this immunosensor configuration was applied to the quantitation of various meaningful biomarkers, namely ALCAM (cancer) [164], CRP (inflammation) [151], and cTnI (myocardial infarct) [144]. In the case of CRP, the use of scFv anti-CRP (30 kDa) as a bioreceptor instead of whole IgG (150 kDa) significantly improved the biosensor response [151]. The mode of grafting of antibody to the AuNR had a significant effect on the performances of the resulting biosensor [148]. Covalent grafting of thiolated anti-human IgG antibody afforded a more sensitive nanoimmunoprobe than electrostatic adsorption to CTAB-capped AuNR sequentially treated with PSS and PAH. In the former configuration, the antibody is preferentially located at the tips of the nanorods where the electric field enhancement is the highest and is comparatively closer to the gold surface. In the latter configuration, the presence of the thick multilayer polymer creates a larger distance between the bioreceptor and the nanorod surface with uniform distribution all over the rods.

Table 5. Strategies, analytes, and analytical performances for solid-phase-based immunosensors.

Analyte	Format	Analytical Performances	Ref.
Human IgG	Direct	LoD = 61 pM DR = 33–233 nM	[133]
	Direct	DR = 10–40 nM	[146,148]
	Direct; visual detection	LoD = 1 ng/mL DR = 1–10 ng/mL	[175]
ALCAM	Direct	LoD = 15 pM DR = 0.05–30 nM	[164]
CRP	Direct	DR = 1–10 ng/mL	[151]
cTnI	Direct *	LoD = 353 pg/mL	[144]

* solid phase = filter paper.

5.3. Aptasensors

Aptasensors are the class of biosensors using an aptamer as bioreceptor [198]. Aptamers are short synthetic nucleic acid (DNA, RNA) or peptide sequences binding to a specific target with high affinity (comparable to that of antibodies). In addition to their target versatility (ranging from small molecules or ions to whole bacteria and viruses), aptamers are particularly well suited as biorecognition elements of LSPR biosensors owing to their small size enabling the binding events to occur at close range of the gold surface. Moreover, binding of small molecule targets to aptamers usually induces a dramatic change of their conformation (structural switch), which may result in a change of the refractive index at the vicinity of the AuNR and in turn convert into a detectable shift of its LSPR band. Such a direct transduction scheme is hardly achievable with antibody bioreceptors that require a competitive format to enable small molecule detection. Table 6 gathers selected AuNRs-based aptasensors.

Table 6. Strategies, analytes, and analytical performances for aptasensors.

Analyte	Format	Analytical Performances	Ref.
Ochratoxin A (OTA)	Direct; glass slide	LoD = 1 nM DR = 0.1 nM–10 μM	[199]
	Direct; optical fiber	LoD = 12 pM DR = 10 pM–100 nM	[137]
	Direct; glass slide	DR = 10 pM–10 μM LoD = 0.56 pM	[139]
	Solution; competitive disassembly	LoD = 0.54 nM DR = 1.2–25 nM	[183]
Aflatoxin B1	Direct; glass slide	DR = 10 pM–10 μM LoD = 0.63 pM	[139]
ATP	Direct; glass slide	DR = 10 pM–10 μM LoD = 0.87 pM	[139]
	Direct; glass slide	DR = 10 pM–10 μM	[140]
MCF-7 cancer cells (mucin-1)	Direct; Cells		[136]
Cytochrome c (apoptosis marker)	Sandwich with MNP-Ab for capture; solution; aggregation	LoD = 0.1 ng/mL	[163]
cTnI	Direct; filter paper	LoD = 35 pg/mL DR = 35 pg/mL–3.5 μg/mL	[144]

A solid-phase aptasensor was built up for the detection of ochratoxin A (OTA), a mycotoxin produced by some *Aspergillus* species and a potential contaminant of foodstuff. Its principle relies on a change of aptamer folding from random and coiled conformation to G-quadruplex structure upon binding of OTA. This and the higher refractive index of OTA induce a red shift of the l-LSPR band proportional to the concentration of analyte in the nanomolar range [199]. Higher sensitivity and wider dynamic range were achieved by co-immobilization of thiol-terminated OTA aptamer and T_3 onto the AuNR and addition of the G-quadruplex binder berberin for signal enhancement [139]. The same group developed another OTA biosensor with AuNR immobilized on an optical fiber using the same surface chemistry combined with an extremely high RI sensitivity (601 nm/RIU) leading to an LoD of 12 pM [137]. An ATP biosensor was developed by the same group following a slightly different design using AuNR coated with a mixture of T_3 and split ATP-specific aptamer and a DNA oligomer comprising the other half of the ATP aptamer flanked on both sides by TAMRA fluorophores. Owing to its ability to absorb the visible light, the TAMRA chromophore can strongly couple with the LSPR of nanoparticles to produce large plasmon band shift [195]. Addition of the target triggered the hybridization of the two DNA strands, bringing the two TAMRA entities close to the gold surface resulting in a large shift of the LSPR band. This configuration noticeably enhanced the sensor response in comparison to the simple immobilization of the full ATP aptamer on the AuNR [140].

Another group set up a different biosensor configuration using two AuNR nanoprobes coated by short DNA sequences and the OTA specific DNA aptamer flanked on both sides by sequences complementary to the nanoprobes. In the absence of target, the three of them form sandwiches leading to nanoparticle aggregation. In the presence of target, dissociation occurred because of preferential binding of OTA to its aptamer that in turn led to blue shift and intensity increase of the l-LSPR band [183].

The release of the apoptosis biomarker cytochrome C upon exposure of cancer cells to the chemotherapeutic agent phenylarsine oxide was measured using a biosensor comprising AuNR coated with specific aptamer as transducer and MNP-Ab for target capture. In a microfluidic cell, the target is successively flown over the MNP-Ab while applying a magnetic field. With the magnetic field off, injection of aptamer@AuNR gave rise to the formation of aggregates resulting in the change of the UV-vis spectrum [163]. An LSPR biosensor was set up to detect the presence of the biomarker mucin-1 at the membrane of certain cancer cells using AuNR coated with a specific aptamer and direct reading by UV-vis spectroscopy [136]. An AuNR-based biosensor was designed to assay the cardiac biomarker cTnI with high sensitivity [144]. It combines a short and highly affine peptide aptamer as bioreceptor and filter paper as a solid support. The short size of the peptide recognition element with respect to the more usual antibody enabled to improve the biosensor LoD by one order of magnitude. The l-LSPR band shift undergoes a distance-dependent decay up to 20 nm away from the gold surface as determined by an LbL experiment. This, and the probably higher density of peptide bioreceptors at the surface of the nanorods, explains why the peptide-based biosensor was more sensitive than the corresponding immunosensor configuration.

5.4. AuNR-Based LSPR Biosensors Using Uncommon Receptors

Although antibodies and aptamers are the most commonly used bioreceptors for AuNR-LSPR biosensors, several less common biomolecules were also employed as summarized in Table 7. AuNR-based biosensors using sugar bioreceptors were developed to assay food allergens [162] or the cancer biomarker galectin-1 [145]. AuNR-based genosensors were reported for the detection of pathogens [141,143]. A peptide nucleic acid (PNA) biosensor was designed for the detection of circulating tumor DNA (ctDNA) point mutation on the KRAS gene in relation to various cancers, including pancreatic cancer [200]. The sequence of the capture PNA probe was carefully designed to preferentially bind to the mutant sequence and less to the wild-type one. Folic acid-coated AuNR were synthesized and shown to bind to HeLa cancer cells overexpressing the folate receptor [172].

Table 7. Strategies, analytes, bioreceptors, and analytical performances for uncommon LSPR-biosensors.

Analyte	Bioreceptor	Format	Analytical Performances	Ref.
SAV	biotin	Direct; glass slide	LoD = 94 pM (5 ng/mL) DR = 2–2000 nM	[160]
			LoD = 25 ng/mL DR = 25–4000 ng/mL	[171]
			DR = 10–100 nM	[181]
Concanavalin A Peanut agglutinin	4-aminophenyl α-D-mannopyranoside 4-aminophenyl b-D-galactopyranoside	Solution; aggregation		[162]
Galectin-1	lactose	Solution; aggregation	DR = 0.1–100 pM LoD = 0.1 pM	[145]
16S rDNA *Serratia marcenscens*	DNA	Sandwich assay; aggregation	DR = 10 pM–10 nM LoD = 5 pM	[141]
ctDNA (KRAS gene mutation)	PNA	Direct	LoD = 2 ng/mL DR = 40–125 ng/mL	[200]
Chlamydia. trachomatis DNA	DNA	Sandwich assay; aggregation	DR = 0.25–20 nM	[143]
Folate receptor	Folic acid	Direct	DR = 100–5000 HeLa cells/mL LoD = 10 cells/mL	[172]

5.5. Single Molecule Plasmonic Biosensors

Since the first reports of the use of single plasmonic nanoparticles as independent unit for biosensing [201,202], considerable progress has been made in this direction in the last decade, especially using AuNR as shown in Table 8. The measurements are often achieved by combining a dark field microscope coupled to a microspectroscopy system. The advantages of these single sensors are unlimited especially regarding sensitivity (close to a single molecule) and miniaturization as a minute amount of analyte are needed and very small coverage of nanoparticles required; interparticle distance is important for dark-field microscope imaging. This format is extremely well suited to AuNRs as their LSPR band is often broader than spherical nanoparticles, and, as widely discussed in this manuscript, this band is very sensitive to the AR, which distribution has been largely improved and mastered during the last years, but nevertheless still suffers from variability. Thus, imaging a single AuNR widens the perspectives in terms of sensitivity while narrowing the noise caused by the unequal distribution of the colloids.

Table 8 summarizes the strategies, analytes, bioreceptors, and analytical performances for single molecule plasmonic biosensors. The applications range from the biosensing of model systems such as Biotin/Streptavidin reaching an LoD as low as 1 nM [203] to relevant bioanalytical targets with LoD down to the aM for PSA detection by an aptamer [142].

Miniaturization is one of the main advantages of single biomolecule biosensing that often include a microfluidic device, sometimes allowing for multiplex detection [186]. We expect an increasing attention to these systems with the 3D printing progress.

Table 8. Strategies, analytes, bioreceptors, and analytical performances for single molecule plasmonic biosensors.

Analyte	Bioreceptor	Format	Analytical Performances	Ref.
SAV	Biotin	Direct	LoD = 1 nM	[203]
Thrombin	Aptamer	Direct	LoD = 10 ng/mL (0.28 nM) DR = 10 ng/mL–100 µg/mL	[204]
			LoD = 0.6 ng/mL (17 pM)	[205]
		Sandwich with Ab	LoD = 1.6 pM DR = 1 ng/mL–10 µg/mL	[206]
NGAL	Ab	Direct	LoD = 8.5 ng/mL (340 pM) DR = 10 ng/mL–1 µg/mL	[165]
PSA	Ab	Direct	DR = 0.1 fM–1 nM LoD = 0.11 fM	[207]
			DR = 1 aM–0.1 nM LoD = 1 aM	[166]
PSA Thrombin IgE	Aptamer	Direct or sandwich with Ab; multiplex; 9-spot array	LoD = 1 ng/mL	[208]
Fibronectin SAV Thrombin IgE	Aptamer	Direct; multiplex	LoD (SAV) = 1 nM DR (SAV) = 1–30 nM	[142]
FtsZ	s1ZipA s2ZipA MinC	Direct; multiplex	DR = 0.2–100 µM	[209]

6. Conclusions and Perspectives

All throughout this manuscript, we have reviewed the recent findings related to AuNRs for LSPR biosensing from their synthesis and coating by silica, to their characterization and further surface functionalization, and up to their bioanalytical applications. AuNR synthesis has been extensively investigated over the last decades; therefore, in the first part of the manuscript, we relied on the existing reviews and updated with the most recent literature on the topic. Then, in a second part of the synthesis section, we discussed the methods applied for AuNRs' coating by a layer of silica sufficiently thin for their potential use as LSPR-biosensors. Although we intentionally highlighted in this section the problems encountered in terms of reproducibility, upscaling, and stability; our main message is that both AuNRs synthesis and their subsequent coating by silica have reached a level of maturity allowing for their extensive use in LSPR biosensing. In the second section of this review, we summarized the experimental techniques allowing for AuNRs characterization at both the microscopic and macroscopic levels. The shape and size, together with the surface charge and the gold content, are measurable data, sometimes applicable for silica shell characterization when needed. In the third part, we comprehensively covered the strategies and methods applied for AuNRs' functionalization to attach the bioreceptors for their further use as LSPR biosensors. We intended in this part to provide the reader with guidance in choosing the adapted protocol to successively attach the bioreceptor and achieve the desired AuNRs bioconjugate. In the last part, we surveyed the applications of AuNRs in relation to LSPR biosensing classified according to the biorecognition element. The superior optical properties of AuNRs make them unmatchable in terms of efficiency and sensitivity for LSPR biosensing. Their analytical performances are extremely competitive regardless of the used bioreceptor. However, it is important to note that most of the reviewed applications related to LSPR biosensing, with the exception of two, did not include silica coating even if, in the mentioned exceptions, it was demonstrated that the silica shell further improved the overall efficiency of the biosensor. The progress made in the synthesis of AuNR@SiO$_2$ highlighted herein promises a booming expansion in their use for LSPR biosensing.

Author Contributions: The manuscript was written through contributions of all authors. All authors have given approval to the final version of the manuscript.

Funding: This research received no external funding.

Acknowledgments: This work was also supported by the ANR grant ChirOptMol (No. ANR-18-CE09-0010) and by the ANR UpPhotoCat (No. ANR- 16-CE07-0031-01).

Conflicts of Interest: The authors declare no conflict of interest.

Abbreviations

AA	L-ascorbic acid
Ab	Antibody
AD	Dehydroascorbic acid
ALCAM	activated leukocyte cell adhesion molecule
APTMS	(3-aminopropyl)trimethoxysilane
AR	aspect ratio
Asc.$^-$	radical ascorbate
Asc$^-$	L-ascorbate
ATP	Adenosine triphosphate
AuNR	Gold nanorod
AuNR@SiO$_2$	silica coated gold nanorod
BDAC	Benzyldimethylhexadecylammonium chloride
CMC	critical micellar concentration
CRP	C-reactive protein
CTAB	cetyltrimethylammonium bromide
CTAC	cetyltrimethylammonium chloride
ctDNA	circulating tumor DNA
cTnI	cardiac troponin I
DA	dehydroascorbic acid
DDLS	depolarised dynamic light scattering
DDT	dodecanethiol
DLS	dynamic light scattering
DNA	Deoxyribonucleic acid
DR	detection range
DTT	Dithiothreitol
EDC	N-(3-Dimethylaminopropyl)-N′-ethylcarbodiimide
FoM	figure of merit
FWHM	full width at half maximum
GBP-SpA	gold-binding polypeptide Staphylococcal Protein A
HBsAg	Hepatitis B surface antigen
hIgG	human Immunoglobulin G
ICP-MS	inductively coupled plasma mass spectroscopy
ICP-OES	inductively coupled plasma optical emission spectrometry
IGEPAL	octylphenoxypolyethoxyethanol
IgG	Immunoglibulin G
KRAS	Kirsten RAt Sarcoma virus
LbL	layer-by-layer
LC-SPDP	N-Succinimidyl-6-(3′-(2-PyridylDithio)-Propionamido)-hexanoate
l-LSPR	longitudinal localized surface plasmon resonance
LoD	Limit of detection

LSPR	localized surface plasmon resonance
MCF-7	Michigan Cancer Foundation-7
MC-LR-OVA	microcystin-LR *ovalbumin*
MHA	6-Mercaptohexanoic acid
MHDA	16-Mercaptohexadecanoic acid
MNP	magnetic nanoparticle
mPEG-SH	(O-[2-(3-mercaptopropionylamino)ethyl] O'-methylpolyethylene glycol
MPTMS	(3-mercaptopropyl)trimethoxysilane
NaOL	Sodium oleate
NGAL	Neutrophil Gelatinase-Associated Lipocalin
NHS	N-Hydroxysuccinimide
NP	Nanoparticle
OTA	Ochratoxin A
PAH	polyallylamine chloride
PEI	Polyethylenimine
PNA	peptide nucleic acid
PSS	polystyrene sulfonate
PVP	Polyvinyl pyrrolidone
PZC	point of zero charge
RI	Refractive index
RNA	Ribonucleic acid
SAM	Self-Assembled Monolayer
SAV	Streptavidin
SAXS	small angle X-ray scattering
scFv	Single-Chain Fragment Variable
SDS	Sodium dodecyl sulfate
SERS	Surface-enhanced Raman spectroscopy
SHE	Standard hydrogen electrode
SPR	Surface Plasmon resonance
ssDNA	single-stranded DNA
TAMRA	5-Carboxytetramethylrhodamine
TEA	triethylamine
TEM	Transmission Electron Microscopy
TEOS	Tetraethyl orthosilicate
t-LSPR	transverse localized surface plasmon resonance
UTSC	ultra-thin silica shell
XANES	X-ray absorption near edge spectroscopy

References

1. Faraday, M.X. The Bakerian Lecture. —Experimental relations of gold (and other metals) to light. *Philos. Trans. R. Soc. Lond.* **1857**, *147*, 145–181. [CrossRef]
2. Mie, G. Beiträge zur Optik trüber Medien, speziell kolloidaler Metallösungen. *Ann. Phys.* **1908**, *330*, 377–445. [CrossRef]
3. Lee, J.-H.; Cho, H.-Y.; Choi, H.K.; Lee, J.-Y.; Choi, J.-W. Application of Gold Nanoparticle to Plasmonic Biosensors. *Int. J. Mol. Sci.* **2018**, *19*, 2021. [CrossRef] [PubMed]
4. Liu, J.; Jalali, M.; Mahshid, S.; Wachsmann-Hogiu, S. Are plasmonic optical biosensors ready for use in point-of-need applications? *Analyst* **2020**, *145*, 364–384. [CrossRef] [PubMed]
5. Carter, T.; Mulholland, P.; Chester, K. Antibody-targeted nanoparticles for cancer treatment. *Immunotherapy* **2016**, *8*, 941–958. [CrossRef]
6. Biju, V. Chemical modifications and bioconjugate reactions of nanomaterials for sensing, imaging, drug delivery and therapy. *Chem. Soc. Rev.* **2014**, *43*, 744–764. [CrossRef]
7. Englebienne, P. Use of colloidal gold surface plasmon resonance peak shift to infer affinity constants from the interactions between protein antigens and antibodies specific for single or multiple epitopes. *Analyst* **1998**, *123*, 1599–1603. [CrossRef]

8. Saha, K.; Agasti, S.S.; Kim, C.; Li, X.; Rotello, V.M. Gold Nanoparticles in Chemical and Biological Sensing. *Chem. Rev.* **2012**, *112*, 2739–2779. [CrossRef]

9. Sepúlveda, B.; Angelomé, P.C.; Lechuga, L.M.; Liz-Marzán, L.M. LSPR-based nanobiosensors. *Nano Today* **2009**, *4*, 244–251. [CrossRef]

10. Tang, L.; Li, J. Plasmon-Based Colorimetric Nanosensors for Ultrasensitive Molecular Diagnostics. *ACS Sens.* **2017**, *2*, 857–875. [CrossRef]

11. Aldewachi, H.; Chalati, T.; Woodroofe, M.N.; Bricklebank, N.; Sharrack, B.; Gardiner, P. Gold nanoparticle-based colorimetric biosensors. *Nanoscale* **2017**, *10*, 18–33. [CrossRef]

12. Loiseau, A.; Zhang, L.; Hu, D.; Salmain, M.; Mazouzi, Y.; Flack, R.; Liedberg, B.; Boujday, S. Core–Shell Gold/Silver Nanoparticles for Localized Surface Plasmon Resonance-Based Naked-Eye Toxin Biosensing. *ACS Appl. Mater. Interfaces* **2019**, *11*, 46462–46471. [CrossRef] [PubMed]

13. Chen, H.; Shao, L.; Li, Q.; Wang, J. Gold nanorods and their plasmonic properties. *Chem Soc Rev* **2013**, *42*, 2679–2724. [CrossRef] [PubMed]

14. Brioude, A.; Jiang, X.C.; Pileni, M.P. Optical Properties of Gold Nanorods: DDA Simulations Supported by Experiments. *J. Phys. Chem. B* **2005**, *109*, 13138–13142. [CrossRef]

15. Hu, M.; Chen, J.; Li, Z.-Y.; Au, L.; Hartland, G.V.; Li, X.; Marquez, M.; Xia, Y. Gold nanostructures: engineering their plasmonic properties for biomedical applications. *Chem. Soc. Rev.* **2006**, *35*, 1084–1094. [CrossRef]

16. Huang, X.; Jain, P.K.; El-Sayed, I.H.; El-Sayed, M.A. Gold nanoparticles: interesting optical properties and recent applications in cancer diagnostics and therapy. *Nanomed.* **2007**, *2*, 681–693. [CrossRef] [PubMed]

17. Huang, X.; Neretina, S.; El-Sayed, M.A. Gold Nanorods: From Synthesis and Properties to Biological and Biomedical Applications. *Adv. Mater.* **2009**, *21*, 4880–4910. [CrossRef]

18. Murphy, C.J.; Gole, A.M.; Stone, J.W.; Sisco, P.N.; Alkilany, A.M.; Goldsmith, E.C.; Baxter, S.C. Gold Nanoparticles in Biology: Beyond Toxicity to Cellular Imaging. *Acc. Chem. Res.* **2008**, *41*, 1721–1730. [CrossRef]

19. Kim, S.; Lee, S.; Lee, H.J. An aptamer-aptamer sandwich assay with nanorod-enhanced surface plasmon resonance for attomolar concentration of norovirus capsid protein. *Sens. Actuators B Chem.* **2018**, *273*, 1029–1036. [CrossRef]

20. Sim, H.R.; Wark, A.W.; Lee, H.J. Attomolar detection of protein biomarkers using biofunctionalized gold nanorods with surface plasmon resonance. *The Analyst* **2010**, *135*, 2528. [CrossRef]

21. Spadavecchia, J.; Casale, S.; Boujday, S.; Pradier, C.-M. Bioconjugated gold nanorods to enhance the sensitivity of FT-SPR-based biosensors. *Colloids Surf. B Biointerfaces* **2012**, *100*, 1–8. [CrossRef] [PubMed]

22. Mei, Z.; Tang, L. Surface-Plasmon-Coupled Fluorescence Enhancement Based on Ordered Gold Nanorod Array Biochip for Ultrasensitive DNA Analysis. *Anal. Chem.* **2017**, *89*, 633–639. [CrossRef]

23. Lio, D.C.S.; Liu, C.; Wiraja, C.; Qiu, B.; Fhu, C.W.; Wang, X.; Xu, C. Molecular Beacon Gold Nanosensors for Leucine-Rich Alpha-2-Glycoprotein-1 Detection in Pathological Angiogenesis. *ACS Sens.* **2018**, *3*, 1647–1655. [CrossRef] [PubMed]

24. Zhang, Z.; Wang, H.; Chen, Z.; Wang, X.; Choo, J.; Chen, L. Plasmonic colorimetric sensors based on etching and growth of noble metal nanoparticles: Strategies and applications. *Biosens. Bioelectron.* **2018**, *114*, 52–65. [CrossRef] [PubMed]

25. Chen, H.; Kou, X.; Yang, Z.; Ni, W.; Wang, J. Shape- and Size-Dependent Refractive Index Sensitivity of Gold Nanoparticles. *Langmuir* **2008**, *24*, 5233–5237. [CrossRef] [PubMed]

26. Lohse, S.E.; Murphy, C.J. The Quest for Shape Control: A History of Gold Nanorod Synthesis. *Chem. Mater.* **2013**, *25*, 1250–1261. [CrossRef]

27. Nikoobakht, B.; El-Sayed, M.A. Evidence for Bilayer Assembly of Cationic Surfactants on the Surface of Gold Nanorods. *Langmuir* **2001**, *17*, 6368–6374. [CrossRef]

28. Jana, N.R.; Gearheart, L.; Murphy, C.J. Wet Chemical Synthesis of High Aspect Ratio Cylindrical Gold Nanorods. *J. Phys. Chem. B* **2001**, *105*, 4065–4067. [CrossRef]

29. Sau, T.K.; Murphy, C.J. Seeded High Yield Synthesis of Short Au Nanorods in Aqueous Solution. *Langmuir* **2004**, *20*, 6414–6420. [CrossRef]

30. Ye, X.; Zheng, C.; Chen, J.; Gao, Y.; Murray, C.B. Using Binary Surfactant Mixtures To Simultaneously Improve the Dimensional Tunability and Monodispersity in the Seeded Growth of Gold Nanorods. *Nano Lett.* **2013**, *13*, 765–771. [CrossRef]

31. Lohse, S.E.; Burrows, N.D.; Scarabelli, L.; Liz-Marzán, L.M.; Murphy, C.J. Anisotropic Noble Metal Nanocrystal Growth: The Role of Halides. *Chem. Mater.* **2014**, *26*, 34–43. [CrossRef]

32. Creutz, C. Complexities of ascorbate as a reducing agent. *Inorg. Chem.* **1981**, *20*, 4449–4452. [CrossRef]

33. Evans, D.H.; Lingane, J.J. Standard potentials of the couples involving AuBr4−, AuBr2− and Au In bromide media. *J. Electroanal. Chem. 1959* **1963**, *6*, 1–10. [CrossRef]

34. Lingane, J.J. Standard potentials of half-reactions involving + 1 and + 3 gold in chloride medium: Equilibrium constant of the reaction AuCl4− + 2Au + 2Cl− = 3AuCl2−. *J. Electroanal. Chem. 1959* **1962**, *4*, 332–342. [CrossRef]

35. Foss Jr, C.A.; Hornyak, G.L.; Stockert, J.A.; Martin, C.R. Template-synthesized nanoscopic gold particles: optical spectra and the effects of particle size and shape. *J. Phys. Chem.* **1994**, *98*, 2963–2971. [CrossRef]

36. Jana, N.R.; Gearheart, L.; Murphy, C.J. Seed-Mediated Growth Approach for Shape-Controlled Synthesis of Spheroidal and Rod-like Gold Nanoparticles Using a Surfactant Template. *Adv. Mater.* **2001**, *13*, 1389–1393. [CrossRef]

37. Nikoobakht, B.; El-Sayed, M.A. Preparation and Growth Mechanism of Gold Nanorods (NRs) Using Seed-Mediated Growth Method. *Chem. Mater.* **2003**, *15*, 1957–1962. [CrossRef]

38. Jana, N.R. Gram-Scale Synthesis of Soluble, Near-Monodisperse Gold Nanorods and Other Anisotropic Nanoparticles. *Small* **2005**, *1*, 875–882. [CrossRef]

39. Ali, M.R.K.; Snyder, B.; El-Sayed, M.A. Synthesis and Optical Properties of Small Au Nanorods Using a Seedless Growth Technique. *Langmuir* **2012**, *28*, 9807–9815. [CrossRef] [PubMed]

40. Burrows, N.D.; Harvey, S.; Idesis, F.A.; Murphy, C.J. Understanding the Seed-Mediated Growth of Gold Nanorods through a Fractional Factorial Design of Experiments. *Langmuir* **2017**, *33*, 1891–1907. [CrossRef]

41. Smith, D.K.; Miller, N.R.; Korgel, B.A. Iodide in CTAB Prevents Gold Nanorod Formation. *Langmuir* **2009**, *25*, 9518–9524. [CrossRef]

42. Scarabelli, L.; Sánchez-Iglesias, A.; Pérez-Juste, J.; Liz-Marzán, L.M. A "Tips and Tricks" Practical Guide to the Synthesis of Gold Nanorods. *J. Phys. Chem. Lett.* **2015**, *6*, 4270–4279. [CrossRef] [PubMed]

43. Reza Hormozi-Nezhad, M.; Robatjazi, H.; Jalali-Heravi, M. Thorough tuning of the aspect ratio of gold nanorods using response surface methodology. *Anal. Chim. Acta* **2013**, *779*, 14–21. [CrossRef] [PubMed]

44. Watt, J.; Hance, B.G.; Anderson, R.S.; Huber, D.L. Effect of Seed Age on Gold Nanorod Formation: A Microfluidic, Real-Time Investigation. *Chem. Mater.* **2015**, *27*, 6442–6449. [CrossRef]

45. Vigderman, L.; Zubarev, E.R. High-Yield Synthesis of Gold Nanorods with Longitudinal SPR Peak Greater than 1200 nm Using Hydroquinone as a Reducing Agent. *Chem. Mater.* **2013**, *25*, 1450–1457. [CrossRef]

46. Orendorff, C.J.; Murphy, C.J. Quantitation of Metal Content in the Silver-Assisted Growth of Gold Nanorods. *J. Phys. Chem. B* **2006**, *110*, 3990–3994. [CrossRef]

47. Xu, D.; Mao, J.; He, Y.; Yeung, E.S. Size-tunable synthesis of high-quality gold nanorods under basic conditions by using H 2 O 2 as the reducing agent. *J. Mater. Chem. C* **2014**, *2*, 4989–4996. [CrossRef]

48. Leng, Y.; Yin, X.; Hu, F.; Zou, Y.; Xing, X.; Li, B.; Guo, Y.; Ye, L.; Lu, Z. High-yield synthesis and fine-tuning aspect ratio of (200) faceted gold nanorods by the pH-adjusting method. *RSC Adv.* **2017**, *7*, 25469–25474. [CrossRef]

49. Canonico-May, S.A.; Beavers, K.R.; Melvin, M.J.; Alkilany, A.M.; Duvall, C.L.; Stone, J.W. High conversion of HAuCl4 into gold nanorods: A re-seeding approach. *J. Colloid Interface Sci.* **2016**, *463*, 229–232. [CrossRef]

50. Kozek, K.A.; Kozek, K.M.; Wu, W.-C.; Mishra, S.R.; Tracy, J.B. Large-Scale Synthesis of Gold Nanorods through Continuous Secondary Growth. *Chem. Mater.* **2013**, *25*, 4537–4544. [CrossRef]

51. Chang, H.-H.; Murphy, C.J. Mini Gold Nanorods with Tunable Plasmonic Peaks beyond 1000 nm. *Chem. Mater.* **2018**, *30*, 1427–1435. [CrossRef]

52. González-Rubio, G.; Scarabelli, L.; Guerrero-Martínez, A.; Liz-Marzán, L.M. Surfactant-Assisted Symmetry Breaking in Colloidal Gold Nanocrystal Growth. *ChemNanoMat* **2020**, *6*, 698–707. [CrossRef]

53. González-Rubio, G.; Kumar, V.; Llombart, P.; Díaz-Núñez, P.; Bladt, E.; Altantzis, T.; Bals, S.; Peña-Rodríguez, O.; Noya, E.G.; MacDowell, L.G.; et al. Disconnecting Symmetry Breaking from Seeded Growth for the Reproducible Synthesis of High Quality Gold Nanorods. *ACS Nano* **2019**, *13*, 4424–4435. [CrossRef] [PubMed]

54. Sharma, V.; Park, K.; Srinivasarao, M. Shape separation of gold nanorods using centrifugation. *Proc. Natl. Acad. Sci. USA* **2009**, *106*, 4981–4985. [CrossRef] [PubMed]

55. Li, S.; Chang, Z.; Liu, J.; Bai, L.; Luo, L.; Sun, X. Separation of gold nanorods using density gradient ultracentrifugation. *Nano Res.* **2011**, *4*, 723–728. [CrossRef]

56. Boksebeld, M.; Blanchard, N.P.; Jaffal, A.; Chevolot, Y.; Monnier, V. Shape-selective purification of gold nanorods with low aspect ratio using a simple centrifugation method. *Gold Bull.* **2017**, *50*, 69–76. [CrossRef]

57. Park, K.; Koerner, H.; Vaia, R.A. Depletion-Induced Shape and Size Selection of Gold Nanoparticles. *Nano Lett.* **2010**, *10*, 1433–1439. [CrossRef]

58. Uson, L.; Sebastian, V.; Arruebo, M.; Santamaria, J. Continuous microfluidic synthesis and functionalization of gold nanorods. *Chem. Eng. J.* **2016**, *285*, 286–292. [CrossRef]

59. Xu, Y.; Zhao, Y.; Chen, L.; Wang, X.; Sun, J.; Wu, H.; Bao, F.; Fan, J.; Zhang, Q. Large-scale, low-cost synthesis of monodispersed gold nanorods using a gemini surfactant. *Nanoscale* **2015**, *7*, 6790–6797. [CrossRef]

60. Kaur, P.; Chudasama, B. Effect of Colloidal Medium on the Shelf-Life and Stability of Gold Nanorods Prepared by Seed-Mediated Synthesis. *J. Nanosci. Nanotechnol.* **2018**, *18*, 1665–1674. [CrossRef]

61. Vassalini, I.; Rotunno, E.; Lazzarini, L.; Alessandri, I. "Stainless" Gold Nanorods: Preserving Shape, Optical Properties, and SERS Activity in Oxidative Environment. *ACS Appl. Mater. Interfaces* **2015**, *7*, 18794–18802. [CrossRef] [PubMed]

62. Lohse, S.E.; Eller, J.R.; Sivapalan, S.T.; Plews, M.R.; Murphy, C.J. A Simple Millifluidic Benchtop Reactor System for the High-Throughput Synthesis and Functionalization of Gold Nanoparticles with Different Sizes and Shapes. *ACS Nano* **2013**, *7*, 4135–4150. [CrossRef] [PubMed]

63. Ma, J.; Li, C.-W. Rapid and continuous parametric screening for the synthesis of gold nanocrystals with different morphologies using a microfluidic device. *Sens. Actuators B Chem.* **2018**, *262*, 236–244. [CrossRef]

64. Park, K.; Hsiao, M.; Yi, Y.-J.; Izor, S.; Koerner, H.; Jawaid, A.; Vaia, R.A. Highly Concentrated Seed-Mediated Synthesis of Monodispersed Gold Nanorods. *ACS Appl. Mater. Interfaces* **2017**, *9*, 26363–26371. [CrossRef]

65. Khanal, B.P.; Zubarev, E.R. Gram-Scale Synthesis of Isolated Monodisperse Gold Nanorods. *Chem. Eur. J.* **2019**, *25*, 1595–1600. [CrossRef]

66. Liz-Marzán, L.M.; Giersig, M.; Mulvaney, P. Synthesis of Nanosized Gold−Silica Core−Shell Particles. *Langmuir* **1996**, *12*, 4329–4335. [CrossRef]

67. Obare, S.O.; Jana, N.R.; Murphy, C.J. Preparation of Polystyrene- and Silica-Coated Gold Nanorods and Their Use as Templates for the Synthesis of Hollow Nanotubes. *Nano Lett.* **2001**, *1*, 601–603. [CrossRef]

68. Pérez-Juste, J.; Correa-Duarte, M.A.; Liz-Marzán, L.M. Silica gels with tailored, gold nanorod-driven optical functionalities. *Appl. Surf. Sci.* **2004**, *226*, 137–143. [CrossRef]

69. Li, C.; Li, Y.; Ling, Y.; Lai, Y.; Wu, C.; Zhao, Y. Exploration of the growth process of ultrathin silica shells on the surface of gold nanorods by the localized surface plasmon resonance. *Nanotechnology* **2014**, *25*, 045704. [CrossRef]

70. Li, J.F.; Tian, X.D.; Li, S.B.; Anema, J.R.; Yang, Z.L.; Ding, Y.; Wu, Y.F.; Zeng, Y.M.; Chen, Q.Z.; Ren, B.; et al. Surface analysis using shell-isolated nanoparticle-enhanced Raman spectroscopy. *Nat. Protoc.* **2013**, *8*, 52–65. [CrossRef]

71. Fernández-López, C.; Mateo-Mateo, C.; Álvarez-Puebla, R.A.; Pérez-Juste, J.; Pastoriza-Santos, I.; Liz-Marzán, L.M. Highly Controlled Silica Coating of PEG-Capped Metal Nanoparticles and Preparation of SERS-Encoded Particles. *Langmuir* **2009**, *25*, 13894–13899. [CrossRef] [PubMed]

72. Wang, J.; Zhang, W.; Li, S.; Miao, D.; Qian, G.; Su, G. Engineering of Porous Silica Coated Gold Nanorods by Surface-Protected Etching and Their Applications in Drug Loading and Combined Cancer Therapy. *Langmuir* **2019**, *35*, 14238–14247. [CrossRef]

73. Kinnear, C.; Dietsch, H.; Clift, M.J.D.; Endes, C.; Rothen-Rutishauser, B.; Petri-Fink, A. Gold Nanorods: Controlling Their Surface Chemistry and Complete Detoxification by a Two-Step Place Exchange. *Angew. Chem. Int. Ed.* **2013**, *52*, 1934–1938. [CrossRef] [PubMed]

74. Graf, C.; Vossen, D.L.J.; Imhof, A.; van Blaaderen, A. A General Method To Coat Colloidal Particles with Silica. *Langmuir* **2003**, *19*, 6693–6700. [CrossRef]

75. Pastoriza-Santos, I.; Pérez-Juste, J.; Liz-Marzán, L.M. Silica-Coating and Hydrophobation of CTAB-Stabilized Gold Nanorods. *Chem. Mater.* **2006**, *18*, 2465–2467. [CrossRef]

76. Pastoriza-Santos, I.; Liz-Marzán, L.M. Reliable Methods for Silica Coating of Au Nanoparticles. In *Nanomaterial Interfaces in Biology*; Bergese, P., Hamad-Schifferli, K., Eds.; Methods in Molecular Biology; Humana Press: Totowa, NJ, USA, 2013; pp. 75–93. ISBN 978-1-62703-461-6.

77. Nallathamby, P.D.; Hopf, J.; Irimata, L.E.; McGinnity, T.L.; Roeder, R.K. Preparation of fluorescent Au–SiO$_2$ core–shell nanoparticles and nanorods with tunable silica shell thickness and surface modification for immunotargeting. *J. Mater. Chem. B* **2016**, *4*, 5418–5428. [CrossRef]

78. Gorelikov, I.; Matsuura, N. Single-Step Coating of Mesoporous Silica on Cetyltrimethyl Ammonium Bromide-Capped Nanoparticles. *Nano Lett.* **2008**, *8*, 369–373. [CrossRef]

79. Cong, H.; Toftegaard, R.; Arnbjerg, J.; Ogilby, P.R. Silica-Coated Gold Nanorods with a Gold Overcoat: Controlling Optical Properties by Controlling the Dimensions of a Gold–Silica–Gold Layered Nanoparticle. *Langmuir* **2010**, *26*, 4188–4195. [CrossRef]

80. Liu, J.; Kan, C.; Cong, B.; Xu, H.; Ni, Y.; Li, Y.; Shi, D. Plasmonic Property and Stability of Core-Shell Au@SiO$_2$ Nanostructures. *Plasmonics* **2014**, *9*, 1007–1014. [CrossRef]

81. Wu, W.-C.; Tracy, J.B. Large-Scale Silica Overcoating of Gold Nanorods with Tunable Shell Thicknesses. *Chem. Mater.* **2015**, *27*, 2888–2894. [CrossRef]

82. Abadeer, N.S.; Brennan, M.R.; Wilson, W.L.; Murphy, C.J. Distance and Plasmon Wavelength Dependent Fluorescence of Molecules Bound to Silica-Coated Gold Nanorods. *ACS Nano* **2014**, *8*, 8392–8406. [CrossRef] [PubMed]

83. Yoon, S.; Lee, B.; Kim, C.; Lee, J.H. Controlled Heterogeneous Nucleation for Synthesis of Uniform Mesoporous Silica-Coated Gold Nanorods with Tailorable Rotational Diffusion and 1 nm-Scale Size Tunability. *Cryst. Growth Des.* **2018**, *18*, 4731–4736. [CrossRef]

84. Rowe, L.R.; Chapman, B.S.; Tracy, J.B. Understanding and Controlling the Morphology of Silica Shells on Gold Nanorods. *Chem. Mater.* **2018**, *30*, 6249–6258. [CrossRef]

85. Wang, M.; Hoff, A.; Doebler, J.E.; Emory, S.R.; Bao, Y. Dumbbell-Like Silica Coated Gold Nanorods and Their Plasmonic Properties. *Langmuir* **2019**, *35*, 16886–16892. [CrossRef] [PubMed]

86. Tian, X.; Guo, J.; Tian, Y.; Tang, H.; Yang, W. Modulated fluorescence properties in fluorophore-containing gold nanorods@mSiO$_2$. *RSC Adv.* **2014**, *4*, 9343–9348. [CrossRef]

87. Wu, C.; Xu, Q.-H. Stable and Functionable Mesoporous Silica-Coated Gold Nanorods as Sensitive Localized Surface Plasmon Resonance (LSPR) Nanosensors. *Langmuir* **2009**, *25*, 9441–9446. [CrossRef]

88. Zhang, R.; Zhou, Y.; Peng, L.; Li, X.; Chen, S.; Feng, X.; Guan, Y.; Huang, W. Influence of SiO 2 shell thickness on power conversion efficiency in plasmonic polymer solar cells with Au nanorod@SiO 2 core-shell structures. *Sci. Rep.* **2016**, *6*, 1–9. [CrossRef]

89. Mohanta, J.; Satapathy, S.; Si, S. Porous Silica-Coated Gold Nanorods: A Highly Active Catalyst for the Reduction of 4-Nitrophenol. *ChemPhysChem* **2016**, *17*, 364–368. [CrossRef]

90. Huang, C.-C.; Huang, C.-H.; Kuo, I.-T.; Chau, L.-K.; Yang, T.-S. Synthesis of silica-coated gold nanorod as Raman tags by modulating cetyltrimethylammonium bromide concentration. *Colloids Surf. Physicochem. Eng. Asp.* **2012**, *409*, 61–68. [CrossRef]

91. Shen, D.; Yang, J.; Li, X.; Zhou, L.; Zhang, R.; Li, W.; Chen, L.; Wang, R.; Zhang, F.; Zhao, D. Biphase Stratification Approach to Three-Dimensional Dendritic Biodegradable Mesoporous Silica Nanospheres. *Nano Lett.* **2014**, *14*, 923–932. [CrossRef]

92. Xu, C.; Chen, F.; Valdovinos, H.F.; Jiang, D.; Goel, S.; Yu, B.; Sun, H.; Barnhart, T.E.; Moon, J.J.; Cai, W. Bacteria-like mesoporous silica-coated gold nanorods for positron emission tomography and photoacoustic imaging-guided chemo-photothermal combined therapy. *Biomaterials* **2018**, *165*, 56–65. [CrossRef]

93. Malik, M.A.; Wani, M.Y.; Hashim, M.A. Microemulsion method: A novel route to synthesize organic and inorganic nanomaterials. *Arab. J. Chem.* **2012**, *5*, 397–417. [CrossRef]

94. Wang, J.; Shah, Z.H.; Zhang, S.; Lu, R. Silica-based nanocomposites via reverse microemulsions: classifications, preparations, and applications. *Nanoscale* **2014**, *6*, 4418. [CrossRef] [PubMed]

95. Han, Y.; Jiang, J.; Lee, S.S.; Ying, J.Y. Reverse Microemulsion-Mediated Synthesis of Silica-Coated Gold and Silver Nanoparticles. *Langmuir* **2008**, *24*, 5842–5848. [CrossRef]

96. Ding, H.L.; Zhang, Y.X.; Wang, S.; Xu, J.M.; Xu, S.C.; Li, G.H. Fe$_3$O$_4$@SiO$_2$ Core/Shell Nanoparticles: The Silica Coating Regulations with a Single Core for Different Core Sizes and Shell Thicknesses. *Chem. Mater.* **2012**, *24*, 4572–4580. [CrossRef]

97. Anderson, B.D.; Wu, W.-C.; Tracy, J.B. Silica Overcoating of CdSe/CdS Core/Shell Quantum Dot Nanorods with Controlled Morphologies. *Chem. Mater.* **2016**, *28*, 4945–4952. [CrossRef]

98. Wang, L.; Guo, S.; Liu, D.; He, J.; Zhou, J.; Zhang, K.; Wei, Y.; Pan, Y.; Gao, C.; Yuan, Z.; et al. Plasmon-Enhanced Blue Upconversion Luminescence by Indium Nanocrystals. *Adv. Funct. Mater.* **2019**, *29*, 1901242. [CrossRef]

99. Feng, J.; Wang, Z.; Shen, B.; Zhang, L.; Yang, X.; He, N. Effects of template removal on both morphology of mesoporous silica-coated gold nanorod and its biomedical application. *RSC Adv.* **2014**, *4*, 28683–28690. [CrossRef]

100. Yu, Y.Y.; Chang, S.S.; Lee, C.L.; Wang, C.C. Gold Nanorods: Electrochemical Synthesis and Optical Properties. *J. Phys. Chem. B* **1997**, *101*, 6661–6664. [CrossRef]

101. Hatakeyama, Y.; Sasaki, K.; Judai, K.; Nishikawa, K.; Hino, K. Growth Behavior of Gold Nanorods Synthesized by the Seed-Mediated Method: Tracking of Reaction Progress by Time-Resolved X-ray Absorption Near-Edge Structure, Small-Angle X-ray Scattering, and Ultraviolet–Visible Spectroscopy. *J. Phys. Chem. C* **2018**, *122*, 7982–7991. [CrossRef]

102. Juvé, V.; Cardinal, M.F.; Lombardi, A.; Crut, A.; Maioli, P.; Pérez-Juste, J.; Liz-Marzán, L.M.; Del Fatti, N.; Vallée, F. Size-Dependent Surface Plasmon Resonance Broadening in Nonspherical Nanoparticles: Single Gold Nanorods. *Nano Lett.* **2013**, *13*, 2234–2240. [CrossRef] [PubMed]

103. Cao, J.; Sun, T.; Grattan, K.T.V. Gold nanorod-based localized surface plasmon resonance biosensors: A review. *Sens. Actuators B Chem.* **2014**, *195*, 332–351. [CrossRef]

104. Al-Sherbini, A.-S.A.-M. Thermal instability of gold nanorods in micellar solution of water/glycerol mixtures. *Colloids Surf. Physicochem. Eng. Asp.* **2004**, *246*, 61–69. [CrossRef]

105. Pérez-Juste, J.; Liz-Marzán, L.M.; Carnie, S.; Chan, D.Y.C.; Mulvaney, P. Electric-Field-Directed Growth of Gold Nanorods in Aqueous Surfactant Solutions. *Adv. Funct. Mater.* **2004**, *14*, 571–579. [CrossRef]

106. Pérez-Juste, J.; Pastoriza-Santos, I.; Liz-Marzán, L.M.; Mulvaney, P. Gold nanorods: Synthesis, characterization and applications. *Coord. Chem. Rev.* **2005**, *249*, 1870–1901. [CrossRef]

107. Grulke, E.A.; Wu, X.; Ji, Y.; Buhr, E.; Yamamoto, K.; Song, N.W.; Stefaniak, A.B.; Schwegler-Berry, D.; Burchett, W.W.; Lambert, J.; et al. Differentiating gold nanorod samples using particle size and shape distributions from transmission electron microscope images. *Metrologia* **2018**, *55*, 254–267. [CrossRef]

108. Sturges, H.A. The Choice of a Class Interval. *J. Am. Stat. Assoc.* **1926**, *21*, 65–66. [CrossRef]

109. Masuda, H.; Gotoh, K. Study on the sample size required for the estimation of mean particle diameter. *Adv. Powder Technol.* **1999**, *10*, 159–173. [CrossRef]

110. Laramy, C.R.; Brown, K.A.; O'Brien, M.N.; Mirkin, C.A. High-Throughput, Algorithmic Determination of Nanoparticle Structure from Electron Microscopy Images. *ACS Nano* **2015**, *9*, 12488–12495. [CrossRef]

111. Pecora, R. Spectrum of Light Scattered from Optically Anisotropic Macromolecules. *J. Chem. Phys.* **1968**, *49*, 1036–1043. [CrossRef]

112. Liu, H.; Pierre-Pierre, N.; Huo, Q. Dynamic light scattering for gold nanorod size characterization and study of nanorod–protein interactions. *Gold Bull.* **2012**, *45*, 187–195. [CrossRef]

113. Glidden, M.; Muschol, M. Characterizing Gold Nanorods in Solution Using Depolarized Dynamic Light Scattering. *J. Phys. Chem. C* **2012**, *116*, 8128–8137. [CrossRef]

114. Nixon-Luke, R.; Bryant, G. A Depolarized Dynamic Light Scattering Method to Calculate Translational and Rotational Diffusion Coefficients of Nanorods. *Part. Part. Syst. Charact.* **2019**, *36*, 1800388. [CrossRef]

115. Hubert, F.; Testard, F.; Thill, A.; Kong, Q.; Tache, O.; Spalla, O. Growth and Overgrowth of Concentrated Gold Nanorods: Time Resolved SAXS and XANES. *Cryst. Growth Des.* **2012**, *12*, 1548–1555. [CrossRef]

116. Guinier, A.; Fournet, G.; Walker, C.B.; Vineyard, G.H. Small-Angle Scattering of X-Rays. *Phys. Today* **1956**, *9*, 38. [CrossRef]

117. Morita, T.; Hatakeyama, Y.; Nishikawa, K.; Tanaka, E.; Shingai, R.; Murai, H.; Nakano, H.; Hino, K. Multiple small-angle X-ray scattering analyses of the structure of gold nanorods with unique end caps. *Chem. Phys.* **2009**, *364*, 14–18. [CrossRef]

118. Schulz, F.; Möller, J.; Lehmkühler, F.; Smith, A.J.; Vossmeyer, T.; Lange, H.; Grübel, G.; Schroer, M.A. Structure and Stability of PEG- and Mixed PEG-Layer-Coated Nanoparticles at High Particle Concentrations Studied In Situ by Small-Angle X-Ray Scattering. *Part. Part. Syst. Charact.* **2018**, *35*, 1700319. [CrossRef]

119. Kuttner, C.; Höller, R.P.M.; Quintanilla, M.; Schnepf, M.J.; Dulle, M.; Fery, A.; Liz-Marzán, L.M. SERS and plasmonic heating efficiency from anisotropic core/satellite superstructures. *Nanoscale* **2019**, *11*, 17655–17663. [CrossRef]

120. Nikoobakht, B.; Wang, J.; El-Sayed, M.A. Surface-enhanced Raman scattering of molecules adsorbed on gold nanorods: off-surface plasmon resonance condition. *Chem. Phys. Lett.* **2002**, *366*, 17–23. [CrossRef]

121. Liao, H.; Hafner, J.H. Gold Nanorod Bioconjugates. *Chem. Mater.* **2005**, *17*, 4636–4641. [CrossRef]

122. Park, K.; Biswas, S.; Kanel, S.; Nepal, D.; Vaia, R.A. Engineering the Optical Properties of Gold Nanorods: Independent Tuning of Surface Plasmon Energy, Extinction Coefficient, and Scattering Cross Section. *J. Phys. Chem. C* **2014**, *118*, 5918–5926. [CrossRef]

123. Scarabelli, L.; Grzelczak, M.; Liz-Marzán, L.M. Tuning Gold Nanorod Synthesis through Prereduction with Salicylic Acid. *Chem. Mater.* **2013**, *25*, 4232–4238. [CrossRef]

124. Rodríguez-Fernández, J.; Pérez-Juste, J.; Mulvaney, P.; Liz-Marzán, L.M. Spatially-Directed Oxidation of Gold Nanoparticles by Au(III)–CTAB Complexes. *J. Phys. Chem. B* **2005**, *109*, 14257–14261. [CrossRef]

125. Hendel, T.; Wuithschick, M.; Kettemann, F.; Birnbaum, A.; Rademann, K.; Polte, J. In Situ Determination of Colloidal Gold Concentrations with UV–Vis Spectroscopy: Limitations and Perspectives. *Anal. Chem.* **2014**, *86*, 11115–11124. [CrossRef]

126. Edgar, J.A.; McDonagh, A.M.; Cortie, M.B. Formation of Gold Nanorods by a Stochastic "Popcorn" Mechanism. *ACS Nano* **2012**, *6*, 1116–1125. [CrossRef] [PubMed]

127. Gao, Z.; Burrows, N.D.; Valley, N.A.; Schatz, G.C.; Murphy, C.J.; Haynes, C.L. In solution SERS sensing using mesoporous silica-coated gold nanorods. *The Analyst* **2016**, *141*, 5088–5095. [CrossRef] [PubMed]

128. Wang, X.; Li, Y.; Wang, H.; Fu, Q.; Peng, J.; Wang, Y.; Du, J.; Zhou, Y.; Zhan, L. Gold nanorod-based localized surface plasmon resonance biosensor for sensitive detection of hepatitis B virus in buffer, blood serum and plasma. *Biosens. Bioelectron.* **2010**, *26*, 404–410. [CrossRef]

129. Zhang, K.; Shen, X. Cancer antigen 125 detection using the plasmon resonance scattering properties of gold nanorods. *The Analyst* **2013**, *138*, 1828. [CrossRef] [PubMed]

130. Wang, C.; Chen, Y.; Wang, T.; Ma, Z.; Su, Z. Biorecognition-Driven Self-Assembly of Gold Nanorods: A Rapid and Sensitive Approach toward Antibody Sensing. *Chem. Mater.* **2007**, *19*, 5809–5811. [CrossRef]

131. Guo, Z.R.; Gu, C.R.; Fan, X.; Bian, Z.P.; Wu, H.F.; Yang, D.; Gu, N.; Zhang, J.N. Fabrication of Anti-human Cardiac Troponin I Immunogold Nanorods for Sensing Acute Myocardial Damage. *Nanoscale Res. Lett.* **2009**, *4*, 1428–1433. [CrossRef]

132. Casas, J.; Venkataramasubramani, M.; Wang, Y.; Tang, L. Replacement of cetyltrimethylammoniumbromide bilayer on gold nanorod by alkanethiol crosslinker for enhanced plasmon resonance sensitivity. *Biosens. Bioelectron.* **2013**, *49*, 525–530. [CrossRef] [PubMed]

133. Wang, Y.; Tang, L. Chemisorption assembly of Au nanorods on mercaptosilanized glass substrate for label-free nanoplasmon biochip. *Anal. Chim. Acta* **2013**, *796*, 122–129. [CrossRef] [PubMed]

134. Boujday, S.; Lambert, J.-F.; Che, M. Amorphous Silica as a Versatile Supermolecular Ligand for NiII Amine Complexes: Toward Interfacial Molecular Recognition. *ChemPhysChem* **2004**, *5*, 1003–1013. [CrossRef]

135. Song, L.; Zhang, L.; Huang, Y.; Chen, L.; Zhang, G.; Shen, Z.; Zhang, J.; Xiao, Z.; Chen, T. Amplifying the signal of localized surface plasmon resonance sensing for the sensitive detection of Escherichia coli O157:H7. *Sci. Rep.* **2017**, *7*, 3288. [CrossRef] [PubMed]

136. Li, Y.; Zhang, Y.; Zhao, M.; Zhou, Q.; Wang, L.; Wang, H.; Wang, X.; Zhan, L. A simple aptamer-functionalized gold nanorods based biosensor for the sensitive detection of MCF-7 breast cancer cells. *Chem. Commun.* **2016**, *52*, 3959–3961. [CrossRef] [PubMed]

137. Lee, B.; Park, J.-H.; Byun, J.-Y.; Kim, J.H.; Kim, M.-G. An optical fiber-based LSPR aptasensor for simple and rapid in-situ detection of ochratoxin A. *Biosens. Bioelectron.* **2018**, *102*, 504–509. [CrossRef] [PubMed]

138. Ha, S.-J.; Park, J.-H.; Lee, B.; Kim, M.-G. Label-Free Direct Detection of Saxitoxin Based on a Localized Surface Plasmon Resonance Aptasensor. *Toxins* **2019**, *11*, 274. [CrossRef]

139. Park, J.-H.; Byun, J.-Y.; Jang, H.; Hong, D.; Kim, M.-G. A highly sensitive and widely adaptable plasmonic aptasensor using berberine for small-molecule detection. *Biosens. Bioelectron.* **2017**, *97*, 292–298. [CrossRef]

140. Park, J.-H.; Byun, J.-Y.; Shim, W.-B.; Kim, S.U.; Kim, M.-G. High-sensitivity detection of ATP using a localized surface plasmon resonance (LSPR) sensor and split aptamers. *Biosens. Bioelectron.* **2015**, *73*, 26–31. [CrossRef]

141. Wang, X.; Li, Y.; Wang, J.; Wang, Q.; Xu, L.; Du, J.; Yan, S.; Zhou, Y.; Fu, Q.; Wang, Y.; et al. A broad-range method to detect genomic DNA of multiple pathogenic bacteria based on the aggregation strategy of gold nanorods. *The Analyst* **2012**, *137*, 4267–4273. [CrossRef]

142. Rosman, C.; Prasad, J.; Neiser, A.; Henkel, A.; Edgar, J.; Sönnichsen, C. Multiplexed Plasmon Sensor for Rapid Label-Free Analyte Detection. *Nano Lett.* **2013**, *13*, 3243–3247. [CrossRef]

143. Parab, H.J.; Jung, C.; Lee, J.-H.; Park, H.G. A gold nanorod-based optical DNA biosensor for the diagnosis of pathogens. *Biosens. Bioelectron.* **2010**, *26*, 667–673. [CrossRef] [PubMed]

144. Tadepalli, S.; Kuang, Z.; Jiang, Q.; Liu, K.-K.; Fisher, M.A.; Morrissey, J.J.; Kharasch, E.D.; Slocik, J.M.; Naik, R.R.; Singamaneni, S. Peptide Functionalized Gold Nanorods for the Sensitive Detection of a Cardiac Biomarker Using Plasmonic Paper Devices. *Sci. Rep.* **2015**, *5*, 16206. [CrossRef] [PubMed]

145. Zhao, Y.; Tong, L.; Li, Y.; Pan, H.; Zhang, W.; Guan, M.; Li, W.; Chen, Y.; Li, Q.; Li, Z.; et al. Lactose-Functionalized Gold Nanorods for Sensitive and Rapid Serological Diagnosis of Cancer. *ACS Appl. Mater. Interfaces* **2016**, *8*, 5813–5820. [CrossRef]

146. Wang, X.; Mei, Z.; Wang, Y.; Tang, L. Comparison of four methods for the biofunctionalization of gold nanorods by the introduction of sulfhydryl groups to antibodies. *Beilstein J. Nanotechnol.* **2017**, *8*, 372–380. [CrossRef]

147. Zhu, L.; Li, G.; Sun, S.; Tan, H.; He, Y. Digital immunoassay of a prostate-specific antigen using gold nanorods and magnetic nanoparticles. *RSC Adv.* **2017**, *7*, 27595–27602. [CrossRef]

148. Wang, X.; Mei, Z.; Wang, Y.; Tang, L. Gold nanorod biochip functionalization by antibody thiolation. *Talanta* **2015**, *136*, 1–8. [CrossRef]

149. Wang, Y.; Tang, L.-J.; Jiang, J.-H. Surface-Enhanced Raman Spectroscopy-Based, Homogeneous, Multiplexed Immunoassay with Antibody-Fragments-Decorated Gold Nanoparticles. *Anal. Chem.* **2013**, *85*, 9213–9220. [CrossRef]

150. Joshi, P.P.; Yoon, S.J.; Hardin, W.G.; Emelianov, S.; Sokolov, K.V. Conjugation of Antibodies to Gold Nanorods through Fc Portion: Synthesis and Molecular Specific Imaging. *Bioconjug. Chem.* **2013**, *24*, 878–888. [CrossRef]

151. Byun, J.-Y.; Shin, Y.-B.; Li, T.; Park, J.-H.; Kim, D.-M.; Choi, D.-H.; Kim, M.-G. The use of an engineered single chain variable fragment in a localized surface plasmon resonance method for analysis of the C-reactive protein. *Chem. Commun.* **2013**, *49*, 9497–9499. [CrossRef]

152. Shams, S.; Bakhshi, B.; Tohidi Moghadam, T.; Behmanesh, M. A sensitive gold-nanorods-based nanobiosensor for specific detection of Campylobacter jejuni and Campylobacter coli. *J. Nanobiotechnology* **2019**, *17*. [CrossRef]

153. Wijaya, A.; Hamad-Schifferli, K. Ligand Customization and DNA Functionalization of Gold Nanorods via Round-Trip Phase Transfer Ligand Exchange. *Langmuir* **2008**, *24*, 9966–9969. [CrossRef]

154. Pekcevik, I.C.; Poon, L.C.H.; Wang, M.C.P.; Gates, B.D. Tunable Loading of Single-Stranded DNA on Gold Nanorods through the Displacement of Polyvinylpyrrolidone. *Anal. Chem.* **2013**, *85*, 9960–9967. [CrossRef] [PubMed]

155. Li, J.; Zhu, B.; Zhu, Z.; Zhang, Y.; Yao, X.; Tu, S.; Liu, R.; Jia, S.; Yang, C.J. Simple and Rapid Functionalization of Gold Nanorods with Oligonucleotides Using an mPEG-SH/Tween 20-Assisted Approach. *Langmuir* **2015**, *31*, 7869–7876. [CrossRef]

156. Cao, J.; Galbraith, E.K.; Sun, T.; Grattan, K.T.V. Effective surface modification of gold nanorods for localized surface plasmon resonance-based biosensors. *Sens. Actuators B Chem.* **2012**, *169*, 360–367. [CrossRef]

157. Xu, X.; Liu, X.; Li, Y.; Ying, Y. A simple and rapid optical biosensor for detection of aflatoxin B1 based on competitive dispersion of gold nanorods. *Biosens. Bioelectron.* **2013**, *47*, 361–367. [CrossRef] [PubMed]

158. Huang, H.; Tang, C.; Zeng, Y.; Yu, X.; Liao, B.; Xia, X.; Yi, P.; Chu, P.K. Label-free optical biosensor based on localized surface plasmon resonance of immobilized gold nanorods. *Colloids Surf. B Biointerfaces* **2009**, *71*, 96–101. [CrossRef]

159. Peixoto, L.P.F.; Santos, J.F.L.; Andrade, G.F.S. Plasmonic nanobiosensor based on Au nanorods with improved sensitivity: A comparative study for two different configurations. *Anal. Chim. Acta* **2019**, *1084*, 71–77. [CrossRef]

160. Marinakos, S.M.; Chen, S.; Chilkoti, A. Plasmonic Detection of a Model Analyte in Serum by a Gold Nanorod Sensor. *Anal. Chem.* **2007**, *79*, 5278–5283. [CrossRef]

161. Mayer, K.M.; Lee, S.; Liao, H.; Rostro, B.C.; Fuentes, A.; Scully, P.T.; Nehl, C.L.; Hafner, J.H. A Label-Free Immunoassay Based Upon Localized Surface Plasmon Resonance of Gold Nanorods. *ACS Nano* **2008**, *2*, 687–692. [CrossRef]

162. Kaushal, S.; Priyadarshi, N.; Pinnaka, A.K.; Soni, S.; Deep, A.; Singhal, N.K. Glycoconjugates coated gold nanorods based novel biosensor for optical detection and photothermal ablation of food borne bacteria. *Sens. Actuators B Chem.* **2019**, *289*, 207–215. [CrossRef]

163. Loo, J.; Lau, P.-M.; Kong, S.-K.; Ho, H.-P. An Assay Using Localized Surface Plasmon Resonance and Gold Nanorods Functionalized with Aptamers to Sense the Cytochrome-c Released from Apoptotic Cancer Cells for Anti-Cancer Drug Effect Determination. *Micromachines* **2017**, *8*, 338. [CrossRef] [PubMed]

164. Pai, J.-H.; Yang, C.-T.; Hsu, H.-Y.; Wedding, A.B.; Thierry, B. Development of a simplified approach for the fabrication of localised surface plasmon resonance sensors based on gold nanorods functionalized using mixed polyethylene glycol layers. *Anal. Chim. Acta* **2017**, *974*, 87–92. [CrossRef]

165. Guo, L.; Huang, Y.; Kikutani, Y.; Tanaka, Y.; Kitamori, T.; Kim, D.-H. In situ assembly, regeneration and plasmonic immunosensing of a Au nanorod monolayer in a closed-surface flow channel. *Lab. Chip* **2011**, *11*, 3299–3304. [CrossRef]

166. Truong, P.L.; Kim, B.W.; Sim, S.J. Rational aspect ratio and suitable antibody coverage of gold nanorod for ultra-sensitive detection of a cancer biomarker. *Lab. Chip* **2012**, *12*, 1102–1109. [CrossRef]

167. Vial, S.; Wenger, J. Single-step homogeneous immunoassay for detecting prostate-specific antigen using dual-color light scattering of metal nanoparticles. *The Analyst* **2017**, *142*, 3484–3491. [CrossRef]

168. Tang, L.; Casas, J.; Venkataramasubramani, M. Magnetic Nanoparticle Mediated Enhancement of Localized Surface Plasmon Resonance for Ultrasensitive Bioanalytical Assay in Human Blood Plasma. *Anal. Chem.* **2013**, *85*, 1431–1439. [CrossRef]

169. Tang, L.; Casas, J. Quantification of cardiac biomarkers using label-free and multiplexed gold nanorod bioprobes for myocardial infarction diagnosis. *Biosens. Bioelectron.* **2014**, *61*, 70–75. [CrossRef] [PubMed]

170. Temur, E.; Zengin, A.; Boyacı, İ.H.; Dudak, F.C.; Torul, H.; Tamer, U. Attomole Sensitivity of Staphylococcal Enterotoxin B Detection Using an Aptamer-Modified Surface-Enhanced Raman Scattering Probe. *Anal. Chem.* **2012**, *84*, 10600–10606. [CrossRef] [PubMed]

171. Chen, C.-D.; Cheng, S.-F.; Chau, L.-K.; Wang, C.R.C. Sensing capability of the localized surface plasmon resonance of gold nanorods. *Biosens. Bioelectron.* **2007**, *22*, 926–932. [CrossRef]

172. Guo, Y.-J.; Sun, G.-M.; Zhang, L.; Tang, Y.-J.; Luo, J.-J.; Yang, P.-H. Multifunctional optical probe based on gold nanorods for detection and identification of cancer cells. *Sens. Actuators B Chem.* **2014**, *191*, 741–749. [CrossRef]

173. Wang, C.; Irudayaraj, J. Gold Nanorod Probes for the Detection of Multiple Pathogens. *Small* **2008**, *4*, 2204–2208. [CrossRef]

174. Wang, Y.; Lee, K.; Irudayaraj, J. SERS aptasensor from nanorod–nanoparticle junction for protein detection. *Chem Commun* **2010**, *46*, 613–615. [CrossRef]

175. Wang, C.; Ma, Z.; Wang, T.; Su, Z. Synthesis, Assembly, and Biofunctionalization of Silica-Coated Gold Nanorods for Colorimetric Biosensing. *Adv. Funct. Mater.* **2006**, *16*, 1673–1678. [CrossRef]

176. Wang, L.; Zhu, Y.; Xu, L.; Chen, W.; Kuang, H.; Liu, L.; Agarwal, A.; Xu, C.; Kotov, N.A. Side-by-Side and End-to-End Gold Nanorod Assemblies for Environmental Toxin Sensing. *Angew. Chem. Int. Ed.* **2010**, *49*, 5472–5475. [CrossRef] [PubMed]

177. Zijlstra, P.; Paulo, P.M.R.; Orrit, M. Optical detection of single non-absorbing molecules using the surface plasmon resonance of a gold nanorod. *Nat. Nanotechnol.* **2012**, *7*, 379–382. [CrossRef] [PubMed]

178. Murphy, C.J.; Sau, T.K.; Gole, A.M.; Orendorff, C.J.; Gao, J.; Gou, L.; Hunyadi, S.E.; Li, T. Anisotropic Metal Nanoparticles: Synthesis, Assembly, and Optical Applications. *J. Phys. Chem. B* **2005**, *109*, 13857–13870. [CrossRef]

179. Caswell, K.K.; Wilson, J.N.; Bunz, U.H.F.; Murphy, C.J. Preferential End-to-End Assembly of Gold Nanorods by Biotin–Streptavidin Connectors. *J. Am. Chem. Soc.* **2003**, *125*, 13914–13915. [CrossRef]

180. Zijlstra, P.; Paulo, P.M.R.; Yu, K.; Xu, Q.-H.; Orrit, M. Chemical Interface Damping in Single Gold Nanorods and Its Near Elimination by Tip-Specific Functionalization. *Angew. Chem. Int. Ed.* **2012**, *51*, 8352–8355. [CrossRef] [PubMed]

181. Paulo, P.M.R.; Zijlstra, P.; Orrit, M.; Garcia-Fernandez, E.; Pace, T.C.S.; Viana, A.S.; Costa, S.M.B. Tip-Specific Functionalization of Gold Nanorods for Plasmonic Biosensing: Effect of Linker Chain Length. *Langmuir* **2017**, *33*, 6503–6510. [CrossRef]

182. Zhen, S.J.; Huang, C.Z.; Wang, J.; Li, Y.F. End-to-End Assembly of Gold Nanorods on the Basis of Aptamer–Protein Recognition. *J. Phys. Chem. C* **2009**, *113*, 21543–21547. [CrossRef]

183. Xu, X.; Xu, C.; Ying, Y. Aptasensor for the simple detection of ochratoxin A based on side-by-side assembly of gold nanorods. *RSC Adv.* **2016**, *6*, 50437–50443. [CrossRef]

184. Yu, C.; Irudayaraj, J. Multiplex Biosensor Using Gold Nanorods. *Anal. Chem.* **2007**, *79*, 572–579. [CrossRef]

185. Chang, J.-Y.; Wu, H.; Chen, H.; Ling, Y.-C.; Tan, W. Oriented assembly of Au nanorods using biorecognition system. *Chem. Commun.* **2005**, 1092–1094. [CrossRef] [PubMed]

186. Zhu, Y.; Kuang, H.; Xu, L.; Ma, W.; Peng, C.; Hua, Y.; Wang, L.; Xu, C. Gold nanorod assembly based approach to toxin detection by SERS. *J Mater Chem* **2012**, *22*, 2387–2391. [CrossRef]

187. Zhu, Y.; Qu, C.; Kuang, H.; Xu, L.; Liu, L.; Hua, Y.; Wang, L.; Xu, C. Simple, rapid and sensitive detection of antibiotics based on the side-by-side assembly of gold nanorod probes. *Biosens. Bioelectron.* **2011**, *26*, 4387–4392. [CrossRef]

188. Park, W.M.; Choi, B.G.; Huh, Y.S.; Hong, W.H.; Lee, S.Y.; Park, T.J. Facile Functionalization of Colloidal Gold Nanorods by the Specific Binding of an Engineered Protein that Is Preferred over CTAB Bilayers. *ChemPlusChem* **2013**, *78*, 48–51. [CrossRef]

189. Ben Haddada, M.; Hu, D.; Salmain, M.; Zhang, L.; Peng, C.; Wang, Y.; Liedberg, B.; Boujday, S. Gold nanoparticle-based localized surface plasmon immunosensor for staphylococcal enterotoxin A (SEA) detection. *Anal. Bioanal. Chem.* **2017**, *409*, 6227–6234. [CrossRef]

190. Wang, Y.; Liu, X.; Chen, P.; Tran, N.T.; Zhang, J.; Chia, W.S.; Boujday, S.; Liedberg, B. Smartphone spectrometer for colorimetric biosensing. *Analyst* **2016**, *141*, 3233–3238. [CrossRef]

191. Zhang, L.; Salmain, M.; Liedberg, B.; Boujday, S. Naked Eye Immunosensing of Food Biotoxins Using Gold Nanoparticle-Antibody Bioconjugates. *ACS Appl. Nano Mater.* **2019**, *2*, 4150–4158. [CrossRef]

192. Mayer, K.M.; Hafner, J.H. Localized Surface Plasmon Resonance Sensors. *Chem. Rev.* **2011**, *111*, 3828–3857. [CrossRef]

193. Nusz, G.J.; Curry, A.C.; Marinakos, S.M.; Wax, A.; Chilkoti, A. Rational Selection of Gold Nanorod Geometry for Label-Free Plasmonic Biosensors. *ACS Nano* **2009**, *3*, 795–806. [CrossRef]

194. Tian, L.; Chen, E.; Gandra, N.; Abbas, A.; Singamaneni, S. Gold Nanorods as Plasmonic Nanotransducers: Distance-Dependent Refractive Index Sensitivity. *Langmuir* **2012**, *28*, 17435–17442. [CrossRef] [PubMed]

195. Anker, J.N.; Hall, W.P.; Lyandres, O.; Shah, N.C.; Zhao, J.; Van Duyne, R.P. Biosensing with plasmonic nanosensors. *Nat. Mater.* **2008**, *7*, 442–453. [CrossRef] [PubMed]

196. Zhang, L.; Mazouzi, Y.; Salmain, M.; Liedberg, B.; Boujday, S. Antibody-Gold Nanoparticle Bioconjugates for Biosensors: Synthesis, Characterization and Selected Applications. *Biosens. Bioelectron.* **2020**, *165*, 112370. [CrossRef] [PubMed]

197. Jackman, J.A.; Ferhan, A.R.; Cho, N.-J. Nanoplasmonic sensors for biointerfacial science. *Chem. Soc. Rev.* **2017**, *46*, 3615–3660. [CrossRef]

198. Ping, J.; Zhou, Y.; Wu, Y.; Papper, V.; Boujday, S.; Marks, R.S.; Steele, T.W.J. Recent advances in aptasensors based on graphene and graphene-like nanomaterials. *Biosens. Bioelectron.* **2015**, *64*, 373–385. [CrossRef]

199. Park, J.-H.; Byun, J.-Y.; Mun, H.; Shim, W.-B.; Shin, Y.-B.; Li, T.; Kim, M.-G. A regeneratable, label-free, localized surface plasmon resonance (LSPR) aptasensor for the detection of ochratoxin A. *Biosens. Bioelectron.* **2014**, *59*, 321–327. [CrossRef]

200. Tadimety, A.; Zhang, Y.; Kready, K.M.; Palinski, T.J.; Tsongalis, G.J.; Zhang, J.X.J. Design of peptide nucleic acid probes on plasmonic gold nanorods for detection of circulating tumor DNA point mutations. *Biosens. Bioelectron.* **2019**, *130*, 236–244. [CrossRef]

201. McFarland, A.D.; Van Duyne, R.P. Single Silver Nanoparticles as Real-Time Optical Sensors with Zeptomole Sensitivity. *Nano Lett.* **2003**, *3*, 1057–1062. [CrossRef]

202. Raschke, G.; Kowarik, S.; Franzl, T.; Sönnichsen, C.; Klar, T.A.; Feldmann, J.; Nichtl, A.; Kürzinger, K. Biomolecular Recognition Based on Single Gold Nanoparticle Light Scattering. *Nano Lett.* **2003**, *3*, 935–938. [CrossRef]

203. Nusz, G.J.; Marinakos, S.M.; Curry, A.C.; Dahlin, A.; Höök, F.; Wax, A.; Chilkoti, A. Label-Free Plasmonic Detection of Biomolecular Binding by a Single Gold Nanorod. *Anal. Chem.* **2008**, *80*, 984–989. [CrossRef] [PubMed]

204. Guo, L.; Zhou, X.; Kim, D.-H. Facile fabrication of distance-tunable Au-nanorod chips for single-nanoparticle plasmonic biosensors. *Biosens. Bioelectron.* **2011**, *26*, 2246–2251. [CrossRef]

205. Guo, L.; Kim, D.-H. Reusable plasmonic aptasensors: using a single nanoparticle to establish a calibration curve and to detect analytes. *Chem. Commun.* **2011**, *47*, 7125–7127. [CrossRef]

206. Guo, L.; Kim, D.-H. LSPR biomolecular assay with high sensitivity induced by aptamer–antigen–antibody sandwich complex. *Biosens. Bioelectron.* **2012**, *31*, 567–570. [CrossRef]

207. Truong, P.L.; Cao, C.; Park, S.; Kim, M.; Sim, S.J. A new method for non-labeling attomolar detection of diseases based on an individual gold nanorod immunosensor. *Lab. Chip* **2011**, *11*, 2591–2597. [CrossRef] [PubMed]

208. Wu, B.; Chen, L.-C.; Huang, Y.; Zhang, Y.; Kang, Y.; Kim, D.-H. Multiplexed Biomolecular Detection Based on Single Nanoparticles Immobilized on Pneumatically Controlled Microfluidic Chip. *Plasmonics* **2014**, *9*, 801–807. [CrossRef]

209. Ahijado-Guzmán, R.; Prasad, J.; Rosman, C.; Henkel, A.; Tome, L.; Schneider, D.; Rivas, G.; Sönnichsen, C. Plasmonic Nanosensors for Simultaneous Quantification of Multiple Protein–Protein Binding Affinities. *Nano Lett.* **2014**, *14*, 5528–5532. [CrossRef] [PubMed]

Publisher's Note: MDPI stays neutral with regard to jurisdictional claims in published maps and institutional affiliations.

Review

An Overview of Artificial Olfaction Systems with a Focus on Surface Plasmon Resonance for the Analysis of Volatile Organic Compounds

Marielle El Kazzy [1], Jonathan S. Weerakkody [1], Charlotte Hurot [1], Raphaël Mathey [1], Arnaud Buhot [1], Natale Scaramozzino [2] and Yanxia Hou [1,*]

[1] Grenoble Alpes University, CEA, CNRS, IRIG-SyMMES, 17 Rue des Martyrs, 38000 Grenoble, France; marielle.elkazzy@cea.fr (M.E.K.); ssweerakkody@gmail.com (J.S.W.); charlotte.hurot@gmail.com (C.H.); raphael.mathey@cea.fr (R.M.); arnaud.buhot@cea.fr (A.B.)

[2] Grenoble Alpes University, CNRS, LIPhy, 38000 Grenoble, France; natale.scaramozzino@univ-grenoble-alpes.fr

* Correspondence: yanxia.hou-broutin@cea.fr; Tel.: +33-43-878-9478

Abstract: The last three decades have witnessed an increasing demand for novel analytical tools for the analysis of gases including odorants and volatile organic compounds (VOCs) in various domains. Traditional techniques such as gas chromatography coupled with mass spectrometry, although very efficient, present several drawbacks. Such a context has incited the research and industrial communities to work on the development of alternative technologies such as artificial olfaction systems, including gas sensors, olfactory biosensors and electronic noses (eNs). A wide variety of these systems have been designed using chemiresistive, electrochemical, acoustic or optical transducers. Among optical transduction systems, surface plasmon resonance (SPR) has been extensively studied thanks to its attractive features (high sensitivity, label free, real-time measurements). In this paper, we present an overview of the advances in the development of artificial olfaction systems with a focus on their development based on propagating SPR with different coupling configurations, including prism coupler, wave guide, and grating.

Keywords: surface plasmon resonance; olfactory sensors; electronic noses; volatile organic compounds; odorants

Citation: El Kazzy, M.; Weerakkody, J.S.; Hurot, C.; Mathey, R.; Buhot, A.; Scaramozzino, N.; Hou, Y. An Overview of Artificial Olfaction Systems with a Focus on Surface Plasmon Resonance for the Analysis of Volatile Organic Compounds. *Biosensors* **2021**, *11*, 244. https://doi.org/10.3390/bios11080244

Received: 17 June 2021
Accepted: 14 July 2021
Published: 23 July 2021

1. Introduction

Over the last few decades, the detection of gases including odorant molecules and volatile organic compounds (VOCs) has attracted great interest and has become increasingly in demand in various field. VOCs constitute a large class of low-molecular-weight (<300 Da) carbon-containing compounds. They can exhibit odorous properties and are characterized by a high vapor pressure ($\geq$0.01 kPa at 20 °C) and a high-to-moderate hydrophobicity [1]. These small volatile molecules have a wide range of sources, both natural (plants, animals, bacteria etc.) and anthropogenic (fossil fuels, automobile exhaust gas etc.). The majority of VOCs have inimical effects on human health such as headaches and nose, eye and throat irritation [2]. Consequently, monitoring the nature and concentration of these compounds in indoor or outdoor environments can be very important, and sometimes, vital. Additionally, they can be considered as chemical messengers. In fact, their analysis has been shown to reveal a considerable amount of information. For instance, studies in medical diagnostics have identified gases associated with different diseases such as rheumatoid arthritis, cancer, and schizophrenia [3]. Furthermore, a recent study showed the possibility of detecting viral infections such as COVID 19 through exhaled breath analysis [4]. VOC and odor analysis can also have applications in the food, beverage and fragrance industries for quality assessments. Finally, gas sensing can be very useful for security applications (detection of drugs, explosives etc.), environmental monitoring or other usages under development such as augmented/virtual reality [5]. Nowadays, the gold standard for VOC detection involves the use of trained human or canine noses or gas chromatography coupled with mass spectrometry (GC-MS). Indeed, to control the quality of raw materials or final food and perfume products, industries often have recourse to human sensory panels. Trained dogs are commonly employed for security controls or even for the detection of diseases such as prostate and breast cancers [6,7]. Although very sensitive and efficient for field studies, the use of the biological nose presents several drawbacks. For instance, human panels may yield biased subjective results and are prone to fatigue. Dogs require expensive training and their application fields are limited and sometimes risky. The second method, namely, GC-MS, is a highly sensitive and accurate analytical technique that allows separating, identifying and quantifying different VOCs in a mixture. However, analyses require skilled operators and are time consuming and expensive [8]. Therefore, there is a need for an affordable, reliable, portable and sensitive device that allows for a rapid analysis of gases including VOCs. Such a context has prompted many researchers to work on the development of alternative technology such as artificial olfaction systems that overcome the various drawbacks mentioned above.

Herein, artificial olfaction systems include gas sensors, olfactory biosensors and an electronic nose (eN). A gas sensor or olfactory biosensor is a single-sensor device which is able to detect gases and that consists of a receptor coupled with a transducer and a data processing system. Olfactory biosensors use biomaterials as receptors. On the other hand, as stated by Julian W. Gardner and Philip N. Bartlett in 1994 [9], an eN is "an instrument, which comprises an array of electronic chemical sensors with partial specificity and an appropriate pattern-recognition system, capable of recognizing simple or complex odours". By its very nature, the eN is, in fact, a biomimetic device that replicates the odor discrimination principle of the mammalian olfactory system. Thanks to considerable research efforts on natural olfaction, and especially, the Nobel prize winning work of Linda B. Buck and Richard Axel (1991) [10], we know that, in order to distinguish among a myriad of odors, the biological nose uses cross-reactive olfactory receptors (ORs) (about 400 different types in the human nose). This particular feature of ORs (i.e., cross reactivity or partial specificity) allows each receptor to interact with different odorant molecules with differential affinities. Therefore, in the same manner as barcodes, odors are encoded by a combination of olfactory receptors, which consequently allows the nose to have this large detection spectrum. Moreover, to transduce an olfactory stimulus, the biological odor sensor uses an extensively studied "molecular switch": the G protein. Indeed, Buck and Axel showed that ORs belong to the large family of G protein coupled receptors (GPCRs). They are located in the plasma membrane of the cilia, i.e., the dendritic extrusions of the olfactory neurons projected into the mucus covering the olfactory epithelium. When a VOC binds to an OR, the G protein transduction cascade is initiated and the binding event is converted into an electrical signal processed by the olfactory bulb and deciphered by the olfactory cortex. Figure 1 shows the analogy between biological and electronic noses.

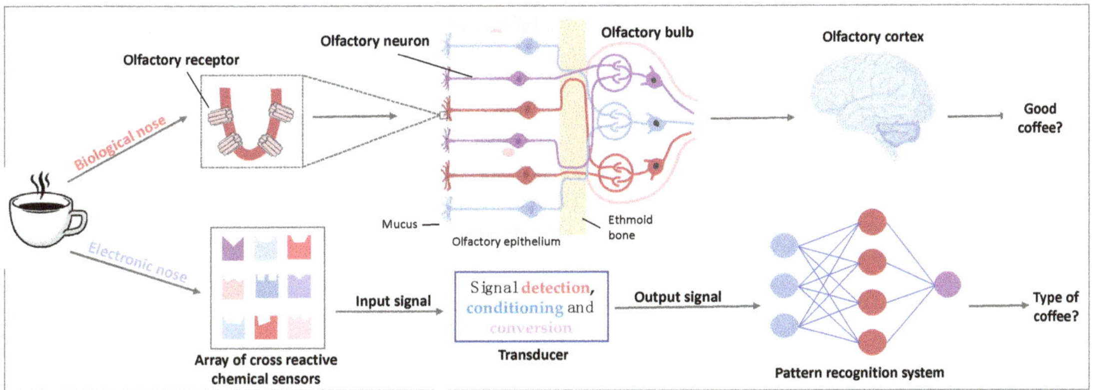

Figure 1. Analogy between the biological and the electronic nose (eN). Figure adapted from [11].

The history of artificial odor detection starts in 1920. In their work on spray electricity and waterfall electricity, Zwaardemaker and Hogewind [12] found that the addition of odorant molecules (e.g., phenol, thymol, citrol) to water markedly raised the spray electricity which could therefore be used to detect these molecules. Subsequently, in 1950, Tanyolac and Eaton [13] attempted to detect air contaminants by measuring variations in the surface tension of a liquid drop. They showed that when contaminated air was in contact with a drop of distilled water, mineral oil or water-stabilized mercury, a considerable change in the surface tension of the drop could be observed. Based on their results, they suggested that an instrument able to classify and measure air contamination at low concentrations could be developed. The first prototype of an electronic device capable of detecting odorants was introduced by Hartman in 1954 [14]. The system was based on polarized microelectrodes as sensing elements. Following this, in 1961, using a thermistor as a transduction device, Moncrieff [15] investigated various coating materials (e.g., polyvinyl chloride, cellulose acetate, milk casein) which interacted differently with odorants. He claimed that using an array of sensors with different coatings could broaden the detection spectrum and, thus, allow for the discrimination of a large number of odors. In 1962, Seiyama et al. [16] developed a gas sensor using semiconductive thin films. The gas detection principle of their system was based on changes in electrical conductivity. A similar study was published in 1965 by Buck et al. [17]. In the same year, Dravnieks and Trotter [18] developed a vapor detector based on the thermal modulation of contact potential. Shaver [19] described a method to enhance the sensitivity of a tungsten oxide gas detector by the addition of a catalytic material such as platinum in 1967. The following year, Taguchi fabricated the first metal oxide semiconductor (MOS) gas sensors for home and industrial usage employing tin oxide as sensitive coating material, which he subsequently patented in 1971 [20]. His company, Figaro Engineering Inc., became the main manufacturer of MOS gas sensors. In 1979, Wohltjen and Dessy [21] introduced the first surface acoustic based gas sensor. However, it was not until 1982, with Persaud and Dodd [22], and then in 1985, with Ikegami and Kaneyasu [23], that the first electronic nose systems based on an array of intelligent chemical sensors emerged. In order to understand the discrimination mechanism of the sense of smell, Persaud and Dodd designed a model of the nose using three Figaro sensors with a differential response spectrum. As a result, their device was able to distinguish among a wide variety of odors, and highlighted the importance of nonspecific interactions in the odor discrimination mechanism. As shown in Figure 2, over the following decades, an exponential number of studies were carried out in order to develop gas sensors and electronic noses. Different sensor systems employing chemiresistive, electrochemical, piezoelectric, and optical transducers [24] have been deployed and assembled in an array to construct eN systems.

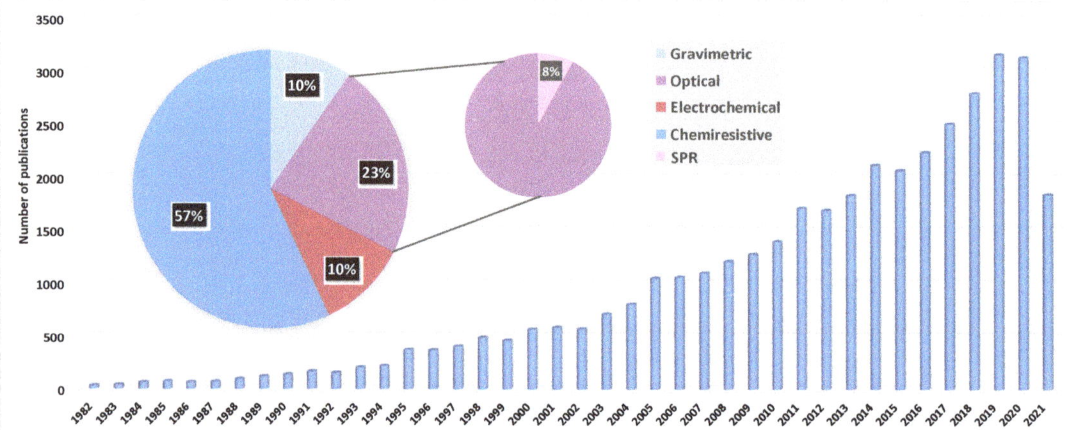

Figure 2. Number of publications on gas sensors and electronic noses since 1982 and the percentage of studies carried out on each type of transduction technique. The data was obtained from Scopus (Keywords used for the histogram: "gas sensor" or "electronic nose" or (gas or vapor or "volatile organic compounds") "sensor array" or multisensor. For the pie chart: "gas sensor" or "electronic nose" or (gas or vapor or "volatile organic compounds") "sensor array" or multisensor and (semiconduct* or chemores* or chemires* or "conducting polymer") or (optical or "surface plasmon resonance" or colorimetric or fluorescen*) or (acoustic or piezoelectric or gravimetric) or (electrochemical).

To date, most eN systems have used chemical layers (metal oxide semiconductor, polymers, etc.) as sensing elements. However, these systems suffer from limited diversity of sensor coatings and poor selectivity. To improve the odor sensing performance, the latest trend consists of using natural biological elements such as ORs and odorant binding proteins (OBPs) or their analogues, such as peptides as sensitive materials [25,26]. Indeed, the sensitivity and selectivity of such receptors have been naturally improved and optimized by millions of years of evolution, making them ideal candidates for odor detection. However, integrating them into an electronic device and maintaining their bioactivity in nonoptimal conditions is very challenging. Promisingly, great improvements have been made in this novel field of olfactory biosensors and electronic noses [25–30].

A large number of reviews have presented the operating principles of the various sensor systems that have been developed so far for VOC and gas detection [8,24,31–36]. In addition, several reviews have focused on the development of gas sensors and eNs based on the main techniques, namely, chemiresistive [11,37–41], gravimetric [42,43], amperometric [44], optical fibers [45], colorimetric and fluorometric [46]. However, to the best of our knowledge, no review has emphasized the development of gas sensors, olfactory biosensors and eNs based on another popular technique, namely, surface plasmon resonance (SPR). Indeed, SPR offers many advantages compared to other techniques, including label free measurement with quantitative and qualitative data, real-time monitoring with information on the affinity and the kinetics of the studied interaction, compatibility with multiplex and high-throughput analyses, reusable sensor chips, and repeatable measurements. Accordingly, in this review, we aim to first give a brief overview of artificial olfaction systems based on various sensor systems, and then a focus on the advances made using SPR.

After this introduction, the second section will review the most common sensing systems currently employed for VOC and gas detection. The third part will be dedicated to advances in SPR-based gas sensors, olfactory biosensors and eNs. It includes a brief description of the theoretical principles of the SPR technique followed by an overview of research works using SPR with different coupling configurations.

2. Gas Sensors and Electronic Noses Based on Various Sensing Systems

As stated, many ingenious systems with different types of sensing materials and transduction techniques have been devised and studied. In the following section, we present a brief overview of the most commonly used sensing platforms for VOC and gas detection. For each system, we will underline the transduction principle, strengths, weaknesses, and present some illustrative examples from the literature.

2.1. Chemiresistive Sensors

This category mainly includes three types of gas sensors, i.e., using MOS, conducting organic polymers (CP) and carbon-based materials [47]. These sensors have a common operating principle whereby the binding of VOCs induces a variation in the electroconductivity. They also have a similar structure that essentially consists of an active layer deposited on a substrate with two electrodes to measure changes in resistance upon exposure to target molecules [39,40,48]. In the following part, popular MOS sensors and CP-based sensors will be discussed more in detail. Gas sensors using carbon material (graphene, carbon-nanotubes, etc.) are not discussed here. More information can be found in recent reviews [49,50].

2.1.1. MOS Sensors

MOS-based sensors are the most commonly used systems for gas and VOC detection among all the sensing technologies [39]. They were first manufactured and marketed by Taguchi in 1968 for gas leak detection [31,35]. These sensors are typically made of a ceramic substrate coated with either n-type (mainly SnO_2, TiO_2, ZnO) or p-type (e.g., NiO) metal-oxide semiconducting film between two electrodes. The ceramic substrate usually contains a heating element that allows the device to reach its operating temperature, generally ranging between 200 and 500 °C [32]. The transduction mechanism of these sensors is based on variations in their conductivity or resistance upon gas molecule binding, which was well addressed in a recently published review [51]. Various factors, such as the bulk resistance, surface effect, grain boundary and contact between the grain interface and the electrode, can affect the electrical properties of gas sensing materials in MOS-based sensors. The detection spectrum and sensitivity of the sensors can be tuned by doping the semiconductor film with noble catalytic metal (e.g., Pt, Pd) [52] or by modifying the working temperature. The grain size, the thickness, and the microstructure and morphology of the coating film can also affect the binding affinity of the device [32,40,53].

These sensors are attractive candidates for eN as they offer high sensitivity with fast response and recovery times. They are also robust and easy to use. Moreover, advances in micro- and nano- fabrication technologies have enabled low-cost production of miniaturized sensor arrays [41,54]. The major drawbacks of these sensors are the lack of selectivity, their susceptibility to humidity and the high operating temperature which leads to high power consumption and reduced lifespan [39,55]. Nevertheless, great efforts have been made to overcome these drawbacks. Low-power microheaters have been designed and new porous structures have been explored [39,54,55]. Moreover, room temperature operating MOS sensors have been developed following different strategies, and involve the use of metal oxide nanostructures such as nanowires, nanotubes and nanobelts [56,57]. MOS sensors and particularly SnO_2-based systems have been extensively studied, miniaturized and combined into arrays for the detection of a large panel of VOCs. Hundreds of outstanding works on experimental and commercial eN systems can be cited. However, this not being the subject of the present review, more details about these systems can be found in the cited reviews [11,34,39,41,54,55,58].

2.1.2. Conducting Organic Polymers Sensors

CP based sensors have received considerable attention since the early 1980s [59]. They are probably the most widely used systems for VOC detection after MOS sensors, and were used in the earlier generations of electronic nose systems [35,36]. CP based sensors are generally composed of a substrate (e.g., glass microscope slide, silicon wafer), on which a film of conducting polymer is deposited between two parallel or interdigitated electrodes [31]. Intrinsic conducting polymers (ICPs) such as polypyrrole, polyaniline, polythiophene and their derivatives have been typically employed for sensor applications [37]. They are usually deposited by electro-polymerization [35]. As for MOS sensors, the transduction principle of these devices relies on variations in the conductivity of the sensors in the presence of VOCs. Several studies have investigated the interaction between the ICPs and the target molecules and suggested different mechanisms [37,38,60]. Reversible modulation of conductance is detected by measuring variations in the current flowing through the polymer when a voltage is applied across the electrodes [31]. The sensing performance of the CPs can be adjusted by modifying the polymer molecular structures, changing the dopants and incorporating a second component into conducting polymers [37]. The addition of a second component gives rise to an original new category of sensing elements called hybrid or composite conducting polymers (CCPs). Further information and examples of CCP-based sensors can be found in the following reviews [37,61].

Unlike MOS sensors, CP-based systems can operate at room temperature, and thus, consume less power. They also exhibit good sensitivity and have short response times [37]. In addition, they are easy to fabricate and resistant to poisoning [8,24]. However, these devices suffer from a lack of selectivity and baseline drift. Moreover, their sensitivity can be affected by humidity and temperature and they can be overloaded by some VOCs resulting in a short lifetime [24,35,54]. Hundreds of papers about CP-based sensors and eNs can be found in the literature [37,38,54]. CP-based gas sensor arrays have been developed for many applications. For instance, Yu et al. have designed a portable array of polypyrrole sensors for the analysis of diabetic patient's breath [62]. Li et al. detected aromatic organic compounds using nanofibers of conducting polyaniline [63]. CP have also been used as sensitive coatings and combined with different sensing platforms such as quartz crystal microbalance [64] and field effect transistors [65].

2.2. Electrochemical Sensors

This family of sensors includes three main categories classified according to their measurement approaches: amperometric, potentiometric and conductimetric/impedimetric sensors [44]. These electroanalytical techniques generally involve monitoring the modulation of an electrical property (current, potential, conductivity or impedance) associated with the interaction of odorant molecules with the working electrode [24]. The working electrode is usually made of gold or platinum and covered with sensing materials, for example, in certain cases, a porous membrane that acts as a transport barrier [35].

These sensors have the advantages of being robust and can function at room temperature [24]. They are also low cost, have low power consumption and can be miniaturized [66], which are all suitable characteristics for eN systems. Additionally, the reactivity of these gas sensors can be customized by adding metal layers, polymers or biological sensing materials to the working electrode surface [34]. However, due to their sensing methodology, some of these sensors have a narrow detection spectrum with a high sensitivity only to a limited number of electrochemically active gases [36]. Several groups have explored the potential of different categories of electrochemical sensors for the detection of VOCs and odorant molecules. For instance, Buttner et al. [67] have demonstrated the usability of an amperometric sensor for in situ detection of explosives in soil. Barou et al. [68] presented a proof of concept for the detection of odorant molecules using square wave voltammetry. Liu et al. [69] designed an olfactory biosensor based on electrochemical impedance spectroscopy (EIS). Also using EIS technique, Hou et al. [70] were able to detect odorant molecules by monitoring the electrical properties of a Langmuir-Blodgett film containing OBPs. In another study [71], employing the same electroanalytical method, the team reported a novel odorant detection strategy using a rat olfactory receptor. As a part of the European project SPOT-NOSED, Akimov et al. [72] worked on the development of nanobiosensors that consist of a single olfactory receptor anchored between nanoelectrodes that detects odorant binding using EIS.

2.3. Field Effect Transistor (FET)

There are several types of FET gas sensors, including thin-film transistor, catalytic metal gate FET, suspended gate FET, capacitively coupled FET and horizontal floating-gate FET. The transduction principle of these devices is mainly based on the modulation of the threshold voltage or the drain source current. Each type of FET sensor has a specific structure, sensing mechanism and characteristics with different advantages and drawbacks. Hong et al. [73] recently published a paper that explains and reviews the operating principle, features and performance of each type of FET sensor.

Many research groups have studied and explored this type of sensor for VOC detection applied to different areas and using various types of sensing materials. For example, Haick's group has extensively worked on the development of silicon nanowire field effect transistors (SiNW FET). The SiNW FET surfaces were modified with different types of organic molecules in order to detect different kinds of VOCs and specially disease biomarkers [74–76]. Park's team developed a highly sensitive FET based bioelectronic noses using single walled carbon nanotubes or polypyrole nanotubes conjugated with human ORs [65,77]. Johnson's group designed and studied VOC sensor arrays using DNA-decorated carbon nanotubes FETs [78–80] and graphene FETs [81]. Kotlowski et al. [82] described an olfactory biosensor employing reduced graphene oxide FET functionalized with OBPs. Liao et al. [83] demonstrated that organic thin-film-transistors are suitable for electronic nose development.

2.4. Gravimetric or Piezoelectric Sensors

Two types of piezoelectric sensors are mainly used for VOC and gas detection: surface acoustic wave (SAW) sensors [8,35] and bulk acoustic wave (BAW) also called quartz crystal microbalance (QCM). A SAW sensor, in delay line configuration, basically consists of two inter-digitated transducers (IDTs) placed on top of a piezoelectric substrate such as quartz or Lithium niobate. To detect target molecules, a sensitive membrane (e.g., conducting polymers, lipids, biomolecules, etc.) is deposited between the IDTs [8]. A QCM sensor comprises a quartz disc coated with two gold electrodes connected to either side of the disc and a layer of sensitive material [35]. Despite their structural differences, both sensors have similar transduction principles. They detect odorant molecules by measuring variations in the resonant frequency caused by a change in mass after VOCs adsorption [8,31,32]. When an alternating voltage is applied across the piezoelectric element, it oscillates at a specific frequency driven by its mechanical properties [31]. This produces 2-dimentional acoustic waves (Rayleigh waves) that propagate along the surface at a frequency between 100 and 400 MHz in SAW sensors. Whereas, in QCM devices, 3-dimentional waves that travel through the bulk at a frequency of 10 to 30 MHz are generated [31].

QCM and SAW sensors have short response time and they are able to work at room temperature. Moreover, the detection spectrum of these devices can be tailored by modifying their sensitive membrane (the sensing materials) [8]. However, they suffer from complex circuitry and limited multiplexing capacity for large sensor array system. Additionally, the coating technologies are poorly controlled resulting in sensors having poor batch-to-batch reproducibility [31]. To tackle this issue, Chevalier et al. [84] showed that diamond nanoparticles can promote homogenous and reproducible coating of SAW sensors. A large number of studies have focused on the development of SAW and QCM based gas sensors and eNs using various sensitive materials. Rapp et al. [85] presented an improved array of eight SAW sensors for the detection of organic gas and an in-built multiplexing technique that allows an easy optimization of signal to noise ratio. They expanded the choice of coatings for the SAW sensors and improved the sensor to sensor reproducibility for a certain coating material. Matatagui et al. [86] recently designed a portable low-cost eN based on SAW sensors and using ferrite nanoparticles as sensing materials for the detection of BTX (benzene, toluene and xylene), which are hazardous gases. Panigrahi et al. [87] worked on the detection of a VOC associated with *Salmonella* contamination in meat using a QCM system coated with synthetic polypeptides. Compagnone et al. reported a QCM sensor array using peptide modified gold nanoparticles for the detection of food aromas [88]. In another study [89], they have investigated the use of metallo porphyrins coated QCM platform for quality control of chocolate. Likewise, Di Natale et al. [90] designed an array of eight QCM sensors coated with metallo porphyrins for the detection of lung cancer. Park's group [91] and Wang's group [92,93] have developed QCM and SAW olfactory biosensors by employing ORs as sensing materials. Furthermore, several studies have explored the performance of QCM based sensors coupled to molecularly imprinted polymers (MIPs) for the detection of VOCs [30].

Other types of gravimetric sensor systems based on film bulk acoustic resonator [94–96], cantilevers [97–101], capacitive micro-machined ultrasonic transducer [102,103] have also been explored and optimized for the detection of VOCs. The following reviews [34,42,43,104] provide more details about these sensors and bring together different research articles that focus on the development of this technique.

2.5. Optical Sensors

This category of sensors detects odorants by measuring variations in the optical properties (e.g., refractive index, fluorescence, absorbance) of the sensing material by monitoring light properties modulation (e.g., wavelength, intensity, phase). They involve the use of a large assortment of techniques including different categories of optical fibers and a diversity of light sources and light-sensitive photodetectors [24]. Depending on the operating principle (i.e., the optical property that is monitored), it is possible to distinguish among several types of optical sensors, each having advantages and drawbacks. It is important to mention that optical spectroscopy (near infrared, infrared, Raman, etc.) is also very promising for gas sensing. Herein, it is not in the scope of this paper and thus will not be considered. More information can be found in a recently published review [105].

The simplest optical sensors effective for electronic nose development are colorimetric sensors. These sensors are based on the measurement of UV−vis absorbance or reflectance and involve the use of chemoresponsive dyes (chromophore) such as metalloporphyrins that will change color upon exposure to VOCs [106]. They have the advantages of being low cost, easy to manufacture and allow real-time multiplexed monitoring of VOCs. However, their main drawback is that they do not offer quantitative measurements [107]. Suslick's group pioneered this technique. They have extensively developed this type of sensors with a large number of published articles where they showed efficient detection of VOCs with very low detection limit for different applications [46]. Hou's group also developed a colorimetric sensor array for the detection of aldehydes and lung cancer biomarkers [108] and for the discrimination of Chinese liquors [109].

Fluorometric or fluorescent sensors are more sensitive than colorimetric sensors and involve the use of fluorophores. They can be categorized into different types based on the fluorescence parameter that is measured (e.g., fluorescence intensity, anisotropy, lifetime, emission and excitation spectra, fluorescence decay, and quantum yield) [24,46,110]. Walt and co-workers pioneered multiplexed fluorescent sensors combined with optical fibers [111]. Indeed, fiber optic platforms are widely used for optical sensor development thanks to their attractive features, including remote and multiplexed sensing capability, biocompatibility, miniaturized structure, light weight, flexibility and immunity to electromagnetic interference [112]. Another main advantage of these systems lies in the temporal response obtained with the kinetic information compared to the equilibrium response obtained with most other sensing technologies. In the field of VOC detection, Walt et al. [111,113] developed an array of optical fibers with a solvatochromic dye (Nile red) encompassed in different polymer matrices with diverse polarity, flexibility, hydrophobicity, pore size and swelling tendency in order to obtain sensors that interact differently with VOCs. The sensitive polymer/dye combinations were deposited at the distal end of the fiber. Changes in the fluorescence intensity at a given wavelength upon the exposure to VOCs were recorded over time thanks to a CCD camera. In another study [114,115], they developed an array of fluorometric fiber optic-based sensors (FOS) where the fluorescent dyes were incorporated into different classes of microbeads. The beads were then immobilized in microwells at the tip of the imaging fiber. Kang's group also developed fiber optic-based fluorometric sensors for VOCs detection [107,116,117]. In particular, the team presented an array of five FOSs using four different types of solvatochromic dyes and two different polymers to form sensitive membranes. The sensing materials were deposited on side-polished optical fibers and pulse width modulations were measured as a response to VOCs [116]. More details and examples about colorimetric and fluorometric sensors can be found in the following reviews [35,46,114,118].

Another important family of optical sensors is based on surface plasmon resonance and involves the excitation of surface plasmons that are extremely sensitive to variations in the refractive index of the sensing materials. In 1982, Nylander et al. [119] investigated the possibility of employing SPR as a transduction technique for gas detection. Using an organic film as a sensing material, their system demonstrated a sensitivity to halothane in the parts per million (ppm) range. Since then, this optical sensing technique has gained substantial popularity. Owing to its prominent attractive features, namely, high sensitivity, label free detection and real time measurements, SPR constitutes a very powerful tool for sensor development comparing to other optical techniques. It has proven to be very useful for monitoring and studying interactions and affinities especially between biological elements (e.g., antibody-antigen, ligand-receptors). Consequently, SPR has been extensively employed for a large number of applications including diseases diagnosis, drug discovery and other bioanalysis [120,121]. Additionally, SPR sensors have been used for the detection of chemical species such as VOCs. Indeed, many research groups have developed efficient gas sensors, olfactory biosensors, and electronic nose systems using SPR as sensing technique. This will be the focus of the following section of the review. To illustrate the progress in this domain, examples of studies with different SPR coupling configurations will be presented and discussed.

3. Propagating SPR-Based Gas Sensors and Electronic Noses

The SPR phenomenon was first observed by Wood [122] in 1902. In his study, he pointed out inexplicable peculiarities in the spectrum of light diffracted by a diffraction grating. To understand this phenomenon, in 1941, Fano [123] re-examined Wood's observations and showed that the anomalous diffraction pattern was caused by the excitation of "polarized quasi-stationary waves" present at the surface of the metallic gratings. In 1952, Pines and Bohm [124] suggested that the energy losses of fast electron passing through foils were caused by the excitation of plasma oscillations or "plasmons" i.e., oscillations of the electronic density in the conducting media. Hereafter, this energy loss and its association with surface plasma oscillations were studied by Ferrell and Stern [125,126], Ritchie [127], Powell [128] and many others. In 1968, Otto [129] presented a method for the excitation of nonradiative surface plasma waves and showed that it resulted in a strong attenuation of the reflected light intensity. Moreover, in the same year, Kretschmann and Raether [130] described another configuration that enabled the excitation of the nonradiative surface plasmons (SPs).

A plasmon corresponds to the collective oscillation of the free electrons in a noble metal [131]. Surface plasmons are collective oscillations of electrons that take place at the interface between two media having dielectric constants of opposite signs typically a metal (e.g., gold, silver) and a dielectric (e.g., air, water) [132]. The SPs are not arbitrary events, they occur upon the excitation or coupling to an electromagnetic photon wave (i.e., light). In fact, when a photon beam interacts with the free electrons of a metal, these electrons will respond by coherently oscillating in resonance with the light wave. This phenomenon is known as surface plasmon resonance and corresponds to the excitation of the SPs.

SPs can be classified into two categories: propagating or localized.

Propagating SPs, also known as surface plasmon polaritons (SPPs) or surface plasma waves (SPWs), are typically produced at the surface of thin metallic layers. SPPs can be considered as electromagnetic waves that propagate along the planar surface of a metal interfacing a dielectric (Figure 3a). The excitation of the SPs in such structures requires the use of coupling elements (e.g., prism, waveguide, gratings) that allow to achieve resonance or matching conditions leading to SPR.

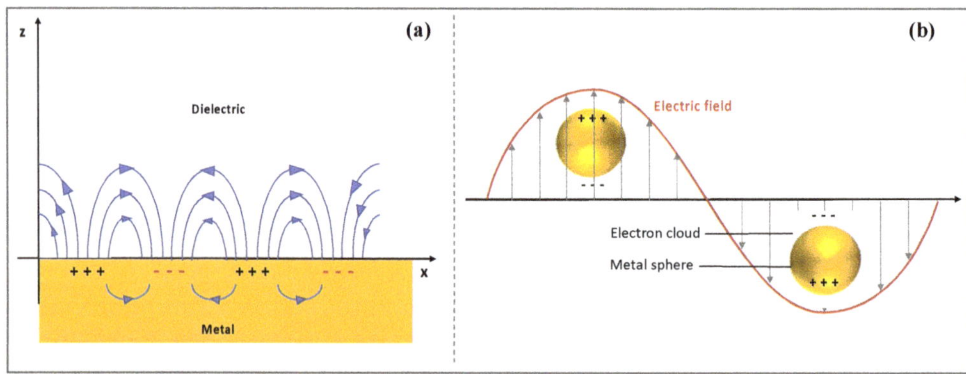

Figure 3. Schematic representation of (**a**) propagating surface plasmon (SP) and (**b**) localized surface plasmon.

On the other hand, localized surface plasmon resonance (LSPR) occurs when light interacts with metallic nanostructures (e.g., gold nanoparticles) that are smaller than the incident wavelength [133,134]. The electric field of the light causes the localized free-electrons in the nanostructure to oscillate with a specific frequency. When the electron cloud is displaced relative to the nuclei a restoring force, generated by the Coulomb attraction between the electrons and the nuclei causes the electron cloud to oscillate relative to the nuclear framework [135] (Figure 3b). This has three main consequences: an enhancement in the local electromagnetic field near the particle's surface and a strong light scattering as well as a sharp spectral absorption with a maximum at the plasmon resonant frequency [136]. For gold nanoparticles (size ranging from few to hundreds of nanometers), a strong absorption pic is observed in the visible light leading to their red color in solution [136]. Unlike the SPR phenomenon, which takes place at the surface of a metallic film, LSPR does not require coupling elements and does not propagate hence its localized character. However, it is likewise sensitive to changes in the local dielectric environment. In particular, the extinction peak (namely the resonance wavelength) is highly affected by the refractive index of the surrounding. Thus, for sensing applications, molecular interactions occurring at the surface of the nanoparticles are typically detected by monitoring shifts in the LSPR wavelength [137]. LSPR sensing platforms consist of either metallic nanoparticles (e.g., nanospheres, nanorods, nanostars), suspended in solution or deposited on a solid support, or micro- and nano- fabricated metallic structures arrays on a solid support (e.g., nanopillar array) [138]. The LSPR peak wavelength can be tuned corresponding to the desired application by modifying the size, shape and material of the nanostructures, which represents an advantage for sensor development [137]. Thanks to the improvement in nanofabrication, various LSPR-based nanosensors have been developed for diverse applications including the detection of various biomolecules such as DNA, disease biomarkers, hormones, amino acids etc. [137,139–141], and different chemical compounds such as inorganic gases [142,143] and VOCs [2,144–146].

In the present review, we will exclusively focus on propagating SPR-based sensor developed for the detection of VOCs. In the literature, a considerable number of reviews and articles that describe and explain the theory behind this phenomenon (propagating SPR) as well as its application for sensor devices can be found [121,147–154].

In the following sections, we will first present a brief theoretical overview of propagating SPR. Then, we will make a comprehensive review of the progress made in the development of gas sensors and electronic noses that employ this technique. In particular, the different systems will be classified based on their coupling configuration.

3.1. The Theory of Propagating SPR

Let us consider a semi-infinite metal with a frequency dependent complex permittivity or dielectric function ε_m and a semi-infinite dielectric with a permittivity ε_d, separated by a planar interface. The solution of Maxwell's equations under appropriate boundary conditions suggests that s-polarized surface oscillations cannot be supported by this type of interface. Consequently, SPWs are transverse-magnetic (TM) or p-polarized waves, i.e., their magnetic field vector is parallel to the interface and perpendicular to the propagation direction [121,154]. Moreover, the existence of surface plasmon requires that the real part of ε_m is negative and its absolute value is greater than ε_d. At optical wavelength (visible and near infrared), this condition is satisfied for various metals including gold which is commonly used for sensor applications [153]. From the analysis of Maxwell's equations, it is also possible to derive the frequency dependent wavevector also called the dispersion relation or propagation constant of the SPW on a smooth surface that is given by [151,155]:

$$k_{SPP} = \frac{\omega}{c} \sqrt{\frac{\varepsilon_d \varepsilon_m}{\varepsilon_d + \varepsilon_m}} \tag{1}$$

where ω/c is the free space wave vector of an optical wave.

The propagation of SPWs along the interface undergoes strong attenuation due to high Ohmic losses in the metal which, consequently, limits the propagation length [149]. This damping is associated with the imaginary part of the wavevector that depends on the metal's permittivity at the oscillation frequency of the SPW [121,149]. The propagation length along the interface is a few microns or even a few tens of microns depending on the metal and the excitation wavelength used [147]. This length can be expressed as [149]:

$$L_{SPP} = \frac{1}{|2Im\{k_{SPP}\}|} \tag{2}$$

Confined at the vicinity of the interface, the electromagnetic field associated with the wave decays evanescently into the metal and the dielectric. However, as shown in Figure 4a the distribution of this field is asymmetric and mostly concentrated in the dielectric [148]. This disparity in the penetration depth is due to the fact that the dielectric constant of the metal is greater than that of the dielectric. The field decay from the surface in the adjacent medium is determined by the dispersion relations of the SPW in the direction perpendicular to the interface (i.e., in the dielectric k_{zd} and in the metal k_{zm}) [155]:

$$k_z = \begin{cases} \dfrac{\omega}{c} \sqrt{\dfrac{\varepsilon_d^2}{\varepsilon_d + \varepsilon_m}} & \text{in dielectric;} \\[2ex] \dfrac{\omega}{c} \sqrt{\dfrac{\varepsilon_m^2}{\varepsilon_d + \varepsilon_m}} & \text{in metal.} \end{cases} \tag{3}$$

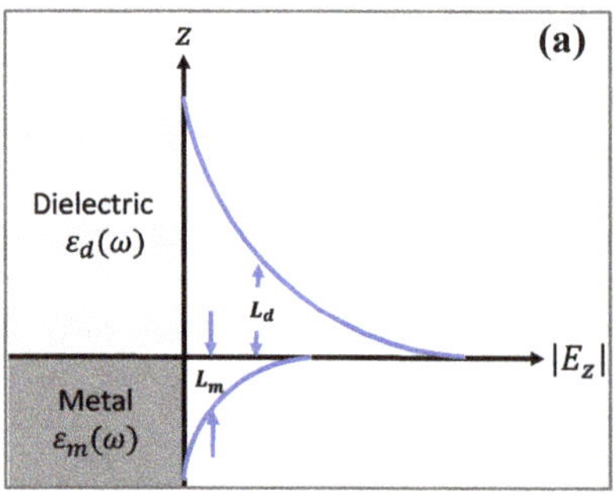 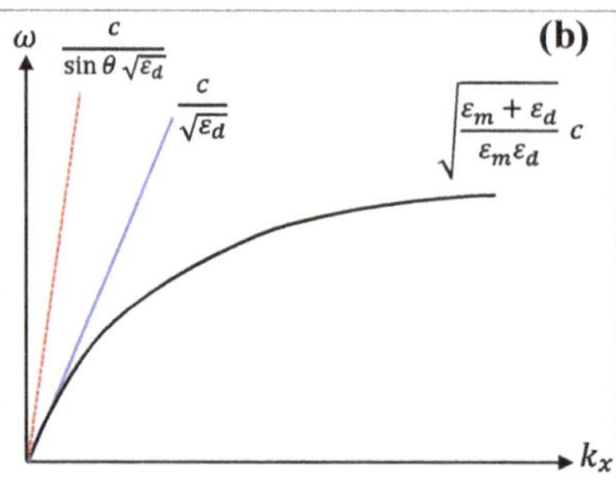

Figure 4. (**a**) Distribution of the electromagnetic field of the surface plasmon polaritons (SPPs) along the z-axis (perpendicular to the surface), the intensity of this field is maximum at the surface and decays exponentially away from it. With L_d and L_m the penetration depth in the dielectric and the metal, respectively. (**b**) Dispersion curve of: free photons propagating in a dielectric (blue solid line), x-component of free photons propagating in a dielectric (red dashed line) and SPPs (black solid line).

The decay length also called penetration depth or skin depth of the SPW in the adjacent medium corresponds to the distance from the interface at which the intensity of the field falls to $1/e$ of its maximum value [148,155]. This value can be expressed as [155]:

$$L_{zi} = \frac{1}{|Im\{k_{zi}\}|} \text{ with } i = \text{metal or dielectric} \qquad (4)$$

To give an order of magnitude, the penetration depth is a few hundred nm (~200 nm) in the dielectric and a few tens of nm (~25 nm) in the metal [147].

The excitation of surface plasmons or the generation of SPWs at the planar interface requires special configurations. Indeed, for the same frequency, the propagation constant (the wavevector) of the surface plasmon at the metal-dielectric interface (black solid line) is higher than the wavevector of photons in the dielectric (blue solid line) (Figure 4b). This mismatch has two consequences. First, the SPPs cannot radiate in light, and are bound to the surface. Second, they cannot be directly coupled or excited by a conventional light illuminating the metal/dielectric interface. Attenuated total reflection (ATR) or diffraction endows the excitation wave with additional momentum to overcome the mismatch and excite SPPs. In practice, this can be achieved using different coupling systems (couplers) such as prim, waveguide and grating couplers [121,147–149]. The excitation of the SPPs manifests itself by a resonant transfer or absorption of the incident light energy resulting in SPR.

As mentioned earlier, SPR is extensively used as transduction technique for optical sensor development and enables the detection of analytes by monitoring changes in the refractive index (n_d) or permittivity (ε_d with $\varepsilon_d = n_d^2$) of the dielectric where the sensing material is deposited. Indeed, since the electromagnetic field of the SPWs is mostly concentrated in this medium, the propagation constant of the wave is strongly affected by its optical properties, namely, its refractive index. The characteristics of the exciting light (i.e., its intensity and phase), are altered upon the interaction with the SPPs and, thus, variations in these parameters can be correlated with changes in the propagation constant of the SPWs and thus the refractive index of the dielectric. In other words, binding-induced modulation in the refractive index at the sensor surface and, consequently, the propagation constant of the SPWs can be detected by measuring changes in the output light properties. Finally, it is worth noting that, since the penetration depth of the field in the dielectric is few hundreds of nm (~200 nm), the SPR can only detect binding events taking place below this limit.

In the following parts, we present the different coupling strategies and review the various studies that employed SPR for the detection of odorant molecules.

3.2. Prism Coupler-Based Sensors

The excitation of SPPs via ATR and prism coupler was first demonstrated by Otto then by Kretschmann and Raether. The Kretschmann configuration is the most commonly used method. This configuration consists of a thin metal film usually gold (about 50 nm thick) deposited on the surface of a prism on top of which the sensitive material is deposited. As shown in Figure 5a, to provoke the coupling, the prism is illuminated with a p-polarized light wave (since the SPW are p-polarized) at an incident angle greater than the critical angle. When the light reaches the prism-metal interface, it is totally internally reflected and an evanescent photon wave is generated at the interface [147]. The high refractive index or permittivity ε_p of the glass prism allows to enhance the momentum or wavevector of the evanescent wave that can thus excite the SPPs [156]. Resonance occurs when the in-plane component of the incident light (photon) wave vector $k_{ph,x}$ (red solid line), which corresponds to the propagation constant of the evanescent wave, matches that of the SPWs. Consequently, a transfer of energy from the incident light to SPWs occurs and is manifested by a sharp dip in the intensity of the reflected light. To satisfy the matching conditions, the angle of incidence or the wavelength of the exiting light can be adjusted since the propagation constant of the evanescent wave is dependent on these parameters. The terms resonance angle and resonance wavelength correspond to values of incident angle and wavelength at which almost 100% efficient coupling and energy transfer are achieved [156]. The resonance condition can be expressed as [121]:

$$k_{ph,x} = \frac{\omega}{c} \sqrt{\varepsilon_p} \sin \theta = Re\{k_{spp}\} \tag{5}$$

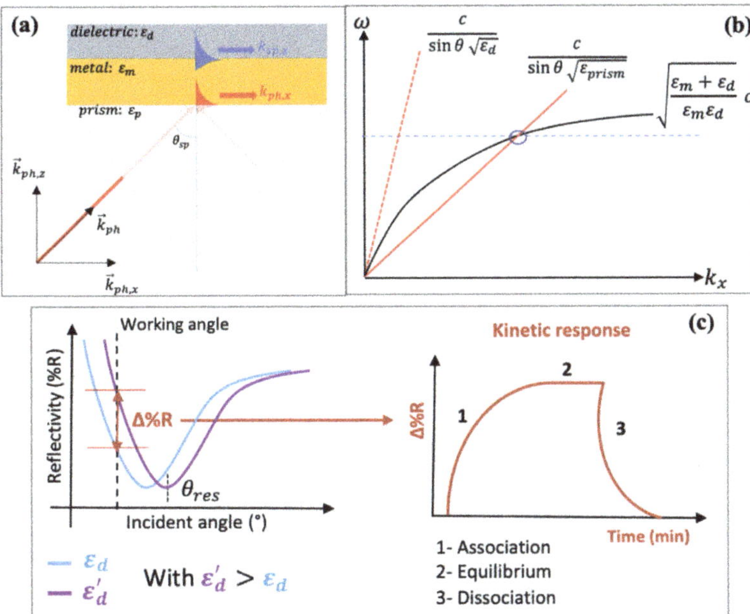

Figure 5. (**a**) Excitation of SPPs by prism coupling in Kretschmann configuration. (**b**) Dispersion curve of: x-component of free photons propagating in a dielectric (red dashed line), x-component of free photons propagating in a prism (red solid line), SPPs (black solid line). (**c**) Intensity interrogation principle.

The same resonant conditions apply for the Otto configuration. The only difference is that, in this configuration the metal film is separated by a small gap from the surface of the prism [129].

In practice, for sensing applications, the sensitive materials are deposited on top of the metal layer, which allows to customize the sensitivity and selectivity of the sensor. Four main measurement methodologies are employed to detect the kinetic interaction of target molecules with the sensitive materials: intensity interrogation, spectral or wavelength interrogation, angular interrogation and finally phase interrogation [157].

In the first method (namely intensity interrogation), variations in the intensity of the reflected light are monitored over time at a fixed wavelength (i.e., using a monochromatic light source) and fixed incident angle (known as the working angle) (Figure 5). The working angle is usually chosen close to the resonance angle (θ_{res}) where small variations in θ_{res} caused by modulation of the surface refractive index will result in large shifts in the intensity of the reflected light. On the other hand, for spectral/wavelength interrogation, a broadband or polychromatic light source is used to excite the SPPs at a fixed incident angle and modulations of the resonance wavelength are monitored. Conversely, in the case of angular interrogation, variations in the resonance angle are measured at a fixed wavelength. Finally, for phase interrogation, shifts in the relative phase difference between p- and s-polarization components are monitored at a specific wavelength and angle. This last interrogation technique offers the highest sensitivity but suffers from a narrow dynamic detection range [158]. The different interrogation methods and especially intensity interrogation allow for simultaneous monitoring of binding events occurring on multiple sensors which is particularly beneficial for electronic nose systems. This multiplexing technique is called surface plasmon resonance imaging (SPR imaging) [159].

Many gas sensors and eN systems can be found in the literature based on this configuration and using a large diversity of sensitive materials including biological elements (e.g., olfactory receptors, odorant binding proteins and peptides) and chemical elements (e.g., polymers and calixarenes). The different systems can be classified into two categories depending on the detection medium, i.e., in the liquid or in gas phase.

3.2.1. Detection of VOCs in Liquid Phase

Prism coupler-based SPR has been widely employed to develop biosensors/biochips for the analysis of large biological molecules. However, it is often considered unsuitable or limited for the analysis of low weight molecules such as VOCs (molecular mass < 300 Da) in the liquid phase. To overcome this limitation, it is essential to couple the optical transduction systems with appropriate sensitive materials in order to generate detectable signals upon their interaction with VOCs. Different biological sensing materials (ORs, OBPs, etc.) have been used for such applications. Very often, signal amplification strategies are needed to obtain reliable SPR signals, which will be highlighted in this review.

Selected and improved by natural evolution, olfactory receptors are very attractive candidates. Since their identification and isolation by Buck and Axel, these proteins have been extensively studied [10]. Great research efforts has been made to deorphanize these receptors [160] and improve their large-scale production that was found to be challenging in some early works [161,162]. The use of OR as sensing materials for the development of olfactory biosensor and eNs presents many assets including high sensitivity and selectivity. In addition, they can be genetically modified to facilitate their purification and immobilization. However, being transmembrane proteins, the presence of a lipid bilayer environment is crucial to maintain their three-dimensional structure and retain their activity when immobilized on sensor chips. This task has been a major drawback and challenge for the development of OR-based olfactory biosensor. Nevertheless, several ingenious strategies have been employed to provide the lipidic environment such as the use of plasma membrane fractions, nanovesicles and nanodiscs [26]. Consequently, OR-based sensors were proven to be effective for the detection of VOCs using different transduction systems including QCM [92], FET [163], electrochemical [71].

SPR platforms have also been associated with ORs. Pajot-Augy's group [164] demonstrated the possibility of using mammalian OR as sensing elements for highly sensitive olfactory biosensors. In their study, they first co-expressed rat ORI7 and human OR17-40 and their associated $G_{\alpha olf}$ subunit in yeast cells. To maintain their structure, the ORs were encompassed in membrane fractions that formed nanosomes with a diameter of approximately 50 nm. The nanosomes were then immobilized on a Biacore sensor chip L1, which consisted of a gold-coated glass support functionalized by a covalently linked carboxylated dextran polymer hydrogel grafted with long alkyl chains (Figure 6). Nanosomes were effectively bound by those alkyl anchors. A BIAcore 3000 was used to perform measurements. This type of setup allows the measurement of resonance angle shifts and consists of a near-infrared LED light source for SPR excitation and a linear array of light sensitive diodes to monitor the reflected light. As reported, no SPR signal was observed when VOCs were injected alone due to poor signal/noise ratio. To solve the problem, an indirect ingenious amplification strategy was designed. It consists of taking advantage of the presence of $G_{\alpha olf}$ anchored to the nanosomes to monitor receptor activation by an odorant ligand, through the desorption of $G_{\alpha olf}$ subunit from the lipidic bilayer. In such a way, when a target odorant binds to the OR, the subunit is activated and then desorbs from the lipidic membrane, resulting in a much stronger SPR signal, as illustrated in Figure 6. To trigger this mechanism, VOCs were injected in the presence of guanosine-5′-triphosphate (GTP). The study demonstrated that ORs retained their functionality in membrane fractions even after immobilization and the obtained olfactory biosensor exhibited high sensitivity and selectivity. The sensor chip kept the same activity level for up to eight injection cycles.

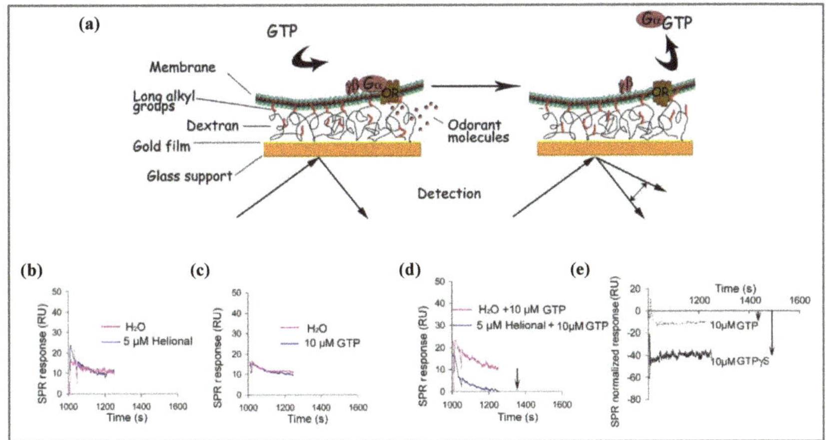

Figure 6. (**a**) BIAcore sensor chip L1 functionalized with nanosomes. No surface plasmon resonance (SPR) response was observed when nanosomes were stimulated either with odorant alone (**b**), or guanosine-5′-triphosphate (GTP) alone (**c**), as compared to the control stimulated with water. The SPR signal was only observed when odorant and GTP were injected simultaneously (**d**). The signal relative to the release of the G_α subunit can be further enhanced four-fold by replacing GTP by GTPγS (**e**) [164].

In a complementary study [165], using this SPR sensing strategy, they investigated the molecular mechanisms underlying odorant detection, in particular, the role of OBPs in the dynamic interactions between OR and odorant ligands. They showed that OBPs play an active role in preserving the conformation and activity of OR especially at high odorant concentration. This finding revealed another role of OBPs in olfaction, in addition to their role in transporting odorants through the olfactory mucus. Importantly, their study showed that SPR-based olfactory biosensors can be used not only for the analysis of odorant molecules, but also for the investigation of basic biologically relevant questions in olfaction.

Furthermore, in collaboration with Jaffrezic-Renault's team, they demonstrated the importance of the surface chemistry on the performance of the system [166]. Human OR17-40 modified with a cmyc tag on the N-terminus and its $G_{\alpha olf}$ subunit were co-expressed in yeast cells (*S. cerevisiae*). The receptors carried by nanosomes were attached to the sensor chip through specific antibody-directed immobilization using Anti-cmyc monoclonal antibodies. Two strategies involving different biofilm architectures were explored: one with controlled antibody orientation and the other with random orientation, as illustrated in Figure 7. A Kretschmann-type SPR spectrometer NanoSPR-6 with two optical channels and a diode light source (650 nm wavelength) was used to perform the study. The response of the system was measured in terms of resonance angle modulation. The setup included a double channel Teflon flow cell that allowed signal acquisition in both custom and differential modes (delta between working and reference channels). They showed that the density of nanosomes and the multilayer bulk thickness are crucial factors for the performance of the olfactory biosensor. The biofilms prepared following the first surface chemistry strategy had higher thickness and nanosome density. However, the corresponding olfactory biosensor exhibited a lower sensitivity for the target odorant molecules compared to the OS based on the second surface chemistry. Indeed, the second strategy provided biofilms with lower thickness and higher porosity that allowed a better accessibility of $G_{\alpha olf}$ to GTPγS, and thus, increased sensitivity.

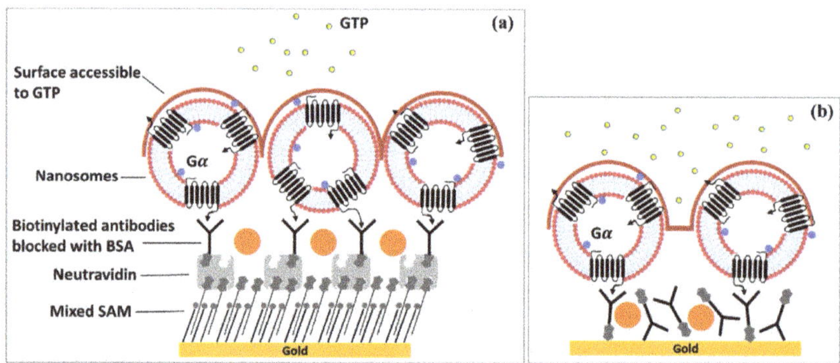

Figure 7. Schemes of the two-surface chemistry employed for the immobilization of olfactory receptors (ORs) in the nanosomes, which were specifically captured via anti-cmyc antibody attached to the gold-coated substrate in an orientated (**a**) or random way (**b**).

Another strategy to exploit the potential of OR for sensing applications is to use so-called artificial olfactory cells, which are genetically modified cells that express olfactory receptors. Park's team developed a sensitive and selective SPR-based olfactory biosensor using whole cells expressing olfactory receptors ORI7 as sensitive materials [167]. The cells were attached to a gold-coated glass slide using poly-D-lysine. The slide was put into optical contact with a prism using a refractive index matching fluid. A p-polarized laser light with a wavelength of 670 nm was used as the probe beam. Thanks to a photodiode detector, variations in the reflected light intensity were monitored as a response to analytes.

In this system, the SPR signal was not directly ascribed to the conformational change of the OR or to the desorption of the G_{α} subunit. In fact, the olfactory receptors expressed on the surface of the cell were not in the detectable range of the SPR (approximately 200 nm above the gold surface), since the size of the cell was several micrometers. However, the G-protein transduction cascade induced by odorant binding generated changes in the intracellular components, mainly with an increase in Ca^{2+} ions. Such changes generated a variation in the local refractive index consequently leading to an SPR signal. In a previous study [168], the group had already demonstrated the feasibility and effectiveness of such a system (i.e., an SPR-based sensor with artificial olfactory cells expressing OR) for the detection of odorants (Figure 8).

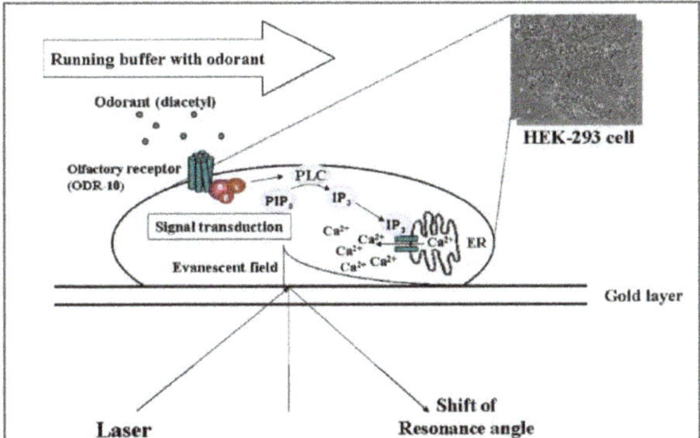

Figure 8. Principle of cell-based measurement of odorant molecules using SPR. An olfactory cell expressing OR was adhered to the gold surface of the sensor chip, and activated by odorant molecule diacetyl. The specific binding of diacetyl to the OR triggered the G protein transduction cascade inside the cell and thus an SPR signal [168].

Although the cell-based olfactory biosensor is very interesting, it is limited by the short lifetime of the sensitive materials. In addition, the system is easily influenced by environmental conditions. Therefore, in another work [169], Park's team explored a different strategy to provide a natural lipidic environment to maintain the stability and biological function of ORs and that is to use liposomes (Figure 9). They controlled the size of the liposome to 40–50 nm, making them fall within the detectable range of the SPR. The liposomes were then immobilized on the poly-D-lysine-coated SPR sensor chip. Their study demonstrated that the reconstituted ORs carried by liposomes were effective sensitive materials for odorant detection.

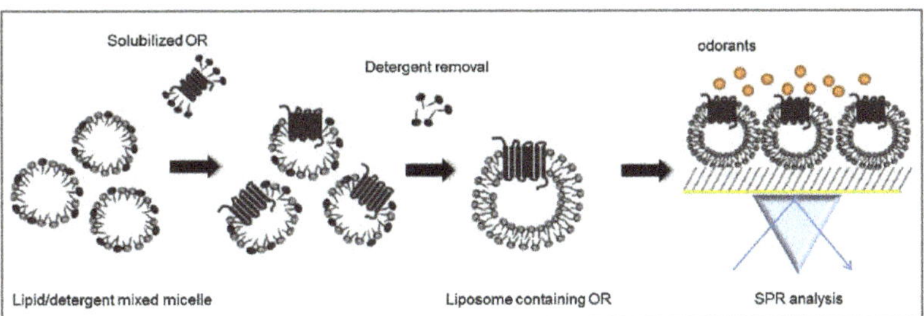

Figure 9. Schematic diagram of reconstitution of OR and SPR analysis. The partially purified OR was reconstituted using lipid/detergent mixed micelle and immobilized on the gold surface of SPR to detect the odorant binding [169].

In a similar work, Sanmartí-Espinal et al. [170] prepared nanovesicles from yeast membranes, with a size of about ~100 nm in diameter, to carry ORs as sensitive materials. Their SPR-based olfactory biosensor had good selectivity. Based on the SPR signal, they even tried to quantify the number of odorants that interacted with a given olfactory receptor.

In addition to ORs, odorant binding proteins also have great potential as sensitive materials in the field of olfactory biosensors. OBPs are small proteins (~20 kDa) highly concentrated in the nasal mucus of vertebrates [171] and in the sensory lymph of insects [172]. Vertebrate OBPs belong to the lipocalin family, characterized by β-barrel structure with eight antiparallel β-sheets that enclose a hydrophobic binding cavity for odorants also known as calyx [173]. Thanks to their binding pocket, OBPs can reversibly bind odorant with micromolar dissociation constant and a broad affinity spectrum (i.e., can interact with different chemical classes) [173]. These proteins are thought to act as shuttles that facilitate the transport and diffusion of hydrophobic odorants across the aqueous mucus to reach the olfactory receptors [174].

Unlike ORs, OBPs are soluble proteins, and thus, do not require a lipidic environment. This also facilitates their large-scale production and purification. They exhibit good stability to high temperature and pH variations, as well as low susceptibility to proteolytic degradation [27]. Moreover, they have a broad specificity and can be genetically modified to tailor their binding properties or facilitate their immobilization. Despite their high stability, maintaining the activity of these proteins over time after their immobilization on the sensor chip and/or after exposure to VOCs is challenging especially in a dry working environment. Nevertheless, many studies have largely investigated the suitability of these sensitive materials for the development of olfactory biosensors and eNs. Indeed, OBPs have been coupled to different transduction platforms (e.g., SAW [175], FET [176]) and their performance were evaluated in both liquid and gas phase [26].

Recently, our team successfully demonstrated the feasibility of a SPR-based OS with OBPs as sensing elements [177]. For that study, three rat OBP3 derivatives with customized binding properties were designed and produced, including OBP3-w, OBP3-a and OBP3-c. The first protein corresponded to the wild type form while the two others were genetically modified mutants. Thanks to site-directed mutagenesis, the binding affinities of the OBPs were customized by varying certain amino acid residues of their binding site. OBP3-a was tuned to have good affinity for aldehydes by introduction of a lysyl residue, while OBP3-c was modified with bulky amino acids to block the binding pocket. Consequently, it could no longer interact with VOCs and was used as negative control. The recombinant proteins were all expressed in *E. coli*. They were immobilized by self-assembly on gold-coated prism by means of a cysteine group that was introduced to their N-terminus, located on the opposite side of the binding cavity. This functionalization strategy allowed easy and orientation-controlled protein immobilization with the OBP at the vicinity of the gold surface. The SPR measurements were performed using a commercial SPR imaging apparatus (SRRiPlex from Horiba). The microarray was illuminated with p-polarized light at 663 nm wavelength. The intensity modulation of the reflected light at a fixed working angle of all the sensors was monitored simultaneously thanks to a CCD camera upon addition of VOCs (Figure 10).

The obtained SPR-based olfactory biosensor had a very low detection limit (DL), e.g., 200 pM for the odorant β-ionone. This result is among the lowest DL reported in the literature. Moreover, the SPR system was able to detect odorants with a molecular weight of 100 g/mol (hexanal) which is lower than DL in mass commonly admitted for commercial SPR imaging, namely, 200 g/mol. Indeed, the intensity of the SPR signal obtained could not be explained solely by the increase in mass after the binding of VOCs on the chip. It is very likely that the binding of VOCs to OBPs induced a conformational change, which led to a variation of the local refractive index with amplified SPR imaging signals. This was possible thanks to our functionalization strategy that enabled the immobilization of the OBPs at the vicinity of the gold surface. Moreover, at low VOC concentration, the olfactory biosensor exhibited an extremely high selectivity with great potential for trace VOC detection.

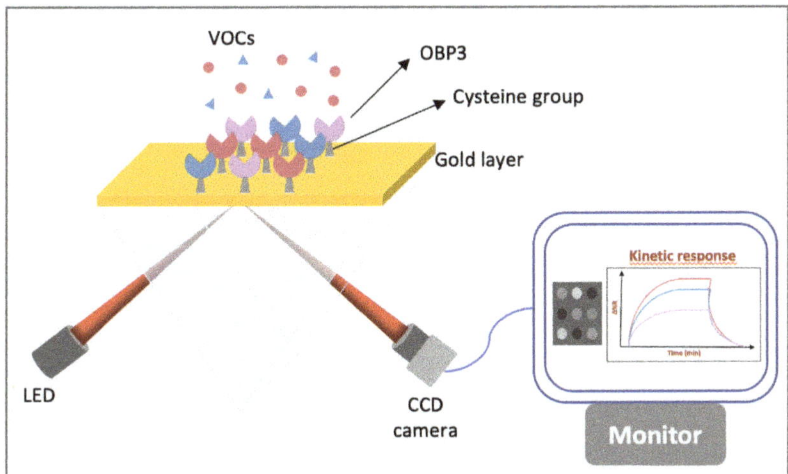

Figure 10. Schematic representation of the SPR-based olfactory biosensor with odorant binding proteins (OBPs) as sensing elements. The three rat OBP3 mutants were immobilized on the gold surface of a prism and their interaction with volatile organic compounds (VOCs) was monitored by SPR imaging.

Biomaterials unrelated to the olfactory system were also used as sensitive materials. Dung et al. developed an efficient SPR-based olfactory biosensor for the detection of toluene using the toluene binding domain (TBD) [178]. TBD belongs to the TodS protein present in the bacterium *Pseudomonas putida*. In this study, a direct immobilization strategy was also employed by introducing three cysteine residues to the N-terminus of the TBD protein. This allows, on the one hand, to control the protein orientation to ensure good accessibility of the binding pocket, on the other hand, to detect SPR signal induced by the conformational change of TBD upon toluene binding. Shifts in reflected light intensity were monitored by a photodiode receptor as a response to analytes. The TBD-based olfactory biosensor showed not only good sensitivity for the target VOC, with DL at 15.62 µM, but also high specificity, with no response for other aromatic hydrocarbons, such as p-xylene and benzene.

The Table 1 summarizes the conditions for VOCs detection of SPR-based olfactory biosensors and their performances in liquid phase.

3.2.2. Detection of VOCs in Gas Phase

The first studies showing the feasibility of prism coupler-based SPR for gas detection date back to the early 1980s [119,179]. However, very few examples were reported in the literature before 2000 [180–184]. These systems were limited in terms of sensitivity and selectivity based on only one or few sensitive chemical layers. Since 2000, an increasing number of articles can be found in the literature using both biological and organic sensitive materials [185–204]. It has been demonstrated that SPR is very effective for sensing VOCs in the gas phase. In fact, when using air as the analysis medium, the detection noise remains relatively low thanks to the low optical index of this medium. Consequently, the binding of the small VOCs can generate reliable SPR signal with very high signal/noise ratio.

For the development of SPR-based olfactory biosensors and eNs for VOC detection in the gas phase, the use of biomolecules such as ORs and OBPs as sensitive materials is limited by their stability under such conditions. Their peptide analogues are particularly interesting alternatives. Indeed, peptides, and in particular, short ones, do not require specific conditions (i.e., humidity, temperature, phospholipidic matrix) to maintain their activity. Moreover, they are much easier to produce and immobilize onto a sensing platform.

Table 1. SPR-based olfactory biosensors in the Kretschmann configuration for the detection of VOCs in liquid phase.

Interrogation	Amplification Strategy	Sensing Material	Performance	Refs.
Resonance angle	Desorption of the $G_{\alpha olf}$ subunit and possible conformation change	Rat ORI7 Human OR17-40 (Carried by nanosomes)	• Conservation of the binding affinity => high selectivity • Repeatability: up to eight activation cycles	[164]
Resonance angle	Desorption of the $G_{\alpha olf}$ subunit and conformational change	Human OR17-40 (Carried by nanosomes)	• Conservation of the binding affinity => high selectivity to helional • Stability: two days	[166]
Reflected light intensity	G-protein transduction cascade	Rat ORI7 (Carried by artificial olfactory cell)	• Conservation of the binding affinity => high selectivity to octanal • Octanal detection limit: 0.1 mM	[167]
Reflected light intensity	Possible conformational change	Three rat OBP-3 mutants	• Very low detection limit in concentration: 200 pM for β-ionone and in molecular weight of VOCs: 100 g/mol for hexanal • Higher selectivity at low concentration of VOCs • Repeatability from measurement to measurement and from chip to chip • Lifespan up to almost two months	[177]
Reflected light intensity	Possible conformational change	Toluene binding domain (TBD)	• High selectivity and sensitivity to toluene (detection limit: 15.62 μM)	[178]

Recently, our group developed an innovative optoelectronic nose using biomimetic peptides based on SPR imaging for the detection of VOCs in the gas phase [185]. For this purpose, a homemade SPR imaging system based on the Kretschmann configuration was constructed, shown in Figure 11. A polarized LED light beam with a 632 nm wavelength was used to excite SPs and a 16-bit CDD camera was used to simultaneously monitor the reflectivity of all the sensors on the chip in real-time. Variation in the reflectivity at a fixed working angle was measured over time upon the exposure of the sensor microarray to VOCs, providing a temporal response.

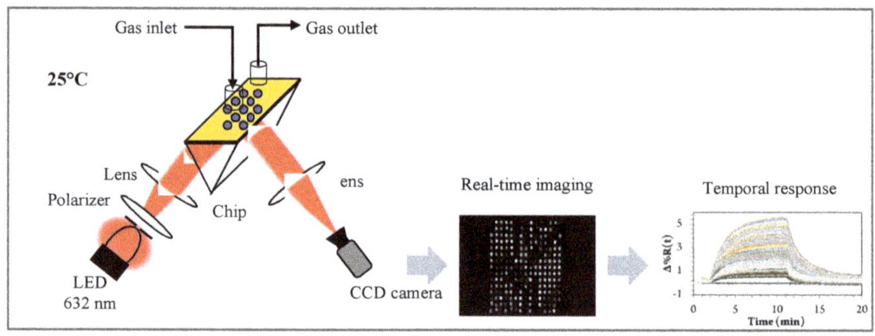

Figure 11. Schematic presentation of the home-made SPR imaging setup.

Such an SPR imaging system is very promising for the development of eN. First, a chip consisting of a large sensor array can be easily prepared and used. The number of sensors is only limited by the resolution of the microarray printing of the sensitive materials. Second, thanks to the imaging mode, the interactions between VOCs and all sensors can be simultaneously monitored using the same instrument. Finally, SPR imaging can provide temporal responses with additional kinetic information compared to a simple equilibrium response obtained with most of the existing eNs.

The peptides were all terminated by a cysteine for their direct immobilization on the gold surface of prism. Thanks to their diverse physicochemical properties and cross-reactivity for VOCs, the obtained eN was found to be very effective in sensing VOCs of different families. In particular, it exhibited extremely high selectivity, capable of discriminating between VOCs differing by a single carbon atom. Additionally, it showed good repeatability and stability under repeated use and prolonged storage.

In order to improve the performance of our eN, in another study [186], we investigated the influence of the wavelength of the LED on the sensitivity of the system by combining numerical simulations with experimental validation. The results showed that the angular sensitivity increased with the wavelength but the angular linearity range decreased due to the narrowing of the plasmon resonance curve at high wavelength. Therefore, a compromise must be made to choose the optimal wavelength depending on the study purposes. Under optimal conditions, the detection limits of our eN reach low parts per billion (ppb) range for VOCs such as 1-butanol.

Furthermore, we investigated the optical contributions to the sensitivity of the SPR imaging [187]. For this, an original characterization method, which was independent of the carrier gas, was established for the SPR prism sensitivity based on pressure jumps [205]. In this work, the impact of different adhesive layer (Cr, Ti) as well as surface topography on the system sensitivity was evaluated. It was found that even though slightly higher sensitivities were theoretically achieved using Ti/Au prism, Cr/Au prisms were more suitable for eN applications since they showed lower sensitivity variabilities, noise, and signal drift due to better adhesive properties. Furthermore, the sensitivity loss due to Au grain-related SPP damping was fully characterized and numerically validated to be free from additional fitting parameters. The adsorption of water vapor was later characterized for such Au surfaces to understand humidity related effects on the eN system. Finally, our study showed that prism sensitivity decreased with increasing temperature [206].

In order to diversify the sensitive materials for eN development, in collaboration with Compagnone's team [191], we tested six novel penta-peptides and nine hairpin DNA selected by virtual screening. Thanks to the complementarity of their binding properties, the obtained eN was able to discriminate not only between VOCs of different chemical families, but also VOCs from the same family with only 1-carbon difference such as 1-butanol and 1-pentanol.

Considering the outstanding potential of our eN system and its great ability to detect and discriminate VOCs, a miniaturized version, called NeOse Pro, was further developed by the company Aryballe. Using the same biomimetic peptide-based chip, Maho et al. [188] demonstrated that NeOse Pro was even able to discriminate between two chiral forms ((R) and (S)) of Carvone and Limonene (Figure 12). Such performance is exceptional for eN system.

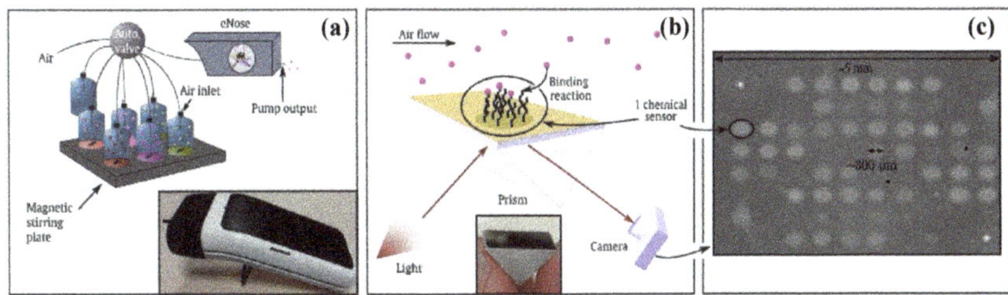

Figure 12. (**a**) Portable NeOse Pro and the experimental set-up for VOC sampling, (**b**) working principal and (**c**) raw image of the prism surface with each spot corresponding to a sensor [188].

NeOse Pro is a very promising tool for field analysis, although, as with most eNs, its use for the headspace analysis of highly humid samples remains a challenge, since its performance may be deteriorated by the presence of a high background signal generated by water vapor from aqueous samples. Slimani et al. [189] have tackled this issue by using a miniaturized silicon preconcentrator packed with hydrophobic adsorbent coupled to the NeOse Pro (Figure 13). As a result, the eN showed not only a great improvement in the detection limit (lowered by 125-fold) for target VOCs, but also an enhancement in the discrimination ability demonstrated by the analysis of eight different flavored waters.

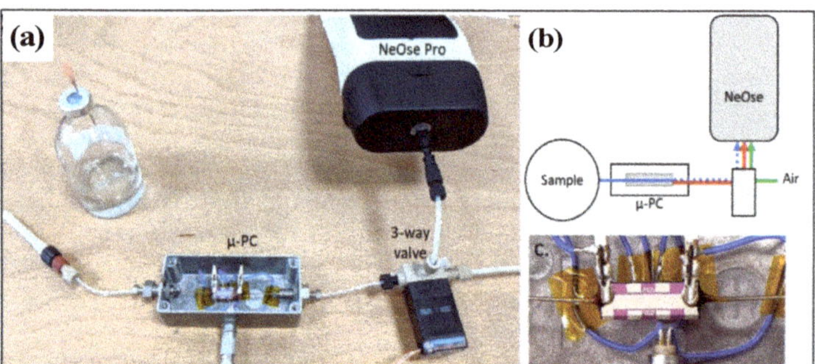

Figure 13. NeOse Pro and μ-preconcentration system coupling. (**a**) Experimental setup, with the sample vial. (**b**) Schematic view of the NeOse Pro/micro preconcentrator (μPC) system and (**c**) View of the preconcentration chip on the metalized side [189].

In a recent study, Fournel et al. [190] compared the performance of the NeOse Pro with human olfaction. They found that the responses of the eN were not a mere reflection of the chemical space of odorants, but rather, that semantic dimensions were also prominent, similar to natural olfaction.

Besides biomolecules, chemical sensitive materials such as cavitands (calixarenes, cyclodextrins) were also used for the detection of gaseous VOCs with prism coupler-based SPR. They are very interesting for trapping VOCs thanks to their molecular structures with cavities, whose sizes, shape and physicochemical properties can be tuned using a wide variety of functional groups.

Daly et al. [192] ingeniously designed new cavitands containing a carboxylic acid group at the upper rim of the cavity for the detection of organophosphorus vapors, and in particular, the sarin nerve gas stimulant dimethylmethylphosphonate (DMMP). The formation of a hydrogen bond between the COOH moieties and P = O group of DMMP was expected. Two different cavitands with four alkyl feet (five carbons long) were produced and studied. Both molecules had similar cavities but with the carboxylic acid group pointing either out of or into the cavity. Their sensitivity to DMMP was compared with that of fluoropolyol, a commonly used polymeric sensing layer for DMMP detection. Cavitands and fluoropolyol layers were deposited on gold-coated glass slide by spin coating and Langmuir-Blodgett technique for comparison. Both techniques allow for the preparation of uniform and homogeneous thin films with a controlled thickness. To perform measurements, a variable wavelength SPR setup in the Kretschmann configuration was used. The interaction between the DMMP and the sensing layers was monitored by measuring the shift in the SPR wavelength at a fixed incident angle upon the exposure to analytes. The results showed that both cavitand layers exhibited almost the same sensitivity and were able to detect ppb levels of DMMP with a rapid and reversible response. The orientation of the COOH group had no effect on DMMP binding, but had strong impact on water uptake. The cavitand-based gas sensor outperformed the fluoropolyol-based one in terms of DMMP sensitivity and with less interference from water vapor and alcohol. Therefore, such a gas sensor is promising for sensitive and specific detection for nerve gas agents. Moreover, the use of cavitands as sensitive materials for SPR based detection of aromatic vapors was also reported by Feresenbet et al. [193].

In a recent study, Şen et al. [194] worked on the development of gas sensor for the detection of VOCs and in particular acetone using synthesized tetranitro-oxacalix[4]arenes. To perform the study, three nitro-substituted heterocalix[4]arenes were synthesized. Thin films of the three sensing materials were deposited on a substrate by spin coating. Their sensing properties for acetone, chloroform, toluene, ethanol and benzene vapors were evaluated by SPR. A BIOSUPLAR 6 Model spectrometer was used to perform SPR measurements. A p-polarized light with a wavelength of 632.8 nm was used to excite the SP. The intensity of the reflected light at a fixed working angle was recorded by a photodetector as a function of time upon the injection of VOCs. The sensing performance of the three films were investigated at room temperature and the VOCs were carried by dry air to avoid the effect of water vapor. As a result, two of the three thin films showed high sensitivity and selectivity to acetone with a detection limit of 3.8 ppm. The system also exhibited a fast and reproducible response with short recovery times (few seconds).

Other chemical sensitive materials such as polymers were also explored and combined with prism coupler-based SPR for the development of gas sensor. Capan et al. [195] investigated the performance of poly(methylmethacrylate) (PMMA) film as a sensitive material for the detection of BTEX (benzene, toluene, ethylbenzene and m-xylene). PMMA films with different thicknesses were deposited onto gold coated glass substrates by spin coating. The different films were obtained by varying the concentration of the polymer solution and the spin speed. The SPR measurements were performed using a Kretschmann type optical setup and a p-polarized monochromatic light at a wavelength of 633 nm was used to excite the SPs. Optical contact between the substrates and a semicylindrical prism was achieved using an index-matching liquid. Two interrogation methods were used to monitor the response of the system upon VOC injection: modulation in the reflection intensity over time at a fixed working angle and shifts in the resonance angle. As a result, among all the BTEX gases, benzene produced the highest SPR response when exposed to PMMA films. Moreover, the response to the other VOCs was very low which indicated that the gas sensor had high selectivity to benzene. Additionally, the team studied other sensitive materials such as calix-4-resorcinarene films [196] and poly[3-(6-methoxyhexyl)thiophene] derivatives films [197] for sensing BTEX and other VOCs using SPR.

Nanto et al. [198] also used synthetic polymer thin films as sensing membranes for the detection of harmful gases such as ammonia and amines with an SPR-based sensor in the Kretschmann configuration. An LED emitting at a wavelength of 660 nm was used as light source and the reflected light was measured by a CCD camera. The response of the system was measured in terms of modulation of the resonance angle as a function of time upon VOC injection. The sensitivity of two types of polymers was investigated: acrylic acid and styrene. A thin film (several tens of nm) of each polymer was deposited on the gold-coated surface of a prism using plasma chemical vapor deposition (CVD). The response of both membranes was tested against eleven harmful gases: ammonia, acetaldehyde, propionaldehyde, xylene, toluene, trimethylamine, triethylamine, dimethyamine, hormaldehyde, acetic acid and butyl acetate. As a result, the gas sensor with the acrylic acid membrane responded only to the basic gases (i.e., ammonia and amines) with high sensitivity and selectivity. In contrast, the OS with the styrene membrane exhibited a 200 times lower sensitivity. The system with the acrylic acid membrane also exhibited a linear response for ammonia in the range of 50–300 ppm and with an estimated detection limit of several ppm. Finally, the study showed that the thickness of the sensing membrane can be optimized to improve the sensitivity. In another study, using a similar system, Nanto's team [199] successfully demonstrated the feasibility of multiplexing with a two-channel odor sensor able to simultaneously detect ammonia and acetic acid with high selectivity. The sensor was based on the same SPR setup but with two sensing membrane, namely, acrylic acid and N,N-dimethylacetamide thin films deposited on one chip by CVD. Two channels of the CCD camera were used to monitor the response to VOCs.

To improve the sensitivity of polymer-based gas sensors, one strategy is to introduce nanoparticles (NPs) such as gold NPs. According to the literature the incorporation of Au NPs in SPR sensors could enhance the sensitivity of the device [207]. Indeed, with a rational design, coupling between the localized surface plasmons of the Au NPs and the propagating surface plasmons of the Au substrate may take place, which can result in a larger plasmon angle shift and changes in reflectivity.

Sih et al. [200] developed an SPR-based gas sensor for the detection of alcohol vapors. In the study, the performance of polythiophene (PT) films as a sensing material was compared with that of Au NPs thin films capped with conjugated oligothiophenes. SPR measurements were performed using a Kretschmann configuration setup and a p-polarized light at a wavelength of 632.8 nm was used to excite the SPs. To prepare the chips, the Au NP/oligothiophene (NPOT) film (~60 nm thickness) was electrodeposited on a gold-coated glass slide and the PT film (~7 nm thickness) was deposited by electropolymerization. The response of the sensors was monitored by measuring the shift of the resonance angle. The performance of the two sensitive materials was tested upon exposure to vapors of five solvents: hexanes, toluene, ethanol, methanol, and water. As a result, the PT layer responded to ethanol, methanol and toluene whereas the NPOT film responded exclusively to alcohols. Therefore, there was an improvement in selectivity in incorporating Au NPs. However, in this study no significant improvement in sensor sensitivity was observed.

Another advantage of nanostructures for gas sensor application is the high surface to volume ratio. Alwahib et al. [201] tested the efficiency of a SPR-based OS with a reduced graphene oxide/maghemite (rGO/γ-Fe$_2$O$_3$) nanocomposite film as sensing layer for hydrocarbon vapor detection. They used a kretschmann-based SPR setup with a helium-neon (He-Ne) laser at 633 nm emission wavelength. A chopper and a polarizer were used to generate the p-polarized excitation beam and a photodetector to monitor the reflected light (Figure 14). Trilayer and bilayer sensing membranes were prepared and compared. The former consisted of a nanocomposite layer (3 nm thick) sandwiched between two gold layers (bottom layer: 37 nm thick, top layer: 2.7 nm thick). For the latter, the rGO/γ-Fe$_2$O$_3$ film (3 nm thick) was deposited on top of a gold layer (49 nm thick). The sensing membranes were placed on microscope glass slides and then brought into contact with a high index prism using an index-matching liquid. The response of the system, upon the exposure to acetone, ethanol, methanol and propanol, was monitored in terms of modulation of the resonance angle. The SPR signal resulted from the adsorption of hydrocarbon vapors that diffused through the pores of the sensing layer inducing a change of the refractive index. As a result, the trilayer-based gas sensor showed higher sensitivity to acetone compared to the other hydrocarbons. Furthermore, it was more stable and had shorter response time comparing to the bilayer-based gas sensor. The authors concluded that this improvement was due to the presence of the third gold layer, which promotes better interactions.

Apart from the improvements of the SPR sensitivity by the optimization of the optical parameters and the use of NPs as described earlier, other approaches have been proposed in the literature based on active plasmonics to add active functionalities to SPR-based devices. For example, Manera et al. [202] reported a study where magnetic field was used to control the SPR. They compared the sensing performance of a magneto-optical SPR (MO-SPR) sensor with that of a traditional SPR sensor for the detection of alcohol. A home-made setup with the Kretschmann configuration was employed to perform the measurements and a p-polarized light with a wavelength of 632.8 nm was used to excite the surface plasmons. To prepare the MO-SPR sensor, a multilayer of Cr/Au/Co/Au was deposited on a glass substrate. Then, a nanoporous TiO$_2$ thin film, used as sensitive material was deposited on top of the multilayer by glancing-angle deposition (GLAD). For comparison, a substrate for classical SPR was also prepared by depositing the TiO$_2$ layer on top of a gold-coated glass substrate. Three VOCs were analyzed, including ethanol, methanol and iso-propanol. The MO-SPR based gas sensor exhibited a significant improvement in sensitivity. Furthermore, its sensitivity was also much higher than that of their previous SPR-based gas sensor using TiO$_2$ thin films [203] and nanometric polyimide films [204].

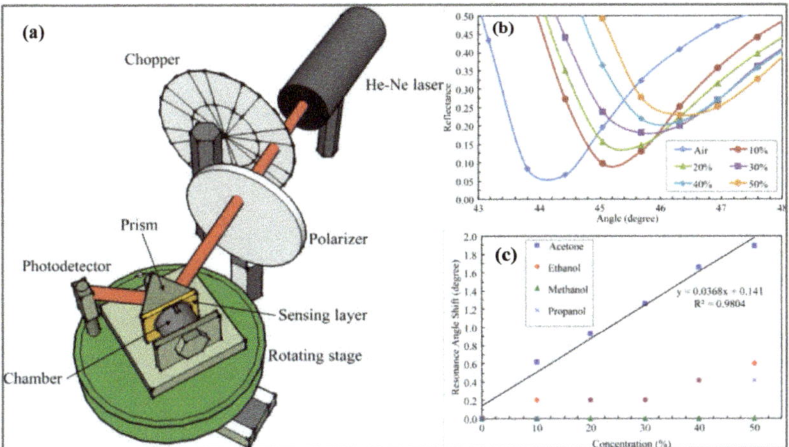

Figure 14. (**a**) SPR setup for detection of hydrocarbon vapor using trilayer Au-rGO/γ-Fe$_2$O$_3$-Au sensor. (**b**) SPR signals of the acetone vapor detection using the reduced graphene oxide/maghemite (rGO/γ-Fe$_2$O$_3$) sensing layer. (**c**) its resonance angle shift for increasing concentrations of different hydrocarbon [201].

The Table 2 summarizes the conditions for VOCs detection of SPR-based artificial olfaction systems and their performances in gas phase.

3.3. Wave Guide Coupler-Based Sensors

The fundamental coupling principle using a waveguide is similar to that of the prism, whereby the excitation of surface plasmons is achieved by an evanescent wave generated by ATR. To clarify the terms, an optical fiber is a special type of waveguide, and one that is widely used. Indeed, fiber optics are less expensive than waveguides and have good flexibility, remote sensing capability and other important features presented earlier. The VOC sensors systems that will be presented in the following section will exclusively involve the use of fiber optic SPR (FO-SPR) sensing platforms.

Fiber optic-based SPR sensors can be elaborated based on either transmission or reflection configuration. A typical fiber optic consists of high refractive index material (the core) sandwiched with a lower refractive index layer (the cladding) which allows light guidance through a succession of total internal reflections (TIRs). In the case of an FO-SPR in transmission configuration, a small region of the optical fiber cladding is removed and replaced by a metal layer where the SPR phenomenon will take place (Figure 15). In the reflection configuration, a thick metal layer deposited at the end of the fiber allows for the SPWs generation and plays the role of a mirror. In both cases, the sensitive materials are deposited on top of the metal layers. As with prism coupling, resonance occurs when the propagation constant of the evanescent wave generated by ATR of the guided mode matches the propagation constant of the SPWs [156]. In SPR-based optical fiber sensors, most of the interrogation methods are based on the detection of loss in the transmitted/reflected light at the resonance. Spectral or wavelength interrogation of the transmitted or the back-reflected light is the most commonly used measurement method. However, fiber-optic sensors based on intensity or phase interrogation have also been reported [208]. Theoretically, the sensitivity of waveguide-based SPR sensors is approximately the same as that of the corresponding ATR configurations [147].

Table 2. SPR-based artificial olfaction systems in the Kretschmann configuration for the detection of VOCs in gas phase.

Artificial Olfaction System	Interrogation	Sensing Material	Performance	Refs.
Electronic nose	Reflected light intensity (Imaging)	Small peptides	• octanol detection limit (DL): below 1 ppm • High discrimination ability (one carbon atom resolution) • Stability: at least three months • Good repeatability	[185]
Electronic nose	Reflected light intensity (Imaging)	Penta-peptides and hairpin DNA	• High discrimination ability (one carbon atom resolution)	[191]
Gas sensor	Resonance wavelength	Cavitands	• High selectivity and sensitivity to DMMP (DL: 16 ppb)	[192]
Gas sensor	Reflected light intensity	Three nitro-substituted heterocalix[4]arenes thin films	• High selectivity and sensitivity to acetone (DL: 3.8 ppm) • Fast and reversible response (few seconds) • Repeatability: from chip to chip and up to four injection cycles	[194]
Gas sensor	Reflected light intensity and resonance angle	Poly(methylmethacrylate) film	• High selectivity and sensitivity to benzene • Fast and reversible response	[195]
Gas sensor	Resonance angle	Acrylic acid and styrene thin film	• Acrylic acid film: good selectivity to ammonia (DL: several ppm) and amines (trimethylamine and trimethylamine • Styrene film: poor selectivity to tested gases	[198]
Gas sensor	Resonance angle	Films of polythiophene (PT) or gold nanoparticles capped with conjugated oligothiophenes (NPOT)	• PT film: responded to alcohol and toluene • NPOT film: responded only to alcohols => high selectivity	[200]
Gas sensor	Resonance angle	Reduced graphene oxide/maghemite nanocomposite film	• High selectivity and sensitivity to acetone	[201]

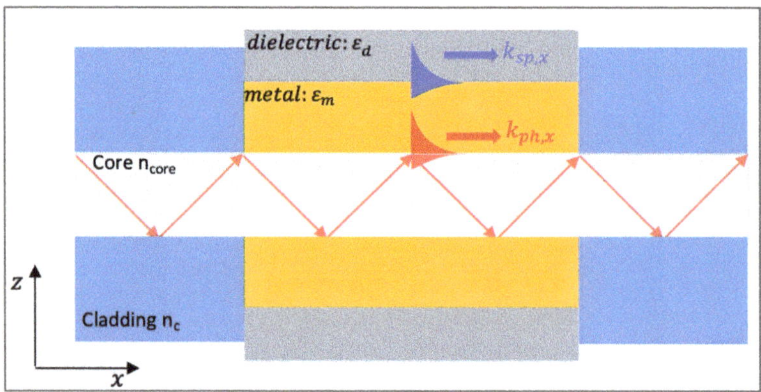

Figure 15. Typical fiber optic SPR (FO-SPR) sensor in transmission configuration.

The first fiber optic based SPR sensor with a conventional geometry (as the one presented in Figure 15) and using spectral interrogation as measurement methodology was proposed by Jorgenson et al. [209] in 1993 for a chemical sensing application. Since then, a large number of studies have experimentally and/or theoretically explored diverse geometry-modified single mode or multimode fibers including side and tip implemented FOS, fiber gratings (e.g., long period fiber gratings and tilted fiber Bragg gratings) and specialty fibers [208] (Figure 16). Different plasmonic coatings (e.g., gold, silver) have also been explored. Moreover, configurations involving the excitation of SPPs on continuous thin metallic layers (i.e., propagating SPPs) as well as those involving LSPR phenomena in metallic nanoparticles at visible and near-infrared wavelengths have been reported and reviewed [210].

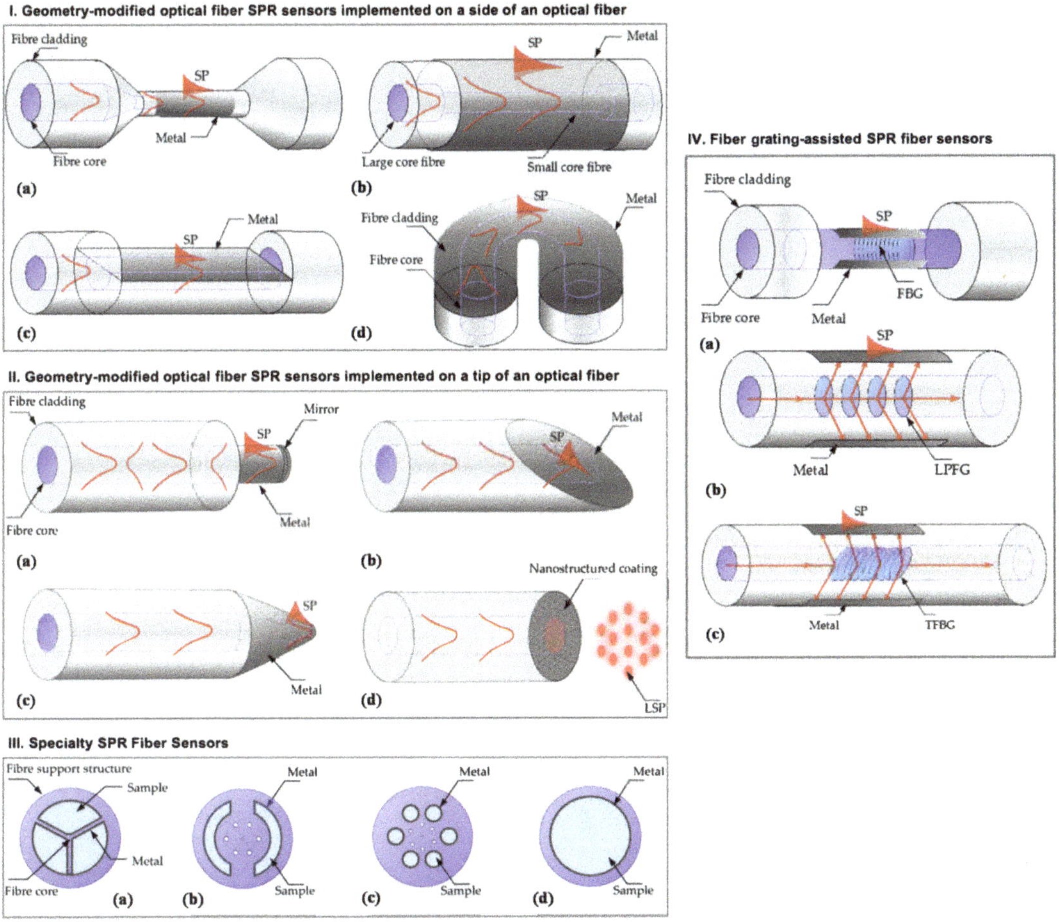

Figure 16. Schematic representation of the different plasmonic fiber-optic sensors **I**: (**a**) Unclad/etched/tapered fiber SPR probe; (**b**) Hetero-core structure; (**c**) Side-polished/D-shaped SPR probe; (**d**) U-shaped SPR probe. **II**: (**a**) Flat fiber tip SPR probe with end mirror; (**b**) Angle polished flat fiber tip SPR sensor; (**c**) Tapered tip SPR probe; (**d**) LSPR fiber tip probe. **III**: (**a**) Wagon-wheel fiber SPR sensor with triangular hole geometry; (**b**) Microstructured optical fiber SPR sensor with crescent-shaped holes; (**c**) Photonic crystal fiber SPR sensor with circular holes; (**d**) Microcapillary fiber SPR sensor geometry. **IV**: (**a**) Etched Fiber Bragg Grating SPR sensor; (**b**) Long Period Fiber Grating SPR sensor; (**c**) Tilted Fiber Bragg Grating SPR sensor [208].

The effectiveness of these sensors has been extensively investigated for physical (e.g., temperature, humidity), chemical (e.g., pH, gas, VOCs) and biological (e.g., DNA, proteins) sensing applications [156,208,210–214]. In the following section, we will focus on the FO-SPR sensors with different configurations developed for the detection of VOCs. Just like prism-based SPR sensor systems, most of the studies on FO-SPR sensors for the detection of VOCs have been reported after the year 2000 [2,215–225]. Only a few papers were published in the 1990s [226,227].

With the aim of achieving simple, low-cost and selective detection of aldehydes (known as cytotoxic and carcinogenic compounds) present in the environmental water, Cennamo et al. [215] developed an SPR sensor using plastic optical fiber (POF). To perform the study, butanal was used as the target VOC and porcine OBP (pOBP) as the sensing material. A plastic optical fiber consisting of a PMMA core of 980 μm and a fluorinated polymer cladding of 20 μm was used to elaborate the sensing platform. For that, the cladding of the POF along half the circumference and about 10 mm in length was removed. The exposed core was then coated with a photoresist buffer (1.5 μm thick) on top of which a 60 nm gold layer was deposited (Figure 17). For signal amplification purposes, a competitive assay was designed. For this, instead of OBP, butanal moieties were immobilized on the gold surface of the POF. Then, to test the detection performance, the sensor was exposed to OBPs pre-incubated with/without butanal. Binding events were detected by monitoring variations in the resonance wavelength. A halogen lamp with a wavelength emission range from 360 nm to 1700 nm was used as light source and the transmitted light spectrum was measured using a spectrum analyzer with a detection range of 200 nm to 850 nm. In a first step, the sensor was subjected to OBP (not pre-incubated with butanal), an increasing response was observed for increasing concentration of OBP. This result confirmed that pOBPs bind to the butanal moieties fixed on the chip, which is a prerequisite for the competitive assay. Next, the sensor was exposed to different concentrations of butanal pre-incubated with a fixed concentration of OBP. The results showed that the lower the concentration of butanal (in the pre-incubation solution), the higher the optical signal obtained. Indeed, the lower concentration of butanal resulted in more free OBPs available to bind to butanal moieties on the sensor surface. The obtained olfactory biosensor was able to detect butanal in aqueous solution for concentrations ranging from 20 μM to 1000 μM.

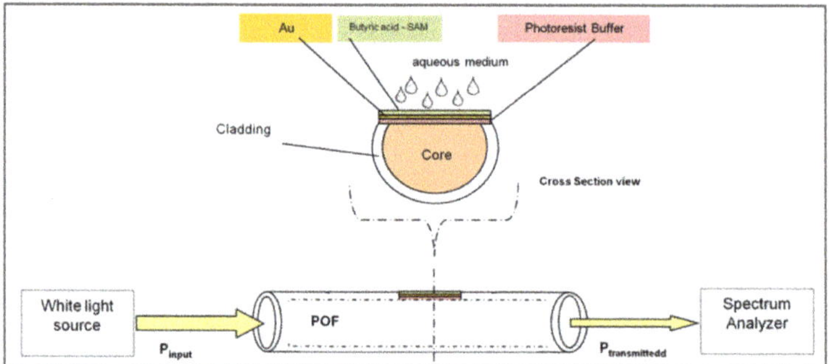

Figure 17. The olfactory biosensor using SPR based plastic optical fiber [215].

In a previous study [216] the team combined the SPR based POF platform with MIP as sensing material to achieve selective sensing of explosives such as 2,3,6-trinitrotoluene (TNT) in aqueous medium. The system exhibited a detection limit of 5.1×10^{-5} M and a sensitivity of 2.7×10^4 nm/M. The authors concluded that despite its limited sensitivity, the sensor was suitable for the detection of TNT with good selectivity. Additionally, the system was easy to prepare and suitable for rapid measurements that did not require any particular skill.

Vandezande et al. [217], designed a FO-SPR sensor for the detection of alcohol vapors. The sensor consisted of an optical fiber with a diameter of 400 μm, from which the inner technology enhanced clad silica (TECS) cladding and the outer protective cladding had been removed from the end. The exposed glass core was then coated with a 39 nm thick gold layer. Metal organic frameworks (MOFs) and more specifically zeolitic imidazolate framework (ZIF) were used as sensing materials and deposited on top of the gold layer Figure 18. MOFs consist of metal ions or a metal oxide cluster interlinked by polydentate linkers into a crystalline 3D framework. These porous materials have large surface area and tunable pore size, which are attractive features for gas and VOC sensing applications [228]. ZIFs were selected among other MOFs because of their high chemical stability and their small pore sizes. In this study, the sensing ability of two ZIF materials: ZIF-8 and ZIF-93, was explored for the detection of different alcohol vapors including methanol, ethanol, isopropanol, and n-butanol. The response of the system was expressed in terms of changes in the refractive index converted from the SPR response. This made it possible not only to monitor the mass and density changes during layer formation of the ZIFs, but also to investigate sorption behavior of VOCs on these layers. The obtained FO-SPR sensors were able to detect VOCs with ppm concentrations and with a detection limit of 2.5 ppm for methanol. However, a significant drift was observed after extended analysis periods. In this study, the authors claimed that the difference in recognition behavior of the hydrophobic ZIF-8 and more hydrophilic ZIF-93 could be exploited to generate qualitative information regarding the vapor composition.

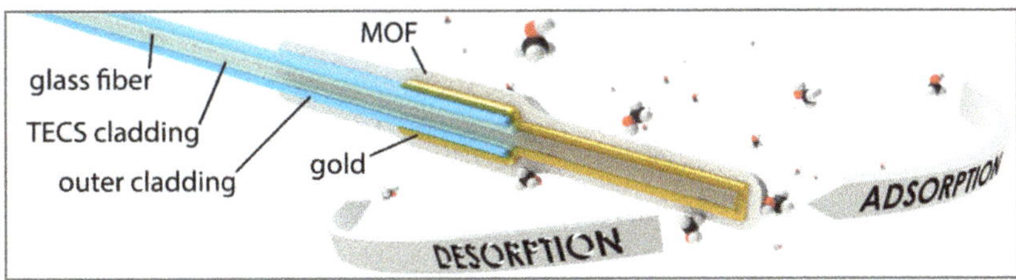

Figure 18. Schematic representation of a metal organic framework FO-SPR probe, not drawn to scale [217].

Gupta's group published several studies on the detection of VOCs and other odorant molecules using FO-SPR sensors [218–220]. In one of these studies [220], they explored the sensing ability of graphene-carbon nanotubes/poly(methyl methacrylate) (GCNT/PMMA) hybrid composites for the detection of methane gas. Their sensitivity and selectivity were compared to that of three other sensing materials including reduced graphene oxide (rGO), carbon nanotubes (CNT), reduced graphene oxide-carbon nanotubes (GCNT). To fabricate different probes, 24 cm long plastic clad silica optical fibers (core diameter 600 μm, numerical aperture 0.4) were used. About 1 cm length of the cladding was removed from the middle portion of the fibers and the uncladded core was coated with a silver layer via thermal evaporation technique. Finally, the sensing materials were deposited on top of the silver by dip coating. To test the performance of the fabricated system, the probe was installed in a gas chamber and a polychromatic light from a tungsten halogen lamp was launched at the input end of the fiber. The spectrum of the transmitted light was recorded with a spectrometer at the other end. The FO-SPR sensors were exposed to different concentrations of methane (ranging from 10 to 100 ppm) and their performance was analyzed in terms of resonance wavelength shift. To evaluate their selectivity, the sensors were exposed to different gases: methane, ammonia, hydrogen sulfide, chlorine, carbon dioxide, hydrogen, and nitrogen. The FO-SPR sensor based on (GCNT/PMMA) hybrid composites showed the best sensitivity and selectivity to methane gas comparing to the three others using rGO, CNT, and GCNT as sensing materials. The authors attributed this performance to the high aspect ratio and the large defect level in the nanocomposite material, which could provide more active sites for VOC adsorption.

Photonic crystal fiber (PCF) is a class of optical fiber characterized by a flexible structure design, which presents a unique light controlling capability with light confinement characteristics not achievable using conventional optical fiber. Combined with SPR, PCF can form a very attractive platform for optical sensing. Accordingly, Lui et al. [221] proposed a novel PCF-SPR sensor to detect mixture of methane and hydrogen. As presented in Figure 19, the PCF-SPR sensor consisted of four ultra-large side-holes symmetrically introduced into the cladding layer. These holes allowed to improve the sensitivity to VOCs since the refractive index variation due to concentration change is usually very low. In practice, the two rows of smaller air-holes along the angle of 45° and 135° surrounding the fiber enabled the introduction of the ultra-large side-holes much closer to the fiber core, which, consequently, led to higher sensitivity. The inner surfaces of the left and top ultra-large air-holes were coated with a gold layer on top of which a film of sensing material was deposited. A film of $Pd-WO_3$ deposited via the sol-gel scheme was used for the detection of hydrogen. The methane-sensitive film consisted of a kind of ultraviolet curable fluoro-siloxane nanofilm with the inclusion of cryptophane A. It was deposited on the gold layer via a capillary dip-coating technique. The sensing performance and response of the system were characterized by analyzing the confinement loss spectra. As a result, the study showed that using polarization filtering, the concentration of methane and hydrogen in a gas mixture could be accurately measured without interfering with each other. The authors suggested that this approach could be broadened to achieve qualitative identification of multiple gases.

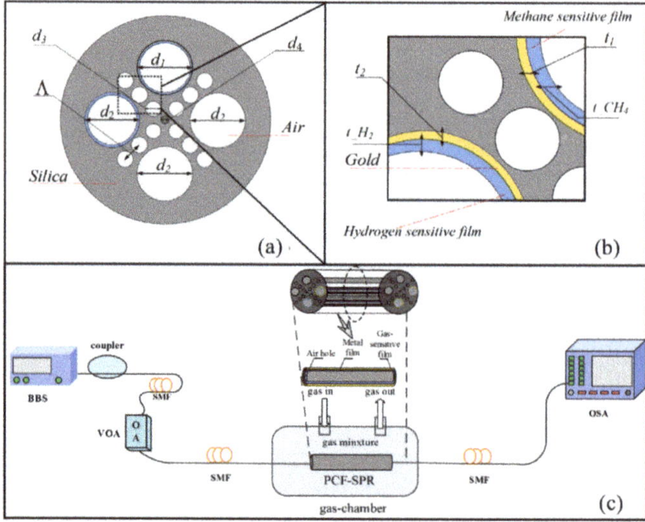

Figure 19. The schematic and cross section of photonic crystal fiber SPR sensor. (**a,b**) structural parameters and (**c**) experimental scheme [221].

Arasu et al. [222] reported a single mode fiber Bragg grating (FBG)-based FO-SPR sensor coated with graphene oxide (GO) layer for ethanol sensing in an aqueous medium. To fabricate the sensor, a standard single mode FBG with a 9 μm core diameter and 125 μm cladding diameter was used. The polymer coating directly over the Bragg grating was removed. Then, a 45 nm thick gold layer was deposited over the grating area without removing the cladding. Finally, a nanostructured GO layer was put on top of the gold surface by drop-casting technique. A tungsten halogen white light source was employed to generate the input signal and the output light was analyzed by a spectrometer. Wavelength interrogation was used to monitor the response of the system upon the addition of different concentrations of ethanol in water.

In order to make sure that the FBG was effective for SPR sensing without the removal of the cladding layer, the team compared the beam profile of a gold coated FBG to that of a standard gold coated single mode fiber (SMF). The results confirmed that, in contrast to the standard SMF, the FBG was able to scatter the light from the fiber core into the cladding, producing TIR at the cladding-air interface and, thus, an evanescent wave that could be exploited for SPR. They also compared the intensity spectrum of a bare FBG, a gold coated FBG and a gold coated FBG with the GO layer, as well as the sensing performance of the last two for ethanol. It was clear that the GO layer enhanced both the sensitivity and accuracy of the FO-SPR sensor thanks to its excellent electrochemical and physical properties.

Wei et al. [223] proposed a long period fiber grating (LPFG) SPR sensor combined with a monolayer of graphene as sensing material. To fabricate the sensor, a single mode fiber with a core diameter of 10 μm, a cladding diameter of 125 μm, and a numerical aperture of 0.22 was used. The long period grating was first inscribed on the fiber core by a CO_2 laser and then the SiO_2 surface of the fiber was coated with an Ag film (50 nm thick) on top of which a monolayer of graphene was deposited by CVD. A schematic representation of the sensor structure is given in Figure 20a,b. To test the performance of the sensor chip, an experimental setup (Figure 20c) comprising a gas flow control system, a wide spectral range light source and a spectrometer was used. The LPFG SPR sensor was exposed to different concentrations of methane carried by a nitrogen gas flow. Wavelength interrogation was employed to monitor changes in the refractive index and thus detect variations in the concentration of VOCs in contact with the sensor. The obtained graphene coated LPFG SPR sensor exhibited a dose dependent linear response to methane and improved sensitivity compared to an uncoated LPFG sensor and an Ag-coated LPFG SPR sensor. The sensor also demonstrated good response repeatability and a baseline recovery (with a recovery time of 65 s). Finally, using finite element simulation, the team showed that the graphene layer enhanced the intensity of the electric field surrounding the sensing layer, which could explain the sensitivity enhancement observed in the presence of this layer.

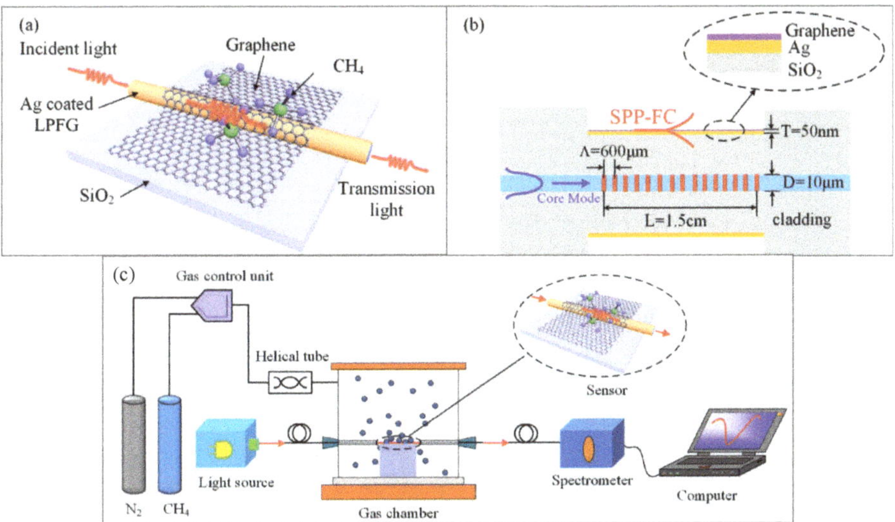

Figure 20. (**a**) Schematic representation of the graphene-based long period fiber grating SPR sensor. (**b**) Longitudinal section of the sensor. (**c**) The experimental setup used [223].

The Table 3 summarizes the conditions for gas or VOCs detection of fiber optic SPR-based artificial olfaction systels and their performances in liquid and gas phase.

Table 3. Fiber optic SPR-based artificial olfaction systems for the detection of VOCs in liquid and gas phase.

Artificial Olfaction System	Fiber Type	Sensing Material	Performance	Sensing Medium	Refs.
Olfactory biosensor	Plastic fiber	Pig odorant binding protein	• High selectivity and to butanal sensitivity (detection limit (DL): 25 μM)	Liquid	[215]
Gas sensor	Glass fiber	Zeolitic imidazolate framework (ZIF-8 and ZIF-93)	• ZIF-8: high sensitivity to methanol (DL: 2.5 ppm)	Gas	[217]
Gas sensor	Plastic clad silica fiber	Graphene-carbon nanotubes/poly(methyl methacrylate) (GCNT/PMMA) hybrid composites, reduced graphene oxide, carbon nanotubes, reduced graphene oxide-carbon nanotubes	• GCNT/PMMA exhibited the highest sensitivity and selectivity to methane compared to the other sensing materials tested • DL: 10 ppm	Gas	[220]
Gas sensor	Photonic crystal fiber	Pd-WO$_3$ film and a kind of ultraviolet curable fluoro-siloxane nanofilm with the inclusion of cryptophane A	• The concentration of methane and hydrogen in a gas mixture could be accurately measured using polarization filtering	Gas	[221]
Gas sensor	Fiber Bragg grating	Graphene oxide (GO)	• The GO layer enhances the sensitivity to ethanol compared to bare gold	Liquid	[222]
Gas sensor	Long Period Fiber Grating (LPFG)	Graphene	• In presence of graphene, the sensitivity to methane is improved 2.96 and 1.31 times with respect to the traditional LPFG sensor and Ag-coated LPFG SPR sensor, respectively • Fast response (50 s) and recovery (65 s) times • Good repeatability	Gas	[223]

3.4. Grating Coupler-Based SPR Sensors

Based on light diffraction effects, the grating coupler is another approach to excite surface plasmons. This method was first observed and described by Wood [122] in 1902. Basically, when a light wave reaches a periodically distorted metal-dielectric interface, it is diffracted into a series of beams that propagate away from the surface at different angles [147] (Figure 21). Coupling occurs when the momentum component along the interface of a scattered order is equal to the propagation constant of the SPPs. The coupling condition can by expressed as [121]:

$$\frac{2\pi}{\lambda}\sqrt{\varepsilon_d}\sin\theta + m\frac{2\pi}{\Lambda} = \pm Re\{k_{spp}\} \tag{6}$$

where λ is the wavelength of the incident p-polarized light, θ the incidence angle, m the diffraction order and Λ the diffraction grating period. To perform measurements using this type of SPR sensors, angular, spectral, phase or intensity interrogation can be employed.

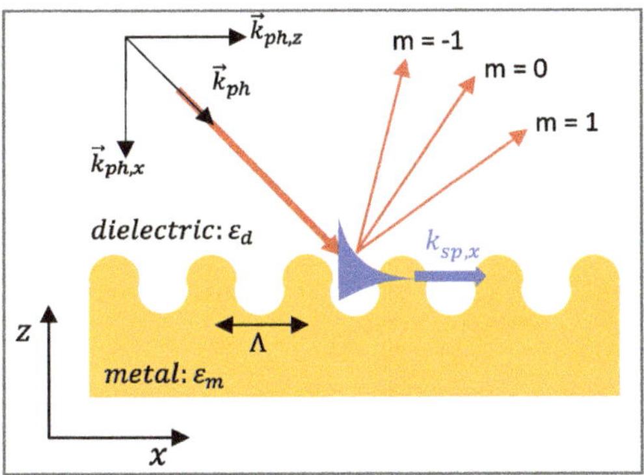

Figure 21. Excitation of SPs by grating coupler.

This category of sensors is much less popular and poorly developed compared to those presented above, because they are generally less sensitive than smooth metal-film coupling based sensors (i.e., prism and optical fiber). Several theoretical and experimental studies have been carried out in attempts to improve the performance of these sensors [132,229,230]. For instance, Nazem et al. [229] recently demonstrated (theoretically and experimentally) the feasibility of a sensitive SPR sensor based on Ag-MgF$_2$ grating. Similarly, Dai et al. [230] experimentally demonstrated a high sensitivity of an SPR sensor with silver rectangular grating coupling. A higher sensitivity than that of a prism-coupled SPR sensor was obtained in the negative order diffraction excitation mode. Borile et al. [231] reported a grating-coupled SPR senor integrated into a microfluidic chamber for label-free monitoring of cell adhesion and cell-surface interaction. Cai et al. [232] worked on the improvement of the sensitivity of grating-based SPR sensors by designing sharp dips of the higher diffraction orders and developing double-dips method. Finally, in the field of VOC sensing, Sambles's group [233,234] presented a prototype gas sensor employing SPR on gratings in the beginning of 1990s. Since then, this field has not been developed much further.

4. Conclusions and Outlook

The reliable analysis of VOCs is of great interest in various fields. To complement traditional analytical methods (GC-MS) and biological noses, great progress has been made in the development of artificial odor detection systems such as gas sensors, olfactory biosensors, and eNs based on diverse sensing technologies. As demonstrated in this paper, propagating SPR with different coupling configurations (prism coupler, wave guide, and grating) is very efficient for such applications. In particular, prism coupler-based gas sensors have been widely studied for sensing VOCs, either in the liquid or gas phase. For VOC analysis in the liquid phase, as highlighted in this review, signal amplification strategies are necessary by selecting appropriate sensitive materials and immobilization techniques to generate reliable SPR signals. In contrast, in the gas phase, the binding of small VOCs on the sensing materials can generate reliable SPR signals with good signal/noise ratios, since the detection noise remains relatively low under such conditions. Moreover, based on SPR imaging mode, a novel generation of eNs with large-scale multiplexed arrays has been developed. Combined with peptides as sensing materials, such eNs offer exceptional performance in terms of sensitivity and selectivity, with the ability to discriminate among chiral forms of VOCs. Regarding wave guide coupler-based gas sensors, most systems use optical fiber in different configurations. They are very interesting thanks to their remote and multiplexed sensing capability, as well as their miniaturized structures. Finally, grating coupler-based gas sensors are much less popular because their sensitivity is still limited.

Although the different systems that we have presented are efficient and sensitive for the detection of VOCs, the current trend is toward the development of more miniaturized sensors. Accordingly, nano plasmonic sensors based on localized SPR are attracting more and more attention, and are being developed for different nanoscale applications including the detection of VOCs. Moreover, the sensing performance of these systems can be optimized by simply varying the size and shape of the nanostructures, which is very advantageous for sensor development. The improvement in nanofabrication processes has made it possible to explore diverse nanostructured geometries to achieve optimal LSPR nanosensors [137].

To further improve the performance (sensitivity, selectivity, and stability) of SPR-based gas sensors, olfactory biosensors, and eNs, it is essential to design novel sensing materials that are able to mimic the binding properties of biomolecules such as ORs and OBPs, but with higher stability. One trend is to use peptides as alternatives. Indeed, peptides are much more robust than proteins, cheaper to synthesize, and could potentially be integrated into industrial devices. On top of that, their selectivity towards target VOCs can be easily tuned through rational designs based on molecular modeling, virtual screening, and phage display. Finally, eNs will benefit greatly from the accelerating growth of artificial intelligence that will allow for more efficient data processing. There is no doubt that novel SPR-based gas sensors and eNs will play a more important role in the field of VOC detection and will find applications in various new domains.

Author Contributions: Conceptualization, M.E.K. and Y.H.; writing—original draft preparation, M.E.K. and Y.H.; revision and validation by M.E.K., Y.H., J.S.W., C.H., R.M., A.B. and N.S.; funding acquisition, Y.H. All authors have read and agreed to the published version of the manuscript.

Funding: French National Research Agency (ANR) and AID/DGA supported this work (ANR-18-CE42-0012).

Institutional Review Board Statement: Not applicable.

Informed Consent Statement: Not applicable.

Data Availability Statement: Not applicable.

Acknowledgments: The authors thank greatly the following organizations for funding the PhD scholarship of M.E.K (ANR and Labex LANEF du Programme d'Investissements d'Avenir: ANR-10-LABX-51-01, of J.S.W. (Fondation Nanosciences) and of C.H. (CEA). The authors also thank ANR and AID/DGA for the financial support (ANR-18-CE42-0012). SyMMES laboratory is part of Labex LANEF, Labex ARCANE and CBH-EUR-GS (ANR-17-EURE-0003).

Conflicts of Interest: Yanxia Hou is scientific counselor and co-founder of the company Aryballe, which manufactures and commercializes the portable device NeOse Pro.

Abbreviations

ATR	attenuated total reflection
BAW	bulk acoustic wave
BTEX	benzene, toluene, ethylbenzene and m-xylene
CCP	composite conducting polymer
CNT	reduced graphene oxide-carbon nanotube
CP	conducting polymer
CVD	chemical vapor deposition
DL	detection limit
DMMP	dimethylmethylphosphonate
EIS	electrochemical impedance spectroscopy
eN	electronic nose
FBG	fiber Bragg grating
FET	field effect transistor
FO-SPR	fiber optic-SPR
FOS	fiber optic sensor
GC-MS	gas chromatography-mass spectrometry

GCNT	reduced graphene oxide-carbon nanotubes
GLAD	glancing-angle deposition
GO	graphene oxide
GPCR	G protein coupled receptors
GTP	guanosine-5′-triphosphate
ICP	intrinsic conducting polymer
IDT	inter-digitated transducer
LPFG	long period fiber grating
LSPR	localized SPR
MIP	molecularly imprinted polymer
MO-SPR	magneto-optical SPR
MOF	metal organic framework
MOS	metal oxide semiconductor
NP	nanoparticles
NPOT	NP/oligothiophene
OBP	odorant binding protein
OR	olfactory receptor
PCF	photonic crystal fiber
PMMA	poly(methylmethacrylate)
POF	plastic optical fiber
ppb	parts per billion
ppm	parts per million
PT	polythiophene
QCM	quartz crystal microbalance
rGO	reduced GO
SAW	surface acoustic wave
SiNW FET	silicon nanowire FET
SMF	single mode fiber
SP	surface plasmon
SPP	surface plasmon polariton
SPR	surface plasmon resonance
SPW	surface plasma wave
TBD	toluene binding domain
TECS	technology enhanced clad silica
TIR	total internal reflection
TM	transverse-magnetic
TNT	2,3,6-trinitrotoluene
VOC	volatile organic compound
ZIF	zeolitic imidazolate framework

References

1. Genva, M.; Kenne Kemene, T.; Deleu, M.; Lins, L.; Fauconnier, M.-L. Is It Possible to Predict the Odor of a Molecule on the Basis of Its Structure? *Int. J. Mol. Sci.* **2019**, *20*, 3018. [CrossRef] [PubMed]
2. He, C.; Liu, L.; Korposh, S.; Correia, R.; Morgan, S.P. Volatile Organic Compound Vapour Measurements Using a Localised Surface Plasmon Resonance Optical Fibre Sensor Decorated with a Metal-Organic Framework. *Sensors* **2021**, *21*, 1420. [CrossRef] [PubMed]
3. Turner, A.P.F.; Magan, N. Electronic Noses and Disease Diagnostics. *Nat. Rev. Microbiol.* **2004**, *2*, 161–166. [CrossRef]
4. Shan, B.; Broza, Y.Y.; Li, W.; Wang, Y.; Wu, S.; Liu, Z.; Wang, J.; Gui, S.; Wang, L.; Zhang, Z.; et al. Multiplexed Nanomaterial-Based Sensor Array for Detection of COVID-19 in Exhaled Breath. *ACS Nano* **2020**, *14*, 12125–12132. [CrossRef] [PubMed]
5. Ward, R.J.; Jjunju, F.P.M.; Griffith, E.J.; Wuerger, S.M.; Marshall, A. Artificial Odour-Vision Syneasthesia via Olfactory Sensory Argumentation. *IEEE Sens. J.* **2021**, *21*, 6784–6792. [CrossRef]
6. Cornu, J.-N.; Cancel-Tassin, G.; Ondet, V.; Girardet, C.; Cussenot, O. Olfactory Detection of Prostate Cancer by Dogs Sniffing Urine: A Step Forward in Early Diagnosis. *Eur. Urol.* **2011**, *59*, 197–201. [CrossRef] [PubMed]
7. Thuleau, A.; Gilbert, C.; Bauër, P.; Alran, S.; Fourchotte, V.; Guillot, E.; Vincent-Salomon, A.; Kerihuel, J.-C.; Dugay, J.; Semetey, V.; et al. A New Transcutaneous Method for Breast Cancer Detection with Dogs. *Oncology* **2019**, *96*, 110–113. [CrossRef]
8. Arshak, K.; Moore, E.; Lyons, G.M.; Harris, J.; Clifford, S. A Review of Gas Sensors Employed in Electronic Nose Applications. *Sens. Rev.* **2004**, *24*, 181–198. [CrossRef]
9. Gardner, J.W.; Bartlett, P.N. A Brief History of Electronic Noses. *Sens. Actuators B Chem.* **1994**, *18*, 210–211. [CrossRef]

10. Buck, L.; Axel, R. A Novel Multigene Family May Encode Odorant Receptors: A Molecular Basis for Odor Recognition. *Cell* **1991**, *65*, 175–187. [CrossRef]
11. Park, S.Y.; Kim, Y.; Kim, T.; Eom, T.H.; Kim, S.Y.; Jang, H.W. Chemoresistive Materials for Electronic Nose: Progress, Perspectives, and Challenges. *InfoMat* **2019**, *1*, 289–316. [CrossRef]
12. Zwaardemaker, H.; Hogewind, F. On Spray-Electricity and Waterfall-Electricity. *KNAW Proc.* **1920**, *22*, 429–437.
13. Tanyolac, N.N.; Eaton, J.R. Study of Odors. *J. Am. Pharm. Assoc. Sci. Ed.* **1950**, *39*, 565–574. [CrossRef] [PubMed]
14. Hartman, J. *A Possible Objective Method for the Rapid Estimation of Flavors in Vegetables*; American Society for Horticultural Science: Alexandria, VA, USA, 1954; Volume 64, pp. 335–342.
15. Moncrieff, R.W. An Instrument for Measuring and Classifying Odors. *J. Appl. Physiol.* **1961**, *16*, 742–749. [CrossRef] [PubMed]
16. Seiyama, T.; Kato, A.; Fujiishi, K.; Nagatani, M. A New Detector for Gaseous Components Using Semiconductive Thin Films. *Anal. Chem.* **1962**, *34*, 1502–1503. [CrossRef]
17. Buck, T.M.; Allen, F.G.; Dalton, J.V. *Detection of Chemical Species by Surface Effects on Metals and Semiconductors*; Bell Telephone Laboratories: Murray Hill, NJ, USA, 1965.
18. Dravnieks, A.; Trotter, P.J. Polar Vapour Detector Based on Thermal Modulation of Contact Potential. *J. Sci. Instrum.* **1965**, *42*, 624–627. [CrossRef]
19. Shaver, P.J. Activated tungsten oxide gas detectors. *Appl. Phys. Lett.* **1967**, *11*, 255–257. [CrossRef]
20. Taguchi, N. Gas Detecting Devices. U.S. Patent 3,631,436, 28 December 1971.
21. Henry, W.; Raymond, D. Surface Acoustic Wave Probe for Chemical Analysis. I. Introduction and Instrument Description. *Anal. Chem.* **1979**, *51*, 1458–1464. [CrossRef]
22. Persaud, K.; Dodd, G. Analysis of Discrimination Mechanisms in the Mammalian Olfactory System Using a Model Nose. *Nature* **1982**, *299*, 352–355. [CrossRef] [PubMed]
23. Ikegami, A.; Kaneyasu, M. Olfactory Detection Using Integrated Sensors. In Proceedings of the 3rd International Conference on Solid-State Sensors and Actuators (Transducers' 85), Philadelphia, PA, USA, 11–14 June 1985; pp. 136–139.
24. Wilson, A.D.; Baietto, M. Applications and Advances in Electronic-Nose Technologies. *Sensors* **2009**, *9*, 5099–5148. [CrossRef] [PubMed]
25. Barbosa, A.J.M.; Oliveira, A.R.; Roque, A.C.A. Protein- and Peptide-Based Biosensors in Artificial Olfaction. *Trends Biotechnol.* **2018**, *36*, 1244–1258. [CrossRef] [PubMed]
26. El kazzy, M.; Hurot, C.; Weerakkody, J.S.; Buhot, A.; Hou, Y. Biomimetic Olfactory Biosensors and Bioelectronic Noses. In *Advances in Biosensors: Reviews*; Yurish, S.Y., Ed.; IFSA Publishing: Barcelona, Spain, 2020; Volume 3, pp. 15–54.
27. Wasilewski, T.; Gębicki, J.; Kamysz, W. Advances in Olfaction-Inspired Biomaterials Applied to Bioelectronic Noses. *Sens. Actuators B Chem.* **2018**, *257*, 511–537. [CrossRef]
28. Sankaran, S.; Khot, L.R.; Panigrahi, S. Biology and Applications of Olfactory Sensing System: A Review. *Sens. Actuators B Chem.* **2012**, *171–172*, 1–17. [CrossRef]
29. Hurot, C.; Scaramozzino, N.; Buhot, A.; Hou, Y. Bio-Inspired Strategies for Improving the Selectivity and Sensitivity of Artificial Noses: A Review. *Sensors* **2020**, *20*, 1803. [CrossRef] [PubMed]
30. Gaggiotti, S.; Della Pelle, F.; Mascini, M.; Cichelli, A.; Compagnone, D. Peptides, DNA and MIPs in Gas Sensing. From the Realization of the Sensors to Sample Analysis. *Sensors* **2020**, *20*, 4433. [CrossRef]
31. Schaller, E.; Bosset, J.O.; Escher, F. 'Electronic Noses' and Their Application to Food. *LWT Food Sci. Technol.* **1998**, *31*, 305–316. [CrossRef]
32. Ampuero, S.; Bosset, J.O. The Electronic Nose Applied to Dairy Products: A Review. *Sens. Actuators B Chem.* **2003**, *94*, 1–12. [CrossRef]
33. Mielle, P. 'Electronic Noses': Towards the Objective Instrumental Characterization of Food Aroma. *Trends Food Sci. Technol.* **1996**, *7*, 432–438. [CrossRef]
34. Albert, K.J.; Lewis, N.S.; Schauer, C.L.; Sotzing, G.A.; Stitzel, S.E.; Vaid, T.P.; Walt, D.R. Cross-Reactive Chemical Sensor Arrays. *Chem. Rev.* **2000**, *100*, 2595–2626. [CrossRef] [PubMed]
35. James, D.; Scott, S.M.; Ali, Z.; O'Hare, W.T. Chemical Sensors for Electronic Nose Systems. *Microchim. Acta* **2005**, *149*, 1–17. [CrossRef]
36. Strike, D.J.; Meijerink, M.G.H.; Koudelka-Hep, M. Electronic Noses—A Mini-Review. *Fresenius' J. Anal. Chem.* **1999**, *364*, 499–505. [CrossRef]
37. Bai, H.; Shi, G. Gas Sensors Based on Conducting Polymers. *Sensors* **2007**, *7*, 267–307. [CrossRef]
38. Wong, Y.C.; Ang, B.C.; Haseeb, A.S.M.A.; Baharuddin, A.A.; Wong, Y.H. Review—Conducting Polymers as Chemiresistive Gas Sensing Materials: A Review. *J. Electrochem. Soc.* **2020**, *167*, 037503. [CrossRef]
39. Berna, A. Metal Oxide Sensors for Electronic Noses and Their Application to Food Analysis. *Sensors* **2010**, *10*, 3882–3910. [CrossRef] [PubMed]
40. Neri, G. First Fifty Years of Chemoresistive Gas Sensors. *Chemosensors* **2015**, *3*, 1–20. [CrossRef]
41. Lin, T.; Lv, X.; Hu, Z.; Xu, A.; Feng, C. Semiconductor Metal Oxides as Chemoresistive Sensors for Detecting Volatile Organic Compounds. *Sensors* **2019**, *19*, 233. [CrossRef]
42. Fanget, S.; Hentz, S.; Puget, P.; Arcamone, J.; Matheron, M.; Colinet, E.; Andreucci, P.; Duraffourg, L.; Myers, E.d.; Roukes, M.L. Gas Sensors Based on Gravimetric Detection—A Review. *Sens. Actuators B Chem.* **2011**, *160*, 804–821. [CrossRef]

43. McGinn, C.K.; Lamport, Z.A.; Kymissis, I. Review of Gravimetric Sensing of Volatile Organic Compounds. *ACS Sens.* **2020**, *5*, 1514–1534. [CrossRef]

44. Stetter, J.R.; Li, J. Amperometric Gas SensorsA Review. *Chem. Rev.* **2008**, *108*, 352–366. [CrossRef]

45. Elosua, C.; Matias, I.; Bariain, C.; Arregui, F. Volatile Organic Compound Optical Fiber Sensors: A Review. *Sensors* **2006**, *6*, 1440–1465. [CrossRef]

46. Li, Z.; Askim, J.R.; Suslick, K.S. The Optoelectronic Nose: Colorimetric and Fluorometric Sensor Arrays. *Chem. Rev.* **2019**, *119*, 231–292. [CrossRef] [PubMed]

47. John, A.T.; Murugappan, K.; Nisbet, D.R.; Tricoli, A. An Outlook of Recent Advances in Chemiresistive Sensor-Based Electronic Nose Systems for Food Quality and Environmental Monitoring. *Sensors* **2021**, *21*, 2271. [CrossRef] [PubMed]

48. Barsan, N.; Koziej, D.; Weimar, U. Metal Oxide-Based Gas Sensor Research: How To? *Sens. Actuators B Chem.* **2007**, *121*, 18–35. [CrossRef]

49. Llobet, E. Gas Sensors Using Carbon Nanomaterials: A Review. *Sens. Actuators B Chem.* **2013**, *179*, 32–45. [CrossRef]

50. Kim, M.I.; Lee, Y.-S. A Comprehensive Review of Gas Sensors Using Carbon Materials. *J. Nanosci. Nanotechnol.* **2016**, *16*, 4310–4319. [CrossRef] [PubMed]

51. Ji, H.; Zeng, W.; Li, Y. Gas Sensing Mechanisms of Metal Oxide Semiconductors: A Focus Review. *Nanoscale* **2019**, *11*, 22664–22684. [CrossRef] [PubMed]

52. Chaturvedi, A.; Mishra, V.N.; Dwivedi, R.; Srivastava, S.K. Response of Oxygen Plasma-Treated Thick Film Tin Oxide Sensor Array for LPG, CCl_4, CO and C_3H_7OH. *Microelectron. J.* **1999**, *30*, 259–264. [CrossRef]

53. Dey, A. Semiconductor Metal Oxide Gas Sensors: A Review. *Mater. Sci. Eng. B* **2018**, *229*, 206–217. [CrossRef]

54. Nazemi, H.; Joseph, A.; Park, J.; Emadi, A. Advanced Micro- and Nano-Gas Sensor Technology: A Review. *Sensors* **2019**, *19*, 1285. [CrossRef] [PubMed]

55. Chiu, S.-W.; Tang, K.-T. Towards a Chemiresistive Sensor-Integrated Electronic Nose: A Review. *Sensors* **2013**, *13*, 14214–14247. [CrossRef] [PubMed]

56. Zhu, L.; Zeng, W. Room-Temperature Gas Sensing of ZnO-Based Gas Sensor: A Review. *Sens. Actuators A Phys.* **2017**, *267*, 242–261. [CrossRef]

57. Joshi, N.; Hayasaka, T.; Liu, Y.; Liu, H.; Oliveira, O.N.; Lin, L. A Review on Chemiresistive Room Temperature Gas Sensors Based on Metal Oxide Nanostructures, Graphene and 2D Transition Metal Dichalcogenides. *Microchim. Acta* **2018**, *185*, 213. [CrossRef] [PubMed]

58. Kim, H.-J.; Lee, J.-H. Highly Sensitive and Selective Gas Sensors Using P-Type Oxide Semiconductors: Overview. *Sens. Actuators B Chem.* **2014**, *192*, 607–627. [CrossRef]

59. Nylander, C.; Armgarth, M.; Lundström, I. An Ammonia Detector Based on a Conducting Polymer. *Anal. Chem. Symp. Ser.* **1983**, *17*, 203–207.

60. Blackwood, D.; Josowicz, M. Work Function and Spectroscopic Studies of Interactions between Conducting Polymers and Organic Vapors. *J. Phys. Chem.* **1991**, *95*, 493–502. [CrossRef]

61. Park, S.; Park, C.; Yoon, H. Chemo-Electrical Gas Sensors Based on Conducting Polymer Hybrids. *Polymers* **2017**, *9*, 155. [CrossRef]

62. Yu, J.-B.; Byun, H.-G.; So, M.-S.; Huh, J.-S. Analysis of Diabetic Patient's Breath with Conducting Polymer Sensor Array. *Sens. Actuators B Chem.* **2005**, *108*, 305–308. [CrossRef]

63. Li, W.; Hoa, N.D.; Cho, Y.; Kim, D.; Kim, J.-S. Nanofibers of Conducting Polyaniline for Aromatic Organic Compound Sensor. *Sens. Actuators B Chem.* **2009**, *143*, 132–138. [CrossRef]

64. Xie, G.; Sun, P.; Yan, X.; Du, X.; Jiang, Y. Fabrication of Methane Gas Sensor by Layer-by-Layer Self-Assembly of Polyaniline/PdO Ultra Thin Films on Quartz Crystal Microbalance. *Sens. Actuators B Chem.* **2010**, *145*, 373–377. [CrossRef]

65. Yoon, H.; Lee, S.H.; Kwon, O.S.; Song, H.S.; Oh, E.H.; Park, T.H.; Jang, J. Polypyrrole Nanotubes Conjugated with Human Olfactory Receptors: High-Performance Transducers for FET-Type Bioelectronic Noses. *Angew. Chem.* **2009**, *121*, 2793–2796. [CrossRef]

66. Cao, Z.; Buttner, W.J.; Stetter, J.R. The Properties and Applications of Amperometric Gas Sensors. *Electroanalysis* **1992**, *4*, 253–266. [CrossRef]

67. Buttner, W.J.; Findlay, M.; Vickers, W.; Davis, W.M.; Cespedes, E.R.; Cooper, S.; Adams, J.W. In Situ Detection of Trinitrotoluene and Other Nitrated Explosives in Soils. *Anal. Chim. Acta* **1997**, *341*, 63–71. [CrossRef]

68. Barou, E.; Sigoillot, M.; Bouvet, M.; Briand, L.; Meunier-Prest, R. Electrochemical Detection of the 2-Isobutyl-3-Methoxypyrazine Model Odorant Based on Odorant-Binding Proteins: The Proof of Concept. *Bioelectrochemistry* **2015**, *101*, 28–34. [CrossRef] [PubMed]

69. Liu, Q.; Wang, H.; Li, H.; Zhang, J.; Zhuang, S.; Zhang, F.; Jimmy Hsia, K.; Wang, P. Impedance Sensing and Molecular Modeling of an Olfactory Biosensor Based on Chemosensory Proteins of Honeybee. *Biosens. Bioelectron.* **2013**, *40*, 174–179. [CrossRef] [PubMed]

70. Hou, Y.; Jaffrezic-Renault, N.; Martelet, C.; Tlili, C.; Zhang, A.; Pernollet, J.-C.; Briand, L.; Gomila, G.; Errachid, A.; Samitier, J.; et al. Study of Langmuir and Langmuir−Blodgett Films of Odorant-Binding Protein/Amphiphile for Odorant Biosensors. *Langmuir* **2005**, *21*, 4058–4065. [CrossRef] [PubMed]

71. Hou, Y.; Jaffrezic-Renault, N.; Martelet, C.; Zhang, A.; Minic-Vidic, J.; Gorojankina, T.; Persuy, M.-A.; Pajot-Augy, E.; Salesse, R.; Akimov, V.; et al. A Novel Detection Strategy for Odorant Molecules Based on Controlled Bioengineering of Rat Olfactory Receptor I7. *Biosens. Bioelectron.* **2007**, *22*, 1550–1555. [CrossRef]
72. Akimov, V.; Alfinito, E.; Bausells, J.; Benilova, I.; Paramo, I.C.; Errachid, A.; Ferrari, G.; Fumagalli, L.; Gomila, G.; Grosclaude, J.; et al. Nanobiosensors Based on Individual Olfactory Receptors. *Analog. Integr. Circ. Signal Process.* **2008**, *57*, 197–203. [CrossRef]
73. Hong, S.; Wu, M.; Hong, Y.; Jeong, Y.; Jung, G.; Shin, W.; Park, J.; Kim, D.; Jang, D.; Lee, J.-H. FET-Type Gas Sensors: A Review. *Sens. Actuators B Chem.* **2021**, *330*, 129240. [CrossRef]
74. Shehada, N.; Brönstrup, G.; Funka, K.; Christiansen, S.; Leja, M.; Haick, H. Ultrasensitive Silicon Nanowire for Real-World Gas Sensing: Noninvasive Diagnosis of Cancer from Breath Volatolome. *Nano Lett.* **2015**, *15*, 1288–1295. [CrossRef]
75. Shehada, N.; Cancilla, J.C.; Torrecilla, J.S.; Pariente, E.S.; Brönstrup, G.; Christiansen, S.; Johnson, D.W.; Leja, M.; Davies, M.P.A.; Liran, O.; et al. Silicon Nanowire Sensors Enable Diagnosis of Patients via Exhaled Breath. *ACS Nano* **2016**, *10*, 7047–7057. [CrossRef]
76. Ermanok, R.; Assad, O.; Zigelboim, K.; Wang, B.; Haick, H. Discriminative Power of Chemically Sensitive Silicon Nanowire Field Effect Transistors to Volatile Organic Compounds. *ACS Appl. Mater. Interfaces* **2013**, *5*, 11172–11183. [CrossRef] [PubMed]
77. Kim, T.H.; Lee, S.H.; Lee, J.; Song, H.S.; Oh, E.H.; Park, T.H.; Hong, S. Single-Carbon-Atomic-Resolution Detection of Odorant Molecules Using a Human Olfactory Receptor-Based Bioelectronic Nose. *Adv. Mater.* **2009**, *21*, 91–94. [CrossRef]
78. Khamis, S.M.; Jones, R.A.; Johnson, A.T.C.; Preti, G.; Kwak, J.; Gelperin, A. DNA-Decorated Carbon Nanotube-Based FETs as Ultrasensitive Chemical Sensors: Discrimination of Homologues, Structural Isomers, and Optical Isomers. *AIP Adv.* **2012**, *2*, 022110. [CrossRef]
79. Kybert, N.J.; Lerner, M.B.; Yodh, J.S.; Preti, G.; Johnson, A.T.C. Differentiation of Complex Vapor Mixtures Using Versatile DNA–Carbon Nanotube Chemical Sensor Arrays. *ACS Nano* **2013**, *7*, 2800–2807. [CrossRef] [PubMed]
80. Staii, C.; Johnson, A.T.C.; Chen, M.; Gelperin, A. DNA-Decorated Carbon Nanotubes for Chemical Sensing. *Nano Lett.* **2005**, *5*, 1774–1778. [CrossRef] [PubMed]
81. Kybert, N.J.; Han, G.H.; Lerner, M.B.; Dattoli, E.N.; Esfandiar, A.; Johnson, A.T.C. Scalable Arrays of Chemical Vapor Sensors Based on DNA-Decorated Graphene. *Nano Res.* **2014**, *7*, 95–103. [CrossRef]
82. Kotlowski, C.; Larisika, M.; Guerin, P.M.; Kleber, C.; Kröber, T.; Mastrogiacomo, R.; Nowak, C.; Pelosi, P.; Schütz, S.; Schwaighofer, A.; et al. Fine Discrimination of Volatile Compounds by Graphene-Immobilized Odorant-Binding Proteins. *Sens. Actuators B Chem.* **2018**, *256*, 564–572. [CrossRef]
83. Liao, F.; Chen, C.; Subramanian, V. Organic TFTs as Gas Sensors for Electronic Nose Applications. *Sens. Actuators B Chem.* **2005**, *107*, 849–855. [CrossRef]
84. Chevallier, E.; Scorsone, E.; Bergonzo, P. New Sensitive Coating Based on Modified Diamond Nanoparticles for Chemical SAW Sensors. *Sens. Actuators B Chem.* **2011**, *154*, 238–244. [CrossRef]
85. Rapp, M.; Reibel, J.; Voigt, A.; Balzer, M.; Bülow, O. New Miniaturized SAW-Sensor Array for Organic Gas Detection Driven by Multiplexed Oscillators. *Sens. Actuators B Chem.* **2000**, *65*, 169–172. [CrossRef]
86. Matatagui, D.; Bahos, F.A.; Gràcia, I.; Horrillo, M.d.C. Portable Low-Cost Electronic Nose Based on Surface Acoustic Wave Sensors for the Detection of BTX Vapors in Air. *Sensors* **2019**, *19*, 5406. [CrossRef] [PubMed]
87. Panigrahi, S.; Sankaran, S.; Mallik, S.; Gaddam, B.; Hanson, A.A. Olfactory Receptor-Based Polypeptide Sensor for Acetic Acid VOC Detection. *Mater. Sci. Eng. C* **2012**, *32*, 1307–1313. [CrossRef]
88. Compagnone, D.; Fusella, G.C.; Del Carlo, M.; Pittia, P.; Martinelli, E.; Tortora, L.; Paolesse, R.; Di Natale, C. Gold Nanoparticles-Peptide Based Gas Sensor Arrays for the Detection of Foodaromas. *Biosens. Bioelectron.* **2013**, *42*, 618–625. [CrossRef] [PubMed]
89. Compagnone, D.; Faieta, M.; Pizzoni, D.; Di Natale, C.; Paolesse, R.; Van Caelenberg, T.; Beheydt, B.; Pittia, P. Quartz Crystal Microbalance Gas Sensor Arrays for the Quality Control of Chocolate. *Sens. Actuators B Chem.* **2015**, *207*, 1114–1120. [CrossRef]
90. Di Natale, C.; Macagnano, A.; Martinelli, E.; Paolesse, R.; D'Arcangelo, G.; Roscioni, C.; Finazzi-Agrò, A.; D'Amico, A. Lung Cancer Identification by the Analysis of Breath by Means of an Array of Non-Selective Gas Sensors. *Biosens. Bioelectron.* **2003**, *18*, 1209–1218. [CrossRef]
91. Sung, J.H.; Ko, H.J.; Park, T.H. Piezoelectric Biosensor Using Olfactory Receptor Protein Expressed in Escherichia Coli. *Biosens. Bioelectron.* **2006**, *21*, 1981–1986. [CrossRef] [PubMed]
92. Du, L.; Wu, C.; Peng, H.; Zou, L.; Zhao, L.; Huang, L.; Wang, P. Piezoelectric Olfactory Receptor Biosensor Prepared by Aptamer-Assisted Immobilization. *Sens. Actuators B Chem.* **2013**, *187*, 481–487. [CrossRef]
93. Wu, C.; Du, L.; Wang, D.; Wang, L.; Zhao, L.; Wang, P. A Novel Surface Acoustic Wave-Based Biosensor for Highly Sensitive Functional Assays of Olfactory Receptors. *Biochem. Biophys. Res. Commun.* **2011**, *407*, 18–22. [CrossRef]
94. Benetti, M.; Cannatà, D.; Di Pietrantonio, F.; Foglietti, V.; Verona, E. Microbalance Chemical Sensor Based on Thin-Film Bulk Acoustic Wave Resonators. *Appl. Phys. Lett.* **2005**, *87*, 173504. [CrossRef]
95. Chen, D.; Xu, Y.; Wang, J.; Zhang, L. Nerve Gas Sensor Using Film Bulk Acoustic Resonator Modified with a Self-Assembled Cu2+/11-Mercaptoundecanoic Acid Bilayer. *Sens. Actuators B Chem.* **2010**, *150*, 483–486. [CrossRef]
96. Chang, Y.; Tang, N.; Qu, H.; Liu, J.; Zhang, D.; Zhang, H.; Pang, W.; Duan, X. Detection of Volatile Organic Compounds by Self-Assembled Monolayer Coated Sensor Array with Concentration-Independent Fingerprints. *Sci. Rep.* **2016**, *6*, 23970. [CrossRef]

97. Korsa, M.T.; Maria Carmona Domingo, J.; Nsubuga, L.; Hvam, J.; Niekiel, F.; Lofink, F.; Rubahn, H.-G.; Adam, J.; Hansen, R.d.O. Optimizing Piezoelectric Cantilever Design for Electronic Nose Applications. *Chemosensors* **2020**, *8*, 114. [CrossRef]

98. Manai, R.; Scorsone, E.; Rousseau, L.; Ghassemi, F.; Possas Abreu, M.; Lissorgues, G.; Tremillon, N.; Ginisty, H.; Arnault, J.-C.; Tuccori, E.; et al. Grafting Odorant Binding Proteins on Diamond Bio-MEMS. *Biosens. Bioelectron.* **2014**, *60*, 311–317. [CrossRef] [PubMed]

99. Yoo, Y.K.; Chae, M.-S.; Kang, J.Y.; Kim, T.S.; Hwang, K.S.; Lee, J.H. Multifunctionalized Cantilever Systems for Electronic Nose Applications. *Anal. Chem.* **2012**, *84*, 8240–8245. [CrossRef]

100. Hwang, K.S.; Lee, M.H.; Lee, J.; Yeo, W.-S.; Lee, J.H.; Kim, K.-M.; Kang, J.Y.; Kim, T.S. Peptide Receptor-Based Selective Dinitrotoluene Detection Using a Microcantilever Sensor. *Biosens. Bioelectron.* **2011**, *30*, 249–254. [CrossRef] [PubMed]

101. Ju, S.; Lee, K.-Y.; Min, S.-J.; Yoo, Y.K.; Hwang, K.S.; Kim, S.K.; Yi, H. Single-Carbon Discrimination by Selected Peptides for Individual Detection of Volatile Organic Compounds. *Sci. Rep.* **2015**, *5*, 9196. [CrossRef] [PubMed]

102. Lee, H.J.; Park, K.K.; Kupnik, M.; Oralkan, Ö.; Khuri-Yakub, B.T. Chemical Vapor Detection Using a Capacitive Micromachined Ultrasonic Transducer. *Anal. Chem.* **2011**, *83*, 9314–9320. [CrossRef] [PubMed]

103. Seok, C.; Mahmud, M.M.; Kumar, M.; Adelegan, O.J.; Yamaner, F.Y.; Oralkan, O. A Low-Power Wireless Multichannel Gas Sensing System Based on a Capacitive Micromachined Ultrasonic Transducer (CMUT) Array. *IEEE Internet Things J.* **2019**, *6*, 831–843. [CrossRef]

104. Grate, J.W. Acoustic Wave Microsensor Arrays for Vapor Sensing. *Chem. Rev.* **2000**, *100*, 2627–2648. [CrossRef]

105. Bogomolov, A. Developing Multisensory Approach to the Optical Spectral Analysis. *Sensors* **2021**, *21*, 3541. [CrossRef]

106. Rakow, N.A.; Suslick, K.S. A Colorimetric Sensor Array for Odour Visualization. *Nature* **2000**, *406*, 710–713. [CrossRef] [PubMed]

107. Khan, M.; Kang, S.-W. A High Sensitivity and Wide Dynamic Range Fiber-Optic Sensor for Low-Concentration VOC Gas Detection. *Sensors* **2014**, *14*, 23321–23336. [CrossRef] [PubMed]

108. Li, J.; Hou, C.-J.; Huo, D.; Yang, M.; Fa, H.; Yang, P. Development of a Colorimetric Sensor Array for the Discrimination of Aldehydes. *Sens. Actuators B Chem.* **2014**, *196*, 10–17. [CrossRef]

109. Li, J.-J.; Song, C.-X.; Hou, C.-J.; Huo, D.-Q.; Shen, C.-H.; Luo, X.-G.; Yang, M.; Fa, H.-B. Development of a Colorimetric Sensor Array for the Discrimination of Chinese Liquors Based on Selected Volatile Markers Determined by GC-MS. *J. Agric. Food Chem.* **2014**, *62*, 10422–10430. [CrossRef] [PubMed]

110. Shin, Y.-H.; Teresa Gutierrez-Wing, M.; Choi, J.-W. Review—Recent Progress in Portable Fluorescence Sensors. *J. Electrochem. Soc.* **2021**, *168*, 017502. [CrossRef]

111. Dickinson, T.A.; White, J.; Kauer, J.S.; Walt, D.R. A Chemical-Detecting System Based on a Cross-Reactive Optical Sensor Array. *Nature* **1996**, *382*, 697–700. [CrossRef]

112. Pawar, D.; Kale, S.N. A Review on Nanomaterial-Modified Optical Fiber Sensors for Gases, Vapors and Ions. *Microchim. Acta* **2019**, *186*, 253. [CrossRef]

113. Johnson, S.R.; Sutter, J.M.; Engelhardt, H.L.; Jurs, P.C.; White, J.; Kauer, J.S.; Dickinson, T.A.; Walt, D.R. Identification of Multiple Analytes Using an Optical Sensor Array and Pattern Recognition Neural Networks. *Anal. Chem.* **1997**, *69*, 4641–4648. [CrossRef]

114. Walt, D.R. Bead-Based Optical Fiber Arrays for Artificial Olfaction. *Curr. Opin. Chem. Biol.* **2010**, *14*, 767–770. [CrossRef] [PubMed]

115. Dickinson, T.A.; Michael, K.L.; Kauer, J.S.; Walt, D.R. Convergent, Self-Encoded Bead Sensor Arrays in the Design of an Artificial Nose. *Anal. Chem.* **1999**, *71*, 2192–2198. [CrossRef] [PubMed]

116. Khan, M.R.R.; Kang, B.-H.; Lee, S.-W.; Kim, S.-H.; Yeom, S.-H.; Lee, S.-H.; Kang, S.-W. Fiber-Optic Multi-Sensor Array for Detection of Low Concentration Volatile Organic Compounds. *Opt. Express* **2013**, *21*, 20119–20130. [CrossRef] [PubMed]

117. Khan, M.R.R.; Kang, B.-H.; Yeom, S.-H.; Kwon, D.-H.; Kang, S.-W. Fiber-Optic Pulse Width Modulation Sensor for Low Concentration VOC Gas. *Sens. Actuators B Chem.* **2013**, *188*, 689–696. [CrossRef]

118. Ma, Y.; Li, Y.; Ma, K.; Wang, Z. Optical Colorimetric Sensor Arrays for Chemical and Biological Analysis. *Sci. China Chem.* **2018**, *61*, 643–655. [CrossRef]

119. Nylander, C.; Liedberg, B.; Lind, T. Gas Detection by Means of Surface Plasmon Resonance. *Sens. Actuators* **1982**, *3*, 79–88. [CrossRef]

120. Nguyen, H.H.; Park, J.; Kang, S.; Kim, M. Surface Plasmon Resonance: A Versatile Technique for Biosensor Applications. *Sensors* **2015**, *15*, 10481–10510. [CrossRef]

121. Homola, J. Surface Plasmon Resonance Sensors for Detection of Chemical and Biological Species. *Chem. Rev.* **2008**, *108*, 462–493. [CrossRef]

122. Wood, R.W. On a Remarkable Case of Uneven Distribution of Light in a Diffraction Grating Spectrum. *Lond. Edinb. Dublin Philos. Mag. J. Sci.* **1902**, *4*, 396–402. [CrossRef]

123. Fano, U. The Theory of Anomalous Diffraction Gratings and of Quasi-Stationary Waves on Metallic Surfaces (Sommerfeld's Waves). *J. Opt. Soc. Am.* **1941**, *31*, 213–222. [CrossRef]

124. Pines, D.; Bohm, D. A Collective Description of Electron Interactions: II. Collective vs Individual Particle Aspects of the Interactions. *Phys. Rev.* **1952**, *85*, 338–353. [CrossRef]

125. Ferrell, R.A. Angular Dependence of the Characteristic Energy Loss of Electrons Passing Through Metal Foils. *Phys. Rev.* **1956**, *101*, 554–563. [CrossRef]

126. Stern, E.A.; Ferrell, R.A. Surface Plasma Oscillations of a Degenerate Electron Gas. *Phys. Rev.* **1960**, *120*, 130–136. [CrossRef]

127. Ritchie, R.H. Plasma Losses by Fast Electrons in Thin Films. *Phys. Rev.* **1957**, *106*, 874–881. [CrossRef]

128. Powell, C.J. The Origin of the Characteristic Electron Energy Losses in Ten Elements. *Proc. Phys. Soc.* **1960**, *76*, 593–610. [CrossRef]

129. Otto, A. Excitation of Nonradiative Surface Plasma Waves in Silver by the Method of Frustrated Total Reflection. *Z. Phys.* **1968**, *216*, 398–410. [CrossRef]

130. Kretschmann, E.; Raether, H. Radiative Decay of Non Radiative Surface Plasmons Excited by Light. *Z. Nat. A* **1968**, *23*, 2135–2136. [CrossRef]

131. Mayer, K.M.; Hafner, J.H. Localized Surface Plasmon Resonance Sensors. *Chem. Rev.* **2011**, *111*, 3828–3857. [CrossRef] [PubMed]

132. Roh, S.; Chung, T.; Lee, B. Overview of the Characteristics of Micro- and Nano-Structured Surface Plasmon Resonance Sensors. *Sensors* **2011**, *11*, 1565–1588. [CrossRef]

133. Willets, K.A.; Van Duyne, R.P. Localized Surface Plasmon Resonance Spectroscopy and Sensing. *Annu. Rev. Phys. Chem.* **2007**, *58*, 267–297. [CrossRef] [PubMed]

134. Petryayeva, E.; Krull, U.J. Localized Surface Plasmon Resonance: Nanostructures, Bioassays and Biosensing—A Review. *Anal. Chim. Acta* **2011**, *706*, 8–24. [CrossRef] [PubMed]

135. Kelly, K.L.; Coronado, E.; Zhao, L.L.; Schatz, G.C. The Optical Properties of Metal Nanoparticles: The Influence of Size, Shape, and Dielectric Environment. *J. Phys. Chem. B* **2003**, *107*, 668–677. [CrossRef]

136. Hutter, E.; Fendler, J.H. Exploitation of Localized Surface Plasmon Resonance. *Adv. Mater.* **2004**, *16*, 1685–1706. [CrossRef]

137. Chung, T.; Lee, S.-Y.; Song, E.Y.; Chun, H.; Lee, B. Plasmonic Nanostructures for Nano-Scale Bio-Sensing. *Sensors* **2011**, *11*, 10907–10929. [CrossRef] [PubMed]

138. Rojalin, T.; Phong, B.; Koster, H.J.; Carney, R.P. Nanoplasmonic Approaches for Sensitive Detection and Molecular Characterization of Extracellular Vesicles. *Front. Chem.* **2019**, *7*, 279. [CrossRef] [PubMed]

139. Sepúlveda, B.; Angelomé, P.C.; Lechuga, L.M.; Liz-Marzán, L.M. LSPR-Based Nanobiosensors. *Nano Today* **2009**, *4*, 244–251. [CrossRef]

140. Haes, A.J.; Chang, L.; Klein, W.L.; Van Duyne, R.P. Detection of a Biomarker for Alzheimer's Disease from Synthetic and Clinical Samples Using a Nanoscale Optical Biosensor. *J. Am. Chem. Soc.* **2005**, *127*, 2264–2271. [CrossRef]

141. Lopez, G.A.; Estevez, M.-C.; Soler, M.; Lechuga, L.M. Recent Advances in Nanoplasmonic Biosensors: Applications and Lab-on-a-Chip Integration. *Nanophotonics* **2017**, *6*, 123–136. [CrossRef]

142. Bingham, J.M.; Anker, J.N.; Kreno, L.E.; Van Duyne, R.P. Gas Sensing with High-Resolution Localized Surface Plasmon Resonance Spectroscopy. *J. Am. Chem. Soc.* **2010**, *132*, 17358–17359. [CrossRef]

143. Kreno, L.E.; Hupp, J.T.; Van Duyne, R.P. Metal−Organic Framework Thin Film for Enhanced Localized Surface Plasmon Resonance Gas Sensing. *Anal. Chem.* **2010**, *82*, 8042–8046. [CrossRef]

144. Shang, L.; Liu, C.; Watanabe, M.; Chen, B.; Hayashi, K. LSPR Sensor Array Based on Molecularly Imprinted Sol-Gels for Pattern Recognition of Volatile Organic Acids. *Sens. Actuators B Chem.* **2017**, *249*, 14–21. [CrossRef]

145. Zhang, D.; Lu, Y.; Zhang, Q.; Yao, Y.; Li, S.; Li, H.; Zhuang, S.; Jiang, J.; Liu, G.L.; Liu, Q. Nanoplasmonic Monitoring of Odorants Binding to Olfactory Proteins from Honeybee as Biosensor for Chemical Detection. *Sens. Actuators B Chem.* **2015**, *221*, 341–349. [CrossRef]

146. Cheng, C.-S.; Chen, Y.-Q.; Lu, C.-J. Organic Vapour Sensing Using Localized Surface Plasmon Resonance Spectrum of Metallic Nanoparticles Self Assemble Monolayer. *Talanta* **2007**, *73*, 358–365. [CrossRef]

147. Homola, J.; Yee, S.S.; Gauglitz, G. Surface Plasmon Resonance Sensors: Review. *Sens. Actuators B Chem.* **1999**, *54*, 3–15. [CrossRef]

148. Homola, J. Present and Future of Surface Plasmon Resonance Biosensors. *Anal. Bioanal. Chem.* **2003**, *377*, 528–539. [CrossRef]

149. Zayats, A.V.; Smolyaninov, I.I. Near-Field Photonics: Surface Plasmon Polaritons and Localized Surface Plasmons. *J. Opt. A Pure Appl. Opt.* **2003**, *5*, S16–S50. [CrossRef]

150. Zayats, A.V.; Smolyaninov, I.I.; Maradudin, A.A. Nano-Optics of Surface Plasmon Polaritons. *Phys. Rep.* **2005**, *408*, 131–314. [CrossRef]

151. Barnes, W.L.; Dereux, A.; Ebbesen, T.W. Surface Plasmon Subwavelength Optics. *Nature* **2003**, *424*, 824–830. [CrossRef] [PubMed]

152. Prabowo, B.; Purwidyantri, A.; Liu, K.-C. Surface Plasmon Resonance Optical Sensor: A Review on Light Source Technology. *Biosensors* **2018**, *8*, 80. [CrossRef] [PubMed]

153. Sambles, J.R.; Bradbery, G.W.; Yang, F. Optical Excitation of Surface Plasmons: An Introduction. *Contemp. Phys.* **1991**, *32*, 173–183. [CrossRef]

154. Pitarke, J.M.; Silkin, V.M.; Chulkov, E.V.; Echenique, P.M. Theory of Surface Plasmons and Surface-Plasmon Polaritons. *Rep. Prog. Phys.* **2007**, *70*, 1–87. [CrossRef]

155. Raether, H. Surface Plasmons on Smooth Surfaces. In *Surface Plasmons on Smooth and Rough Surfaces and on Gratings*; Springer: Berlin/Heidelberg, Germany, 1988; pp. 4–39.

156. Gupta, B.D.; Verma, R.K. Surface Plasmon Resonance-Based Fiber Optic Sensors: Principle, Probe Designs, and Some Applications. *J. Sens.* **2009**, *2009*, 979761. [CrossRef]

157. Gupta, B.; Shrivastav, A.; Usha, S. Surface Plasmon Resonance-Based Fiber Optic Sensors Utilizing Molecular Imprinting. *Sensors* **2016**, *16*, 1381. [CrossRef]

158. Zeng, Y.; Wang, X.; Zhou, J.; Miyan, R.; Qu, J.; Ho, H.-P.; Zhou, K.; Gao, B.Z.; Shao, Y. Phase Interrogation SPR Sensing Based on White Light Polarized Interference for Wide Dynamic Detection Range. *Opt. Express* **2020**, *28*, 3442–3450. [CrossRef]

159. Wong, C.L.; Olivo, M. Surface Plasmon Resonance Imaging Sensors: A Review. *Plasmonics* **2014**, *9*, 809–824. [CrossRef]

160. Glatz, R.; Bailey-Hill, K. Mimicking Nature's Noses: From Receptor Deorphaning to Olfactory Biosensing. *Prog. Neurobiol.* **2011**, *93*, 270–296. [CrossRef] [PubMed]

161. Firestein, S. How the Olfactory System Makes Sense of Scents. *Nature* **2001**, *413*, 211–218. [CrossRef] [PubMed]

162. Cook, B.L.; Steuerwald, D.; Kaiser, L.; Graveland-Bikker, J.; Vanberghem, M.; Berke, A.P.; Herlihy, K.; Pick, H.; Vogel, H.; Zhang, S. Large-Scale Production and Study of a Synthetic G Protein-Coupled Receptor: Human Olfactory Receptor 17-4. *Proc. Natl. Acad. Sci. USA* **2009**, *106*, 11925–11930. [CrossRef] [PubMed]

163. Jin, H.J.; Lee, S.H.; Kim, T.H.; Park, J.; Song, H.S.; Park, T.H.; Hong, S. Nanovesicle-Based Bioelectronic Nose Platform Mimicking Human Olfactory Signal Transduction. *Biosens. Bioelectron.* **2012**, *35*, 335–341. [CrossRef]

164. Vidic, J.M.; Grosclaude, J.; Persuy, M.-A.; Aioun, J.; Salesse, R.; Pajot-Augy, E. Quantitative Assessment of Olfactory Receptors Activity in Immobilized Nanosomes: A Novel Concept for Bioelectronic Nose. *Lab Chip* **2006**, *6*, 1026–1032. [CrossRef]

165. Vidic, J.; Grosclaude, J.; Monnerie, R.; Persuy, M.-A.; Badonnel, K.; Baly, C.; Caillol, M.; Briand, L.; Salesse, R.; Pajot-Augy, E. On a Chip Demonstration of a Functional Role for Odorant Binding Protein in the Preservation of Olfactory Receptor Activity at High Odorant Concentration. *Lab Chip* **2008**, *8*, 678–688. [CrossRef] [PubMed]

166. Benilova, I.; Chegel, V.I.; Ushenin, Y.V.; Vidic, J.; Soldatkin, A.P.; Martelet, C.; Pajot, E.; Jaffrezic-Renault, N. Stimulation of Human Olfactory Receptor 17–40 with Odorants Probed by Surface Plasmon Resonance. *Eur. Biophys. J.* **2008**, *37*, 807–814. [CrossRef]

167. Lee, S.H.; Ko, H.J.; Park, T.H. Real-Time Monitoring of Odorant-Induced Cellular Reactions Using Surface Plasmon Resonance. *Biosens. Bioelectron.* **2009**, *25*, 55–60. [CrossRef] [PubMed]

168. Lee, J.Y.; Ko, H.J.; Lee, S.H.; Park, T.H. Cell-Based Measurement of Odorant Molecules Using Surface Plasmon Resonance. *Enzym. Microb. Technol.* **2006**, *39*, 375–380. [CrossRef]

169. Oh, E.H.; Lee, S.H.; Ko, H.J.; Park, T.H. Odorant Detection Using Liposome Containing Olfactory Receptor in the SPR System. *Sens. Actuators B Chem.* **2014**, *198*, 188–193. [CrossRef]

170. Sanmartí-Espinal, M.; Iavicoli, P.; Calò, A.; Taulés, M.; Galve, R.; Marco, M.P.; Samitier, J. Quantification of Interacting Cognate Odorants with Olfactory Receptors in Nanovesicles. *Sci. Rep.* **2017**, *7*, 17483. [CrossRef]

171. Briand, L.; Eloit, C.; Nespoulous, C.; Bézirard, V.; Huet, J.-C.; Henry, C.; Blon, F.; Trotier, D.; Pernollet, J.-C. Evidence of an Odorant-Binding Protein in the Human Olfactory Mucus: Location, Structural Characterization, and Odorant-Binding Properties. *Biochemistry* **2002**, *41*, 7241–7252. [CrossRef] [PubMed]

172. Briand, L.; Nespoulous, C.; Huet, J.-C.; Takahashi, M.; Pernollet, J.-C. Ligand Binding and Physico-Chemical Properties of ASP2, a Recombinant Odorant-Binding Protein from Honeybee (*Apis Mellifera* L.): Odorant Binding by a Honeybee OBP. *Eur. J. Biochem.* **2001**, *268*, 752–760. [CrossRef] [PubMed]

173. Pelosi, P.; Mastrogiacomo, R.; Iovinella, I.; Tuccori, E.; Persaud, K.C. Structure and Biotechnological Applications of Odorant-Binding Proteins. *Appl. Microbiol. Biotechnol.* **2014**, *98*, 61–70. [CrossRef]

174. Briand, L.; Nespoulous, C.; Perez, V.; Rémy, J.-J.; Huet, J.-C.; Pernollet, J.-C. Ligand-Binding Properties and Structural Characterization of a Novel Rat Odorant-Binding Protein Variant: Ligand Binding and Characterization of Rat OBP-1F. *Eur. J. Biochem.* **2000**, *267*, 3079–3089. [CrossRef]

175. Di Pietrantonio, F.; Benetti, M.; Cannatà, D.; Verona, E.; Palla-Papavlu, A.; Fernández-Pradas, J.M.; Serra, P.; Staiano, M.; Varriale, A.; D'Auria, S. A Surface Acoustic Wave Bio-Electronic Nose for Detection of Volatile Odorant Molecules. *Biosens. Bioelectron.* **2015**, *67*, 516–523. [CrossRef]

176. Larisika, M.; Kotlowski, C.; Steininger, C.; Mastrogiacomo, R.; Pelosi, P.; Schütz, S.; Peteu, S.F.; Kleber, C.; Reiner-Rozman, C.; Nowak, C.; et al. Electronic Olfactory Sensor Based on *A. Mellifera* Odorant-Binding Protein 14 on a Reduced Graphene Oxide Field-Effect Transistor. *Angew. Chem.* **2015**, *127*, 13245–13248. [CrossRef]

177. Hurot, C.; Brenet, S.; Buhot, A.; Barou, E.; Belloir, C.; Briand, L.; Hou, Y. Highly Sensitive Olfactory Biosensors for the Detection of Volatile Organic Compounds by Surface Plasmon Resonance Imaging. *Biosens. Bioelectron.* **2019**, *123*, 230–236. [CrossRef]

178. Dung, T.T.; Kim, M. A Surface Plasmon Resonance Sensor for Detection of Toluene ($C_6H_5CH_3$). *Appl. Sci. Converg. Technol.* **2018**, *27*, 184–188. [CrossRef]

179. Liedberg, B.; Nylander, C.; Lunström, I. Surface Plasmon Resonance for Gas Detection and Biosensing. *Sens. Actuators* **1983**, *4*, 299–304. [CrossRef]

180. Miwa, S.; Arakawa, T. Selective Gas Detection by Means of Surface Plasmon Resonance Sensors. *Thin Solid Films* **1996**, *281*, 466–468. [CrossRef]

181. Granito, C.; Wilde, J.N.; Petty, M.C.; Houghton, S.; Iredale, P.J. Toluene Vapour Sensing Using Copper and Nickel Phthalocyanine Langmuir-Blodgett Films. *Thin Solid Films* **1996**, *284*, 98–101. [CrossRef]

182. Wilde, J.N.; Petty, M.C.; Saffell, J.; Tempore, A.; Valli, L. Surface Plasmon Resonance Imaging for Gas Sensing. *Meas. Control.* **1997**, *30*, 269–272. [CrossRef]

183. Podgorsek, R.P.; Sterkenburgh, T.; Wolters, J.; Ehrenreich, T.; Nischwitz, S.; Franke, H. Optical Gas Sensing by Evaluating ATR Leaky Mode Spectra. *Sens. Actuators B Chem.* **1997**, *39*, 349–352. [CrossRef]

184. Wilde, J.N.; Nagel, J.; Petty, M.C. Optical Sensing of Aromatic Hydrocarbons Using Langmuir–Blodgett Films of a Schiff Base Co-Ordination Polymer. *Thin Solid Films* **1998**, *327*, 726–729. [CrossRef]

185. Brenet, S.; John-Herpin, A.; Gallat, F.-X.; Musnier, B.; Buhot, A.; Herrier, C.; Rousselle, T.; Livache, T.; Hou, Y. Highly-Selective Optoelectronic Nose Based on Surface Plasmon Resonance Imaging for Sensing Volatile Organic Compounds. *Anal. Chem.* **2018**, *90*, 9879–9887. [CrossRef]

186. Brenet, S.; Weerakkody, J.S.; Buhot, A.; Gallat, F.-X.; Mathey, R.; Leroy, L.; Livache, T.; Herrier, C.; Hou, Y. Improvement of Sensitivity of Surface Plasmon Resonance Imaging for the Gas-Phase Detection of Volatile Organic Compounds. *Talanta* **2020**, *212*, 120777. [CrossRef]

187. Weerakkody, J.S.; Brenet, S.; Livache, T.; Herrier, C.; Hou, Y.; Buhot, A. Optical Index Prism Sensitivity of Surface Plasmon Resonance Imaging in Gas Phase: Experiment versus Theory. *J. Phys. Chem. C* **2020**, *124*, 3756–3767. [CrossRef]

188. Maho, P.; Herrier, C.; Livache, T.; Rolland, G.; Comon, P.; Barthelmé, S. Reliable Chiral Recognition with an Optoelectronic Nose. *Biosens. Bioelectron.* **2020**, *159*, 112183. [CrossRef]

189. Slimani, S.; Bultel, E.; Cubizolle, T.; Herrier, C.; Rousselle, T.; Livache, T. Opto-Electronic Nose Coupled to a Silicon Micro Pre-Concentrator Device for Selective Sensing of Flavored Waters. *Chemosensors* **2020**, *8*, 60. [CrossRef]

190. Fournel, A.; Mantel, M.; Pinger, M.; Manesse, C.; Dubreuil, R.; Herrier, C.; Rousselle, T.; Livache, T.; Bensafi, M. An Experimental Investigation Comparing a Surface Plasmon Resonance Imaging-Based Artificial Nose with Natural Olfaction. *Sens. Actuators B Chem.* **2020**, *320*, 128342. [CrossRef]

191. Gaggiotti, S.; Hurot, C.; Weerakkody, J.S.; Mathey, R.; Buhot, A.; Mascini, M.; Hou, Y.; Compagnone, D. Development of an Optoelectronic Nose Based on Surface Plasmon Resonance Imaging with Peptide and Hairpin DNA for Sensing Volatile Organic Compounds. *Sens. Actuators B Chem.* **2020**, *303*, 127188. [CrossRef]

192. Daly, S.M.; Grassi, M.; Shenoy, D.K.; Ugozzoli, F.; Dalcanale, E. Supramolecular Surface Plasmon Resonance (SPR) Sensors for Organophosphorus Vapor Detection. *J. Mater. Chem.* **2007**, *17*, 1809–1818. [CrossRef]

193. Feresenbet, E.B.; Dalcanale, E.; Dulcey, C.; Shenoy, D.K. Optical Sensing of the Selective Interaction of Aromatic Vapors with Cavitands. *Sens. Actuators B Chem.* **2004**, *97*, 211–220. [CrossRef]

194. Şen, S.; Cömert, Ö.F.; Çapan, R.; Ay, M. A Room Temperature Acetone Sensor Based on Synthesized Tetranitro-Oxacalix[4]Arenes: Thin Film Fabrication and Sensing Properties. *Sens. Actuators A Phys.* **2020**, *315*, 112308. [CrossRef]

195. Capan, R.; Ray, A.K.; Hassan, A.K.; Tanrisever, T. Poly (Methyl Methacrylate) Films for Organic Vapour Sensing. *J. Phys. D Appl. Phys.* **2003**, *36*, 1115. [CrossRef]

196. Hassan, A.K.; Ray, A.K.; Nabok, A.V.; Wilkop, T. Kinetic Studies of BTEX Vapour Adsorption onto Surfaces of Calix-4-Resorcinarene Films. *Appl. Surf. Sci.* **2001**, *182*, 49–54. [CrossRef]

197. Chaure, S.; Yang, B.; Hassan, A.K.; Ray, A.K.; Bolognesi, A. Interaction Behaviour of Spun Films of Poly[3-(6-Methoxyhexyl)Thiophene] Derivatives with Ambient Gases. *J. Phys. D Appl. Phys.* **2004**, *37*, 1558–1562. [CrossRef]

198. Nanto, H.; Kitade, Y.; Takei, Y.; Kubota, N. Odor Sensor Utilizing Surface Plasmon Resonance. *Sens. Mater.* **2005**, *17*, 405–412.

199. Nanto, H.; Yagi, F.; Hasunuma, H.; Takei, Y.; Koyama, S.; Oyabu, T.; Mihara, T. Multichannel Odor Sensor Utilizing Surface Plasmon Resonance. *Sens. Mater.* **2009**, *21*, 201–208.

200. Sih, B.C.; Wolf, M.O.; Jarvis, D.; Young, J.F. Surface-Plasmon Resonance Sensing of Alcohol with Electrodeposited Polythiophene and Gold Nanoparticle-Oligothiophene Films. *J. Appl. Phys.* **2005**, *98*, 114314. [CrossRef]

201. Alwahib, A.A.; Sadrolhosseini, A.R.; An'amt, M.N.; Lim, H.N.; Yaacob, M.H.; Abu Bakar, M.H.; Ming, H.N.; Mahdi, M.A. Reduced Graphene Oxide/Maghemite Nanocomposite for Detection of Hydrocarbon Vapor Using Surface Plasmon Resonance. *IEEE Photonics J.* **2016**, *8*, 1–9. [CrossRef]

202. Manera, M.G.; Montagna, G.; Ferreiro-Vila, E.; González-García, L.; Sánchez-Valencia, J.R.; González-Elipe, A.R.; Cebollada, A.; Garcia-Martin, J.M.; Garcia-Martin, A.; Armelles, G.; et al. Enhanced Gas Sensing Performance of TiO2 Functionalized Magneto-Optical SPR Sensors. *J. Mater. Chem.* **2011**, *21*, 16049–16056. [CrossRef]

203. Manera, M.G.; Leo, G.; Curri, M.L.; Cozzoli, P.D.; Rella, R.; Siciliano, P.; Agostiano, A.; Vasanelli, L. Investigation on Alcohol Vapours/TiO$_2$ Nanocrystal Thin Films Interaction by SPR Technique for Sensing Application. *Sens. Actuators B Chem.* **2004**, *100*, 75–80. [CrossRef]

204. Manera, M.G.; de Julián Fernández, C.; Maggioni, G.; Mattei, G.; Carturan, S.; Quaranta, A.; Della Mea, G.; Rella, R.; Vasanelli, L.; Mazzoldi, P. Surface Plasmon Resonance Study on the Optical Sensing Properties of Nanometric Polyimide Films to Volatile Organic Vapours. *Sens. Actuators B Chem.* **2007**, *120*, 712–718. [CrossRef]

205. Livache, T.; Gallat, F.-X.; Hou-Broutin, Y.; Herrier, C.; Rousselle, T. Method for Calibrating an Electronic Nose. International Patent Application No. PCT/EP2018/055233, 7 September 2018.

206. Weerakkody, J.S.; Hurot, C.; Brenet, S.; Mathey, R.; Raillon, C.; Livache, T.; Buhot, A.; Hou, Y. Opto-Electronic Nose—Temperature and VOC Concentration Effects on the Equilibrium Response. In Proceedings of the 2019 IEEE International Symposium on Olfaction and Electronic Nose (ISOEN), Fukuoka, Japan, 26–29 May 2019; pp. 1–3. [CrossRef]

207. Hutter, E.; Cha, S.; Liu, J.-F.; Park, J.; Yi, J.; Fendler, J.H.; Roy, D. Role of Substrate Metal in Gold Nanoparticle Enhanced Surface Plasmon Resonance Imaging. *J. Phys. Chem. B* **2001**, *105*, 8–12. [CrossRef]

208. Klantsataya, E.; Jia, P.; Ebendorff-Heidepriem, H.; Monro, T.; François, A. Plasmonic Fiber Optic Refractometric Sensors: From Conventional Architectures to Recent Design Trends. *Sensors* **2016**, *17*, 12. [CrossRef] [PubMed]

209. Jorgenson, R.C.; Yee, S.S. A Fiber-Optic Chemical Sensor Based on Surface Plasmon Resonance. *Sens. Actuators B Chem.* **1993**, *12*, 213–220. [CrossRef]

210. Caucheteur, C.; Guo, T.; Albert, J. Review of Plasmonic Fiber Optic Biochemical Sensors: Improving the Limit of Detection. *Anal. Bioanal. Chem.* **2015**, *407*, 3883–3897. [CrossRef]

211. Vindas, K.; Leroy, L.; Garrigue, P.; Voci, S.; Livache, T.; Arbault, S.; Sojic, N.; Buhot, A.; Engel, E. Highly Parallel Remote SPR Detection of DNA Hybridization by Micropillar Optical Arrays. *Anal. Bioanal. Chem.* **2019**, *411*, 2249–2259. [CrossRef]

212. Sharma, A.K.; Pandey, A.K.; Kaur, B. A Review of Advancements (2007–2017) in Plasmonics-Based Optical Fiber Sensors. *Opt. Fiber Technol.* **2018**, *43*, 20–34. [CrossRef]
213. Sharma, A.K.; Jha, R.; Gupta, B.D. Fiber-Optic Sensors Based on Surface Plasmon Resonance: A Comprehensive Review. *IEEE Sens. J.* **2007**, *7*, 1118–1129. [CrossRef]
214. Desmet, C.; Vindas, K.; Alvarado Meza, R.; Garrigue, P.; Voci, S.; Sojic, N.; Maziz, A.; Courson, R.; Malaquin, L.; Leichle, T.; et al. Multiplexed Remote SPR Detection of Biological Interactions through Optical Fiber Bundles. *Sensors* **2020**, *20*, 511. [CrossRef] [PubMed]
215. Cennamo, N.; Di Giovanni, S.; Varriale, A.; Staiano, M.; Di Pietrantonio, F.; Notargiacomo, A.; Zeni, L.; D'Auria, S. Easy to Use Plastic Optical Fiber-Based Biosensor for Detection of Butanal. *PLoS ONE* **2015**, *10*, e0116770. [CrossRef] [PubMed]
216. Cennamo, N.; D'Agostino, G.; Galatus, R.; Bibbò, L.; Pesavento, M.; Zeni, L. Sensors Based on Surface Plasmon Resonance in a Plastic Optical Fiber for the Detection of Trinitrotoluene. *Sens. Actuators B Chem.* **2013**, *188*, 221–226. [CrossRef]
217. Vandezande, W.; Janssen, K.P.F.; Delport, F.; Ameloot, R.; De Vos, D.E.; Lammertyn, J.; Roeffaers, M.B.J. Parts per Million Detection of Alcohol Vapors via Metal Organic Framework Functionalized Surface Plasmon Resonance Sensors. *Anal. Chem.* **2017**, *89*, 4480–4487. [CrossRef]
218. Semwal, V.; Shrivastav, A.M.; Verma, R.; Gupta, B.D. Surface Plasmon Resonance Based Fiber Optic Ethanol Sensor Using Layers of Silver/Silicon/Hydrogel Entrapped with ADH/NAD. *Sens. Actuators B Chem.* **2016**, *230*, 485–492. [CrossRef]
219. Mishra, S.K.; Rani, S.; Gupta, B.D. Surface Plasmon Resonance Based Fiber Optic Hydrogen Sulphide Gas Sensor Utilizing Nickel Oxide Doped ITO Thin Film. *Sens. Actuators B Chem.* **2014**, *195*, 215–222. [CrossRef]
220. Mishra, S.K.; Tripathi, S.N.; Choudhary, V.; Gupta, B.D. Surface Plasmon Resonance-Based Fiber Optic Methane Gas Sensor Utilizing Graphene-Carbon Nanotubes-Poly(Methyl Methacrylate) Hybrid Nanocomposite. *Plasmonics* **2015**, *10*, 1147–1157. [CrossRef]
221. Liu, H.; Wang, M.; Wang, Q.; Li, H.; Ding, Y.; Zhu, C. Simultaneous Measurement of Hydrogen and Methane Based on PCF-SPR Structure with Compound Film-Coated Side-Holes. *Opt. Fiber Technol.* **2018**, *45*, 1–7. [CrossRef]
222. Arasu, P.T.; Noor, A.S.M.; Shabaneh, A.A.; Yaacob, M.H.; Lim, H.N.; Mahdi, M.A. Fiber Bragg Grating Assisted Surface Plasmon Resonance Sensor with Graphene Oxide Sensing Layer. *Opt. Commun.* **2016**, *380*, 260–266. [CrossRef]
223. Wei, W.; Nong, J.; Zhang, G.; Tang, L.; Jiang, X.; Chen, N.; Luo, S.; Lan, G.; Zhu, Y. Graphene-Based Long-Period Fiber Grating Surface Plasmon Resonance Sensor for High-Sensitivity Gas Sensing. *Sensors* **2017**, *17*, 2. [CrossRef] [PubMed]
224. Bharadwaj, R.; Mukherji, S. Gold Nanoparticle Coated U-Bend Fibre Optic Probe for Localized Surface Plasmon Resonance Based Detection of Explosive Vapours. *Sens. Actuators B Chem.* **2014**, *192*, 804–811. [CrossRef]
225. Paul, D.; Dutta, S.; Saha, D.; Biswas, R. LSPR Based Ultra-Sensitive Low Cost U-Bent Optical Fiber for Volatile Liquid Sensing. *Sens. Actuators B Chem.* **2017**, *250*, 198–207. [CrossRef]
226. Niggemann, M.; Katerkamp, A.; Pellmann, M.; Bolsmann, P.; Reinbold, J.; Cammann, K. Remote Sensing of Tetrachloroethene with a Micro-Fibre Optical Gas Sensor Based on Surface Plasmon Resonance Spectroscopy. *Sens. Actuators B Chem.* **1996**, *34*, 328–333. [CrossRef]
227. Abdelghani, A.; Chovelon, J.M.; Jaffrezic-Renault, N.; Veilla, C.; Gagnaire, H. Chemical Vapour Sensing by Surface Plasmon Resonance Optical Fibre Sensor Coated with Fluoropolymer. *Anal. Chim. Acta* **1997**, *337*, 225–232. [CrossRef]
228. Li, Y.; Xiao, A.-S.; Zou, B.; Zhang, H.-X.; Yan, K.-L.; Lin, Y. Advances of Metal–Organic Frameworks for Gas Sensing. *Polyhedron* **2018**, *154*, 83–97. [CrossRef]
229. Nazem, S.; Malekmohammad, M.; Soltanolkotabi, M. Theoretical and Experimental Study of a Surface Plasmon Sensor Based on Ag-MgF2 Grating Coupler. *Appl. Phys. B* **2020**, *126*, 96. [CrossRef]
230. Dai, Y.; Xu, H.; Wang, H.; Lu, Y.; Wang, P. Experimental Demonstration of High Sensitivity for Silver Rectangular Grating-Coupled Surface Plasmon Resonance (SPR) Sensing. *Opt. Commun.* **2018**, *416*, 66–70. [CrossRef]
231. Borile, G.; Rossi, S.; Filippi, A.; Gazzola, E.; Capaldo, P.; Tregnago, C.; Pigazzi, M.; Romanato, F. Label-Free, Real-Time on-Chip Sensing of Living Cells via Grating-Coupled Surface Plasmon Resonance. *Biophys. Chem.* **2019**, *254*, 106262. [CrossRef] [PubMed]
232. Cai, D.; Lu, Y.; Lin, K.; Wang, P.; Ming, H. Improving the Sensitivity of SPR Sensors Based on Gratings by Double-Dips Method (DDM). *Opt. Express* **2008**, *16*, 14597. [CrossRef] [PubMed]
233. Vukusic, P.S.; Bryan-Brown, G.P.; Sambles, J.R. Surface Plasmon Resonance on Gratings as a Novel Means for Gas Sensing. *Sens. Actuators B Chem.* **1992**, *8*, 155–160. [CrossRef]
234. Jory, M.J.; Vukusic, P.S.; Sambles, J.R. Development of a Prototype Gas Sensor Using Surface Plasmon Resonance on Gratings. *Sens. Actuators B Chem.* **1994**, *17*, 203–209. [CrossRef]

biosensors

Review

Plasmonic Metal Nanoparticles Hybridized with 2D Nanomaterials for SERS Detection: A Review

Caterina Serafinelli [1,2,3,4,*], Alessandro Fantoni [1,3], Elisabete C. B. A. Alegria [1,2] and Manuela Vieira [1,3,4]

[1] Instituto Superior de Engenharia de Lisboa—Instituto Politécnico de Lisboa, 1949-014 Lisboa, Portugal; afantoni@deetc.isel.ipl.pt (A.F.); ebastos@deq.isel.ipl.pt (E.C.B.A.A.); mv@isel.ipl.pt (M.V.)

[2] Centro de Química Estrutural, Institute of Molecular Sciences, Instituto Superior Técnico, Universidade de Lisboa, 1049-001 Lisboa, Portugal

[3] CTS—Centre of Technology and Systems, Caparica, 2829-516 Almada, Portugal

[4] Department of Electrotechnical and Computer Engineering, Faculty of Science and Technology, Universidade NOVA de Lisboa, DEE-FCT-UNL, Caparica, 2829-516 Almada, Portugal

[*] Correspondence: c.serafinelli@campus.fct.unl.pt

Abstract: In SERS analysis, the specificity of molecular fingerprints is combined with potential single-molecule sensitivity so that is an attractive tool to detect molecules in trace amounts. Although several substrates have been widely used from early on, there are still some problems such as the difficulties to bind some molecules to the substrate. With the development of nanotechnology, an increasing interest has been focused on plasmonic metal nanoparticles hybridized with (2D) nanomaterials due to their unique properties. More frequently, the excellent properties of the hybrids compounds have been used to improve the drawbacks of the SERS platforms in order to create a system with outstanding properties. In this review, the physics and working principles of SERS will be provided along with the properties of differently shaped metal nanoparticles. After that, an overview on how the hybrid compounds can be engineered to obtain the SERS platform with unique properties will be given.

Keywords: SERS analysis; plasmonic metal nanoparticles; hotspots; hybrid materials

Citation: Serafinelli, C.; Fantoni, A.; Alegria, E.C.B.A.; Vieira, M. Plasmonic Metal Nanoparticles Hybridized with 2D Nanomaterials for SERS Detection: A Review. *Biosensors* **2022**, *12*, 225. https://doi.org/10.3390/bios12040225

Received: 2 March 2022
Accepted: 6 April 2022
Published: 9 April 2022

Publisher's Note: MDPI stays neutral with regard to jurisdictional claims in published maps and institutional affiliations.

1. Introduction

The SERS technology fits very well in the scenario of the rapidly emerging new technologies, realizing the ultimate goal of analytical chemistry: the detection, analysis, and manipulation of single molecules, namely the single-molecule detection. For example, a growing interest is focused on groundbreaking "single-molecule electrical approaches" methods that translate chemical or physical processes into detectable electrical signals at the single-event level on the platform of single-molecule electronic devices [1]. Among the different strategies developed in recent years, graphene—molecule—graphene single-molecule junctions (GMG-SMJs) are particularly attractive, showing great potential for the routine applications [2]. Other approaches to reach single-molecule sensitivity are focused on mechanical strategies such as Atomic Force Microscopy (AFM). Owing to the atomically well-defined tip apex and its mechanical flexibility, AFM has revealed an intriguing potential to directly characterize the molecular structure in real space with an outstanding single-molecule sensitivity, and on these bases, a new generation of AFM, the single-molecule AFM (sm-AFM), has been developed for the characterization and manipulation of single molecules [3]. Other approaches are based on optical strategies: for example, new fluorescence microscopy techniques surpassing the diffraction limit of the traditional optical microscopes are receiving a great interest. The development of super-resolved fluorescence microscopy to achieve "super-resolution" led to wide applications in many scientific fields and was recognized by the Nobel Prize in Chemistry in 2014. Other approaches are based on LSPR of plasmonic nanoparticles [4]. A small overview has been depicted on the techniques aiming to reach the single-molecule sensitivity, but there is still a lot of work to do for their massive use as commercial biosensing platforms. Due to its great potential to be implemented in commercial use, in recent years a growing interest has been concentrated in SERS analysis, so that on the basis of extensive research being produced to go deep inside the mechanisms and in its applications, numerous reviews have been published with the aim to help the researchers in the next steps of their studies [5,6]. However, despite the large number of works published covering a wide range of topics, only a few works have been focused on the use of hybrid structures composed of plasmonic metal nanoparticles and 2D nanomaterials in SERS analysis [7]. Stimulated by the enormous progress made on the knowledge and applications of the hybrid compounds, in this review we highlight their use in SERS analysis in order to fill the gap in this direction. The following work is organized as follows: in Section 1, an introduction illustrating the scenario of SERS analysis has been presented, whereas in Section 2, physics and working principles of the SERS technique are provided. In Section 3, the work on the early use of hybrid structures is addressed, focusing attention on spherical plasmonic metal noble nanoparticles and graphene or its derivates in SERS analysis. In Section 4, anisotropy in plasmonic metal nanoparticles is introduced and it is explained how the performances in SERS analysis are enhanced. The use of anisotropic metal nanoparticles in hybrid compounds to improve the performance of SERS analysis is depicted in Section 5. After discussing the effect of particle anisotropy in hybrid compounds, in the following sections, the different approaches to improve the performance of the SERS technique will be highlighted. In Section 6, the engineering of the bidimensional material is discussed, while considering how the hybrid compounds are evolving towards three-dimensional structures with the aim of improving the SERS properties even more in Section 7. The use of bidimensional material as a nano spacer is presented in Section 8, taking into account the use of plasmonic metal nanoparticles veiled with bidimensional material in Section 9. Considering different directions in which the research of SERS is focused, a general overview on the future of SERS will be provided in Section 10, along with the conclusion in Section 11.

2. Physics and Working Principles

Raman spectroscopy is a valuable technique to study chemical and intramolecular bonds by producing and examining inelastic scattering generated from molecules, thus providing a vibrational fingerprint, unique for each molecule, becoming a powerful tool to identify chemical species, supplying both qualitative and quantitative molecular information from any sample. Unfortunately, this is a very weak process, but it has been found that when the molecules are located near a rough metal surface or metal nanoparticles (NP), a Raman scattering boost occurs, greatly enhancing the signal intensity [8], thus opening the way to a new emerging research field: the surface-enhanced Raman scattering (SERS) spectroscopy, that has proved to be a powerful technique for non-invasive, rapid and reliable sensing of chemicals and biomolecules [9–11]. The Raman enhancement in SERS spectroscopy occurs when an incident electromagnetic field interacts with surface plasmon resonance at a metal surface (SERS effect) [12]. The unique optical properties of metal nanostructures are related to the presence of strong localized plasmon resonances (LSPR), excitation of coherent, collective oscillations of delocalized electrons in the conduction band by an external electromagnetic (EM) field as the driving force [13]. When metal nanostructures interact with a light beam, part of the incident photons are absorbed, and part are scattered in different directions: both absorption and scattering are greatly enhanced when the LSPR is excited [14]. The LSPR generated by metal nanostructures produces extremely intense and highly confined electromagnetic fields within the gaps between metallic nanostructures termed "hotspots" [15] that have been claimed to provide extraordinary enhancements of up to 10^{15} orders of magnitude to the surface-enhanced Raman scattering (SERS) signal [16]. Generally accepted mechanisms of SERS enhancement are attributed to two effects: the first one is an electromagnetic mechanism (EM) related to the striking increase in the local electromagnetic field near the metal nanoparticle surface, whereas the second one is a chemical mechanism (CM) involving a charge transfer [17]. The EM is a process originated from the electromagnetic interaction between the metal nanoparticles and the molecules, implicating two mechanisms: the first is the result of the interaction of the metal nanoparticles with the incident beam, whereas the second one is a re-radiation phenomenon [18]. When a metal nanoparticle interacts with an incident field E_0 at a wavelength λ_0 surface plasmon resonance (SPR), oscillating dipoles are induced and the resulting polarization generates large fields around the particle. The enhanced local field E_{loc} around the metal nanoparticles generated by the interaction with incident light is proportional to the incident field E_0 and to a factor called local enhancement factor M_{loc} (λ_0) and can be quantified as:

$$E_{loc} = E_0 \, M_{loc} \, (\lambda_0) \tag{1}$$

The enhanced local field E_{loc} excites the molecule inducing a dipole so that the molecule scatters the Raman signal in all directions radiating an enhanced scattered field E_{scat} at a Raman wavelength λ_r shifted from λ_0 with an intensity proportional to the molecule polarizability α and the enhanced local field E_{loc}.

$$E_{scat} = \alpha \, E_{loc} = \alpha \, E_0 \, M_{loc} \, (\lambda_0) \tag{2}$$

In these circumstances, the Raman signal is already enhanced with respect to the case in which molecules do not undergo the presence of the nanoparticle at its vicinity. On the second step, the scattered field E_{scat} interacts with the metal nanoparticle and as a result, it is enhanced by a re-radiation process that is expressed by:

$$E_{rad} = E_{scat} \, M_{rad} \, (\lambda_r) = \alpha \, M_{rad} \, (\lambda_r) \, M_{loc} \, (\lambda_0) \, E_0 \tag{3}$$

The chemical enhancement is a process originated from a change in the molecule polarizability (along with the Raman cross-section of its vibrational modes), resulting from the physicochemical interaction between the substrate and the molecule. When the molecule interacts with the substrate by means of physisorption or chemisorption, its geometrical and electronic structures are modified, so that the Raman cross-section of its vibrational modes is different compared to that of the free molecule. The chemical enhancement arises from two different mechanisms. The first one is the *non-resonant chemical effect*: when the molecular orbitals have energies far from the Fermi level of the metal, new electronic states are not formed but the electronic and geometrical structures of the molecule are transformed, resulting in the modification of the Raman shifts and the intensity of the vibrational modes. The second mechanism is a *resonant charge transfer chemical effect*: the interaction between the metal and the molecule leads to a metal–molecule charge transfer state (CT) and in the case in which the laser source is in resonance with the CT state, the Raman modes are strongly enhanced. In addition, the chemical effect can also originate from a temporary electron transfer between the molecule and the metal, a "transient" charge transfer [19]. Although amplified by both EM and CM, SERS efficacy greatly benefits from EM enhancements, with a contribution of 10^8 or more, whereas the contribution of the chemical effect would not exceed a factor 100, thus suggesting that the EM enhancement is the dominant contribution to the SERS sensitivity [20]. Figure 1 presents a schematic showing the SERS signal arising from a molecule localized in the hotspot formed in the junction between two gold nanoparticles.

A key parameter in quantifying the overall signal increasing is the SERS enhancement factor (*EF*) that is experimentally evaluated by means of SERS intensity measurements for the adsorbed molecule on the metal surface, relative to the normal Raman intensity of the same, "free" molecule in the solution. One of the most used equations to calculate *EF* is expressed by:

$$EF = \frac{I_{SERS}/N_{Surf}}{I_{RS}/N_{Vol}} \tag{4}$$

where $N_{Vol} = c_{RS} V$ is the average number of molecules in the scattering volume (V) for the Raman (non-SERS) measurement, and N_{Surf} is the average number of adsorbed molecules in the scattering volume for the SERS experiments while I_{SERS} and I_{RS} are the intensities of the same band for the SERS and bulk spectra [21].

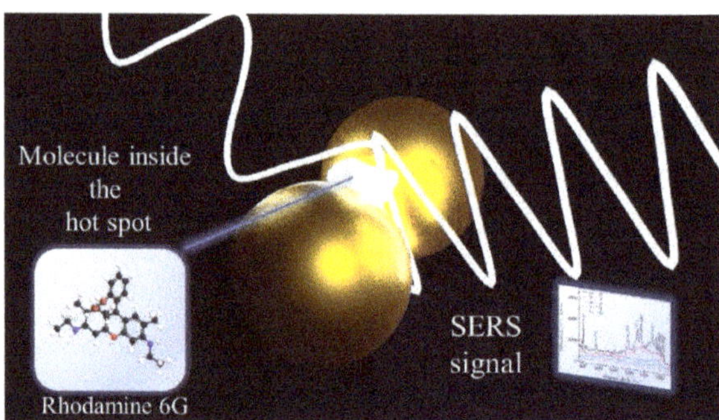

Figure 1. Schematics showing the SERS signal arising from a molecule localized inside a hotspot created in the space between two plasmonic metal nanoparticles.

3. Hybrid Nanocomposites

Plasmonic nanoparticles, in condition of resonance, are able to generate a strong electromagnetic field [22] and in the past they have been exploited as excellent SERS substrates, mainly due to their huge enhancement induced by the EM effect. Despite the huge potential of SERS analysis, the low affinity of non-thiolated molecules to metals, such as gold or silver, is still a challenge because the affinity determines the retention of analytes. To bypass this complication, several approaches have been developed. For example, core-shell colloidal material comprising gold nanoparticles coated with a thermally responsive poly-(N-isopropylacrylamide) (pNIPAM) (Au@pNIPAM) has been developed as a SERS substrate [23]: the plasmonic metal core provides the enhancing properties, whereas the pNIPAM shell is exploited to trap and get the molecules close to the metal core to obtain the SERS signal. Other applications include other coatings such as calixarenes [24]. In this context, bidimensional (2D) nanomaterials such as graphene and its derivatives, 2D metallic oxides, hexagonal boron nitride (h-BN)) etc., have attracted great interest and have been investigated as SERS substrates for their intriguing properties such as an enhanced photogeneration rate, the plasmon-induced "hot electrons" and improved conductivity [25]. Between 2D nanomaterials, transition metal dichalcogenides (TMDCs) is a large family of materials of MX_2 type, where M is a transition metal element from group IV, V or VI, for example, Mo or W and X is a chalcogen (S, Se or Te). They possess a layered structure, and each layer includes three atomic planes with an arrangement of X-M-X type, where a hexagonally packed plane of transition metal atoms M is enclosed within two atomic planes of chalcogen X resulting in a thickness of 6–7 Å for each single layer [26]. Inside the TMDCs family, MoS_2 and Tungsten disulfide (WS_2) received great interest because of their properties and have been exploited to produce advanced SERS substrates. Graphene and its derivative such as graphene oxide (GO) or reduced graphene oxide (r-GO) have been the most extensively investigated and have emerged as a material for SERS substrates, due to their intriguing properties. Firstly, its unique structure is favorable for interactions with analytes via π-π stacking and hydrophobic interactions thus facilitating the adsorption of non-thiolated molecules, resulting in an enhancement of the SERS signal. Secondly, it has been shown that it contributes to the SERS signal enhancement with a magnitude depending on the degree of GO chemical reduction [27]. The signal SERS enhancement is dominated by chemical mechanism [28] rather than the electromagnetic mechanism, although the EF is not dramatic as in the case of metal substrates. Moreover, graphene offers the additional advantage of fluorescence quenching, improving the SERS efficacy.

Larger cross-sections for the fluorescent signal, compared to Raman signal, are observed, so that Raman characteristics are often interfered with, or even submerged, by the intense fluorescence background, thus lowering the quality of SERS analysis. A fluorescence quenching effect of fluorescent dyes rhodamine 6G (R6G) and protoporphyrin IX (PPP) adsorbed on graphene was first reported by Zhang [29]. The approximate evaluated quenching factor has been found on the order of 10^3 and the quenching effect has been assigned to a resonance energy transfer process according to the results obtained by Swathi [30]. The chemical enhancement on three types of different 2D nanomaterials, graphene, hexagonal boron nitride (h-BN) and molybdenum disulfide (MoS_2) (each having different electronic properties) has been investigated by Ling et al. [31] by means of the copper phthalocyanine (CuPc) molecule as a probe. It has been observed that different vibrational modes showed different enhancement factors depending strongly on the substrates. These inconsistencies have been related to three different enhancement mechanisms determined by the distinct electronic properties and chemical bonds exhibited by the three substrates. Graphene is zero-gap semiconductor and has a nonpolar C-C bond [32], so that the Raman enhancement is assigned to the strong charge transfer with the CuPc molecule and not to the weak dipole-dipole interactions due to the nonpolar nature of graphene. Differently, h-BN is highly polar (due to the strong B-N bond) and insulating with a large band gap (5.9 eV) [33] so that the dipole interactions are dominant while the charge transfer interactions are negligible, and the signal amplification results from dipole-dipole interactions between h-BN and CuPc molecules. In addition, MoS_2 is semiconducting [34] and less polar compared to h-BN so that, although both dipole-dipole interactions and charge transfer occur, they are weaker thus resulting in a weaker signal enhancement. In view of the huge progress in designing and developing plasmonic metal nanoparticles as SERS substrates originating from their EM enhancement, and in the use of graphene and bidimensional nanomaterials as effective Raman enhancement substrates due to their CM, further advances in reaching greater performances in SERS analysis have been obtained by the combination of plasmonic metal nanoparticles with 2D nanomaterials. The obtained hybrid composites displayed advanced enhancement properties, arising from the synergistic effect of the EM (originating from the high local electric field at hotspots formed in metallic NPs) and CM (deriving from the charge transfer between 2D nanomaterials and probe molecule). In the early works, hybrid composites have been produced by growing spherical metal nanoparticles on graphene nanosheets. A hybrid system (GO/PDDA/AgNPs) has been developed, self-assembling Ag spherical nanoparticles on the surface of graphene oxide by means of poly (diallyldimethyl ammonium chloride) (PDDA) and successively tested as SERS platform sensing for folic acid detection, showing strong SERS activity. The SERS spectra of p-aminothiophenol (p-ATP) collected using Ag colloid and GO/PDDA/AgNPs showed a greater enhancement signal for spectrum obtained with Ag colloid due to the fact that GO did not assist the adsorption of p-ATP on AgNPs of GO/PDDA/AgNPs. In the case of folic acid detection, the SERS signals on GO/PDDA/AgNPs were much stronger than that on Ag nanoparticles [35]. Stimulated by the striking properties arising from the synergistic effect of EM and CM in the hybrid nanocomposites, Chen [36] developed a SERS platform based on p-aminothiophenol (PATP)-functionalized silver nanoparticles supported on graphene nanosheets (Ag/GNs) for the sensitive and selective detection of 2,4,6-trinitrotoluene (TNT). As a first step, graphene nanosheets were decorated with silver nanoparticles by reducing silver nitrate with sodium citrate and then the Ag/GNs composites have been functionalized with p-aminothiophenol (PATP) obtaining the PATP-Ag/GNs composites. In presence of TNT, π-π conjugated structures between TNT and PATP are created, promoting the effective charge transfer from the electron-rich PATP to the electron-poor TNT that leads to the enhanced Raman signals. With a LOD of 5.0×10^{-16} M, the PATP-Ag/GNs hybrid displayed a great sensitivity towards TNT detection. A SERS substrate has been prepared according to the one-step strategy in the work of Wei [37]. Hybrid structures have been produced by simultaneous reduction of GO and $HAuCl_4$ with sodium citrate and ammonia, and then an RGO/AuNP film has been deposited on a silicon wafer and on a poly (ethylene terephthalate) (PET) substrate to create the SERS platform. To explore the potential of the created RGO/AuNP composite, 4-aminothiophenol (4-ATP) has been exploited as probe molecule in SERS analysis, finding a value of 5.6×10^5 for the enhancement factor (EF), such that the RGO/AuNP hybrid displayed remarkable performances in 4-ATP detection. On the basis of the results obtained, the composite has been exploited as a SERS substrate to detect 2-thiouracil (2-TU) with a low concentration to 1 μM, thus confirming the great ability as a biodetection platform. However, due to their attractive properties, not only graphene has been exploited to create hybrid structures with plasmonic metal nanoparticles, but several types of 2D nanomaterials have also been used to produce composites to be utilized as SERS substrates. A SERS active substrate has been constructed by Chao [38] by growing Au nanoparticles on Molybdenum disulfide (MoS_2) nanosheets. The gold precursor has been directly reduced on the MoS_2 surface in the presence of carboxymethyl cellulose (CMC) as a stabilizer in an aqueous solution to create the $AuNPs@MoS_2$ nanocomposite. Several $AuNPs@MoS_2$ nanocomposites have been prepared with different amounts of Au nanoparticles and the one with the best performance has been selected by means of the standard probe rhodamine 6G (R6G), finding a value of 8.2×10^{-5} for the enhancement factor (EF). The amplification of the SERS signal has been assigned to the hot spots generated by the little aggregation and closeness of Au nanoparticles. An approach to exploit the unique properties of a nanohybrid formed by gold nanoparticles (AuNPs) deposited onto exfoliated nanosheets of tungsten disulfide (WS_2) was presented in the work of Sabherwal [39]: an active SERS platform based on an Au NPs/WS_2 nanohybrid has been developed for the label-free detection of Myoglobin, a cardiac biomarker. The AuNPs/WS_2 nanohybrid has been prepared by the in situ reduction of gold salt precursor, and then the surface was functionalized with specific aptamers to impart high selectivity towards Myoglobin. The prepared nanohybrid has been tested for SERS detection. The obtained results showed the synergistic use of the unique properties of chemical and electromagnetic enhancement of both WS_2 and AuNPs for a many fold increase in the SERS signal intensity. The AuNPs/WS_2 nanohybrid system allowed the Myoglobin detection with a LOD of 10^{-2} pg mL^{-1}, considerably lower with respect to that measured in other works. In Table 1, the results obtained for SERS analysis using hybrids of plasmonic metal nanoparticles and graphene or other 2D nanomaterials are listed.

Table 1. Performances of the SERS platform based on hybrids of plasmonic metal nanoparticles and graphene or other 2D nanomaterials.

NANOPARTICLES DEPOSITED ON GRAPHENE						
System	Molecule Used to Calculate LOD	Limit of Detection (LOD)	Enhancement Factor (EF)	Molecule Used to Calculate EF	Equation Used to Calculate EF	Reference
Graphene oxide/Ag nanoparticle hybrids (GO/AgNPs)	Acid folic	9 nM	Not calculated			[35]
Graphene nanosheets/Ag nanoparticle hybrids (Ag/GNs)	2,4,6-trinitrotoluene (TNT)	5×10^{-16} M	Not calculated			[36]
Reduced graphene oxide/Au nanoparticle hybrids (RGO/AuNPs)	2-thiouracil (2-TU)	1 μM	5.6×10^5	4-aminothiophenol (4-ATP)	$(I_{SERS}/I_{bulk}) \times (M_{bulk}/M_{ads})$	[37]
NANOPARTICLES GROWN ON DIFFERENT 2D NANOMATERIALS						
System	Molecule used to calculate LOD	Limit of Detection (LOD)	Enhancement Factor (EF)	Molecule used to calculate EF	Equation used to calculate EF	Reference
AuNP-decorated MoS_2 nanosheets ($AuNPs@MoS_2$)	Rhodamine 6G (R6G)	10^{-6} M	8.2×10^5	Rhodamine 6G (R6G)	$(I_{SERS}/I_{bulk}) \times (N_{bulk}/N_{SERS})$	[38]
AuNP-decorated tungsten disulfide (WS_2) nanosheets (Au/WS_2)	Myoglobin (Mb)	10^{-2} pg mL^{-1}	6.78×10^6	Rhodamine 6G (R6G)	(I_{SERS}/I_{bulk}) (C_{bulk}/C_{SERS}) (P_{bulk}/P_{SERS})	[39]

4. Effect of Nanoparticle Shape

Plasmon resonances in spherical nanoparticles can be tuned in a limited range of wavelengths by changing the particle diameter, while introducing anisotropy in the particle shape provides a fine control of plasmon resonance thus broadening the range of wavelengths from the visible through the mid IR, by varying the aspect ratio (AR) of the NPs [40]. On these presuppositions, the early attempts to further improve the SERS efficacy of the hybrid composites exploited the engineering of particle morphology tailoring the particle anisotropy along with the plasmonic properties and intrinsic electromagnetic "hotspots".

4.1. Au Nanorods

Rod-shaped nanoparticles are nanostructures where one dimension is longer than the other two, so that the term nanorods indicates elongated nanoparticles. They exhibit two plasmon resonances bands: the first one is longitudinal mode parallel to the long rod axis, whereas the second one is a transverse mode perpendicular to the long axis of the rod [41]. Different from the transverse band and characterized by a low intensity and independence from the aspect ratio (AR), the longitudinal mode is much more intense and strongly independent from AR, so that when controlling the AR of AuNRs, it is possible to tune the plasmon resonance across the UV-visible region of the spectrum [42]. From theoretical discrete dipole approximation (DDA) and experimental electron energy loss spectroscopy (EELS), near-field maps illustrated an intense EM field in the proximity of the AuNRs originating from plasmon modes [43,44]. From the calculations, results show a high electromagnetic (EM) field enhancement at the Au nanorods' tips (hotspots) under longitudinal excitation, whereas the enhancement is moderate on the NR lateral sides for transverse excitation. In addition, the AR also controls the SERS efficacy of the AuNRs: exploiting crystal violet (CV) as a molecule probe, it has been shown that the EF increases when the AR values become greater, so that controlling the AuNRs' aspect ratio enables a fine tuning of plasmon resonance and the EF for an optimized SERS analysis [45]. For the sake of having a more complete view on the properties of elongated particles, it is worth mentioning that, with a tight and precise control over the synthesis conditions, other elongated structures have been obtained, such as gold nano bipyramids. Depending on their sharp nanotips, gold nano bipyramids (AuNBPs) revealed a stronger local field enhancement compared to AuNRs [46]. Due to the sharp tips, in AuNBs the radiation can be centralized into a strong local electric field and enhancing at the same time the local density of photonic states [47], resulting in the beneficial application for spectroscopy, photocatalysis, detection and biomedicine [48].

4.2. Au Nanotriangles

The term nanoplates indicates nanoparticles in which one dimension is much smaller than the other two, and the specific case in which the base is triangular, the nanoparticles are referred as nanotriangles (NTs). Nanotriangles are characterized by intriguing properties originating from the morphology: the EM fields can be confined by their intrinsic sharp corners and edges, thus showing a strong enhancement. EES mapping indicates a strong EM field enhancement near the tips corresponding to a dipolar mode, whereas for increasing energies other modes start to appear presenting the highest enhancement at the edges and at the center of the NT [49]. It is possible to select the wavelengths of plasmon resonance by changing the aspect ratio (edge/thickness) of the AuNTs so that, as in the case of AuNRs, the LSPR of AuNTs are characterized by a fine tunability. Furthermore, depending on the synthetic procedure, AuNTs display a different grade of truncation that enable the dipolar LSPR tuning, leading to a blue-shift in the spectrum when increasing the snip size of the missing corner [50]. As in the case of AuNRs, the ability to generate great EM field enhancement and the plasmon resonance tunability by means of morphology renders AuNTs an intriguing substrate for SERS analysis. As demonstrated by Tan [51], there is a strong relationship between the SERS enhancement and the LSPR of the corresponding substrate: when laser excitation wavelengths match the LSPR, the SERS signal is found about two orders stronger compared to the case in which the laser excitation wavelength is far away from the LSPR band.

4.3. Au Nanostars

Gold nanostars are particles with a star-like morphology comprising a spherical central core from which radial, acute tips branch out, and over the last few years have raised particular attention thanks to their features such as the ease of synthesis for large scale production, the high surface-to-volume ratio useful for improving drug loading efficiency, and above all, thanks to their unusual optical and plasmonic properties [52,53] which pave the way to a great potential for nanomedicine applications. Despite the irregularity of the particle star-like morphology, the extinction spectra show well-defined localized surface plasmon features. The UV-Visible spectra of Au nanostars display an intense band centered at ca. 650–900 nm and a weaker band localized at ca. 500–600 nm [54]. Assigning the LSPR of Au nanostar bands by means of theoretical models for the resolution of the Maxwell's equations, such as Finite-Difference Time-Domain (FDTD) [52], Discrete Dipole Approximation (DDA) [55] or Boundary Element Method (BEM) [54], produced the same results: the band at lower energy is assigned to dipolar resonances localized at the individual tip, whereas the band localized at higher energy is assigned to dipolar resonances localized at the central core. Interestingly, FDTD calculations have demonstrated, despite the general observation that the UV-Visible spectra are dominated by the LSPR band associated with tip oscillations, that the plasmon modes of Au nanostars arise from the hybridization of resonances associated with the core and the tips generating bonding and antibonding nanostar plasmon. Thus, a contribution from the core plasmon to the tip plasmon is proposed. The core plasmons have larger frequencies than the tip plasmons, so that the conduction electron of the core structure can adiabatically follow the lower frequency tip plasmon oscillations. This results in an "antenna effect" which is responsible for an increase in the extinction cross-section (a factor of 4-fold with respect to the individual tip plasmons) as well as in the electric field enhancement. In Au nanostars, the central sphere acts as an electron reservoir. Regarding the influence of the specific morphological details, it has been found that the aperture angle and the roundness of the tip are of major importance in affecting the energy of the LSPR tip mode and thus its position in the UV-Visible spectra, since small changes lead to a significant shift in the main band. BEM calculations have been confirmed when investigating the spatial distribution of the plasmon modes of gold nanostars by electron energy-loss spectroscopy (EELS) mapping performed on a single particle in a scanning transmission electron microscope (STEM) [56], which also showed a high localization near the tips. Similar to Au nanorods, the plasmon resonance in AuNSs can be tuned through modifications in the aspect ratio and/or in the tip sharpness. When the aspect ratio increases, it is possible to observe a red-shift in LSPR, and larger Au nanostars present an increased number and longer tips along with a red-shift in plasmon resonances at an increase in AuNS size. Despite the great potential, understanding how the morphology can impact the efficacy of AuNSs as a field enhancer is a controversial task. In an early work, Vo-Dinh [57] investigated the properties of AuNSs as SERS substrates and compared with respect to the size. Changing the ratio between seeds and $HAuCl_4$, PVP-functionalized AuNs with sizes in the range from 45 nm to 116 nm have been synthesized. Even though the morphology of AuNSs with distinct sizes is different, exploiting p-mercaptobenzoic acid (p-MBA) as a probe molecule, EF around 5×10^3 has been achieved, thus revealing no great differences among the different sizes. Nevertheless, from the results obtained by Ganesh [58] it has been found that when the size and length of the spikes of Au NS have been changed, the intensity of the SERS signal is significantly altered. Exploiting AuNSs with two different morphologies, short spike (SSNS) and long spike (LSNS) Au nanostars, from the SERS experiment demonstrated that SSNS exhibit a higher SERS enhancement compared to LSNS, and it has been justified with a size effect. According to the work of Hong and Li [59], the optimized size of Au nanoparticles for obtaining maximum enhancement in Raman signal is around 50 nm and the core size of SSNS falls within this regime exhibiting higher intense Raman signals. Regardless of the controversy in correlating the AuNSs morphology to SERS activity, their own optical properties such as the strong near field enhancement and plasmon resonance tunability enable Au nanostars to be employed in SERS analysis with great potential.

In Figure 2, the electromagnetic field around a gold nanoparticle with a different morphology shown.

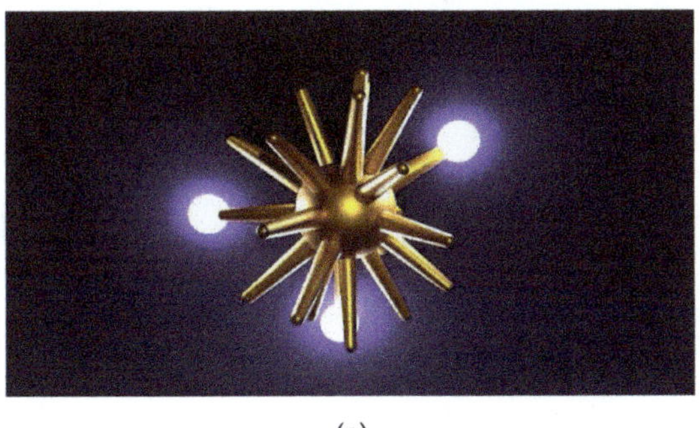

(a)

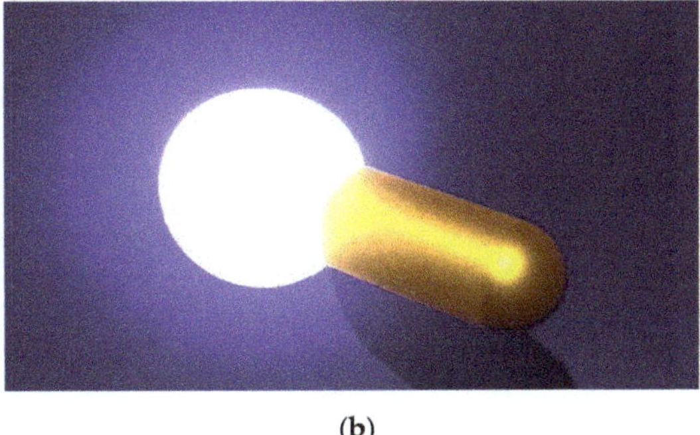

(b)

Figure 2. (a) Electromagnetic (EM) field (bright spot) around a gold nanostars' (AuNS) spike and (b) EM field around a gold nanorod (AuNR) tip.

With the aim to evaluate how the nanoparticles' morphology can affect the SERS activity, Kundu, in his work [60], used rhodamine 6 G (R6G) as probe molecule exploiting nanoparticles with different shapes (Au nanospheres, Au nanorods, Au nanowires and Au nanoprisms) as SERS substrates. Calculating the enhancement factor (EF) according to Equation (1), it has been observed that Au nanoprisms showed much higher value of EF compared to other shapes as follows: nanoprisms > nanowires > nanorods > nanospheres. The enhanced SERS activity for nanoprisms has been related with the presence of a greater number of edges: the largest electric field is localized near the sharpest surface or at the sharp ends of the NPs so that Raman enhancement reaches its maximum value at the sharpest surface and, in addition, edges can interact strongly with R6G compared to the smooth surface of the nanospheres. The synergistic effect of EM field enhancement and strong interactions with probe molecules results in the highest EF values. Similar results have been obtained by Wu [61]. The effect of particle shape has been investigated choosing three types of Au nanoparticles (nanospheres, nanorods and nanostars) as SERS substrates and using malachite green isothiocyanate (MGITC) molecules as probes. From the spectra collected under excitation from the 532 nm and 785 nm lasers, it was shown that Au nanostars displayed the highest SERS enhancement factor (EF) while the nanospheres possessed the lowest SERS activity under excitation with 532 and 785 nm lasers. The experimental results have been combined with theoretical calculations in order to provide a deep insight into the relationship between the particle morphology and its SERS activity. The field distribution has been investigated by means of finite-difference time-domain (FDTD) simulation, revealing considerable differences in the distribution of the EM field around the diverse nanoparticle types induced by their localized surface plasmon resonance (LSPR) under both 532 and 785 nm incident lasers. From FDTD simulations, it can be seen that the maximum electric field intensity is concentrated around the sharp tip in the anisotropic structure generating 'hot spots' thus leading to the dominant contribution to the SERS intensity. According to the particle morphology and anisotropy, the Au nanoparticles follow the order nanostars > nanorods > nanospheres in terms of enhancement factors. The size and morphology of plasmonic metal nanostructures have been discussed as key factors in determining the formation of the hotspots' network in a SERS substrate, but other factors, such as their spatial arrangement and nanogap distances need to be controlled in order to obtain optimal plasmonic properties and maximal Raman signal amplification. When plasmonic metal nanostructures are connected, a direct exchange of free electrons is possible producing different types of plasmons, such as the longitudinal antenna plasmon mode (LAP), bonding dipole plasmon mode (BDP) and charge transfer plasmon mode (CTP) [62]. The type of plasmon created (BDP, CTP or LAP) will affect the performances of the SERS substrate. The particle surface morphology is a further key factor to be considered to control the performances of the substrate sensing. In addition, the shape of plasmonic metal nanostructures has been engineered to obtain an enhanced EM field that is beneficial to the SERS analysis, but it is worth mentioning other approaches aiming to obtain promising results in this direction and that are based on the combination of LSPRs with propagating surface plasmons (PSPs). Often, surface plasmons (SPs) are classified in two categories: PSPs, that are running surface waves, and localized surface plasmons (LSPs), collective surface charge oscillations in the form of the standing waves confined in a metal nanoparticle [63]. The incident light cannot easily excite the PSPs and in order to achieve SPRs' excitation, a specific configuration is required and needs to be assembled [64]. As opposed, the LSPRs are excited by directly shining the light on an assembly of plasmonic metal nanoparticles. Very recently, it has been found that the enhancement factors in SERS analysis are greatly enhanced when the excitation of LSPs is produced by PSPs generated as surface EM waves, and excited from a special coupling medium (prism, waveguide, fiber, or grating) [65], thus paving the way to the development of new devices.

5. Effect of Plasmonic Metal Particle Morphology in Hybrid Compounds on SERS

Taking into account the effect of particle size on the SERS efficacy, it has been easy to combine anisotropic nanoparticles with bidimensional material to further improve the SERS efficacy of the hybrid composites thus exploiting the engineering of particle morphology by tailoring the particle anisotropy along with the plasmonic properties and intrinsic electromagnetic "hotspots". In Figure 3, a schematic of a hybrid compound formed by a graphene layer and an Au nanorods with SERS signal is represented.

Figure 3. Au nanorods under the illumination of a laser beam and the resulting SERS signal.

In the work of Liu [66], for the first time, the fabrication of a hybrid composed of AuNRs and GO for SERS (GO–AuNR) analysis has been reported. The GO–AuNR composite material has been developed exploiting an electrostatic self-assembly strategy and then tested as a SERS substrate. To test the SERS efficacy of the composite system, the Raman signals of four dye molecules (crystal violet (CV), neutral red (NR), trypan blue (TB) and ponceau S (PS)) on different substrates (SiO_2/Si, AuNRs, GO and GO–AuNR) have been compared. From the results obtained, it has been possible to observe that strong SERS activity is observed when CV and NR are deposited on the GO–AuNRs substrate whereas no SERS effect has been detected for TB and PS dye molecules. In the case of cationic dye (CV and NR), the molecules are electrostatically attracted by the negatively charged GO–AuNRs substrate, whereas the anionic dye TB and PS due to their negative charge are prevented to interact with the substrate, thus weaking the Raman signal. Although exploiting hybrid composites containing AuNRs has revealed improved performances in SERS analysis, the substrate is still not properly adequate for detecting real samples, that can be negatively charged. Further improvement has been reported in successive works. For example, Jiang et al. [67] fabricated a composite containing aggregated Ag nanorods and GO showing good stability and used as a SERS platform to detect Rh6G molecules and trace I^- ions in the solution exploiting the SERS quenching due to the formation of the Rh6G-I complex thus finding a LOD of 0.2 nmol/L for Rh6G and a LOD of 0.004 µmol/L. A different strategy exploiting core–shell Au@Ag nanorods has been developed by Gao et al. [68] to obtain an advanced SERS substrate. Using Au@Ag core–shell nanorods hybridized with reduced graphene oxide (GO-Au@AgNRs) Rh6G molecules have been detected with an enhancement factor (EF) up to $(5.0 \pm 0.2) \times 10^8$, 4-fold higher compared to that obtained with rGO-AuNRs and pesticide thiram with a limit of detection (LOD) of 5.12×10^{-3} µM has been revealed. An early hybrid system containing Au nanotriangles has been developed by Jiang [69]. Reduced graphene oxide/silver nanotriangle (rGO/AgNT) composite sol was prepared by the reduction of silver ions with sodium borohydride in the presence of H_2O_2 and sodium citrate and exploited for SERS detection of Dopamine (DA). The detection is based on the competitive adsorption occurred between DA and the molecular probe acridine red (AR) onto the reduced graphene oxide (GO) nanosheets. Depending on the formation of multiple hydrogen bonding and π-π stacking, DA molecules display a much stronger affinity towards the GO nanosheets compared to AR molecules, so that, when added to the system, DA molecules competed with AR for similar adsorption sites on the rGO surface thus leading to the desorption of AR molecules from the rGO surface. The desorbed AR molecules can be successively adsorbed on the surface of Au nanotriangles thus enhancing the SERS signal: when increasing the DA concentration, the amount of AR molecules adsorbed onto AuNTs rises with SERS intensity responding linearly with DA concentration. Stimulated by intense local electric field enhancements of silver nanoplatelets (Ag-NPs) caused by the anisotropic morphology and sharp corners compared with other morphologies, such as nanospheres, Meng [70] developed hybrid systems containing AgNPs and graphene nanosheets (Ag-NP@GH) to be exploited as SERS substrates taking advantage of their unique properties. The SERS enhancement capability of the Ag-NP@GH composite has been tested calculating the enhancement factor (EF) using rhodamine 6G (R6G) as a probe molecule and a value of 4.7×10^8 for EF has been found, thus confirming the good performance of Ag-NP@GH in SERS analysis. After testing, the composite has been exploited for organic pesticide detection, including thiram and methyl parathion (MP), and their mixtures finding a LOD of 40 nM for thiram and 600 nM for MP. The improved sensitivity of the Ag-NPs@GH results from the combination of the EM effect of the AgNPs and the CM effect of the graphene. Hotspots are generated by AgNPs staying close to each other, whereas graphene nanosheets contribute to SERS efficacy with a chemical enhancement (CM) due to the strong adsorption capability by means of π-π interactions. Aiming to exploit the unique plasmonic properties of gold nanostars (AuNSs), in the work of Krishnan [71] hybrid systems containing graphene oxide (GO) and AuNSs have been produced and used as SERS-active substrates. A simple and eco-friendly synthetic route based on a deep eutectic solvent (DES) has been developed and evaluated as a SERS substrate using crystal violet (CV) as a probe molecule and a value of 1.7×10^5 for the enhancement factor EF and a limit of detection (LOD) of 10^{-11} M have been found. The improved performances of the composite as a SERS substrate have been explained with the large number of nanogaps between two contiguous AuNSs generating the SERS hotspots and to its morphology, permitting to the CV analyte molecules to diffuse inside the structure. One of the early hybrid structures comprising branched Au nanoparticles was developed by Ray in his work [72] in a four-step process, binding the Au-branched nanoparticles, (termed Au nanopopcorns) by means of Cysteamine molecules on graphene oxide (GO) nanosheets. To evaluate the SERS enhancement capability of the composites, SERS spectra were collected using Rh6G dyes as probe molecules and GO, Au nanopocorns and hybrid nanostructures as substrates. Calculating the enhancement factor (EF), it has been found that the higher value of EF is for the composites, followed by Au nanopopcorns and GO. The signal enhancement in the case of GO has been explained with the chemical mechanism, whereas the effect of nanoparticles has been further investigated. When exploiting hybrid composites containing differently shaped (spherical, cage and popcorns) Au nanoparticles, the highest EF for the popcorn shape has been found, depending on the presence of sharp tips. As well as in combination with GO, the differently shaped Au nanoparticles showed the highest EF for nanopopcorns. The hybrid structures containing Au nanopopcorns revealed the best performances in the HIV-1 gag-gene DNA and Staphylococcus aureus (MRSA). Animated by the intriguing properties of hybrid structures, Li [73] developed a SERS substrate for bilirubin detection integrating composites containing AuNS-decorated graphene oxide (GO) nanosheets on common filter paper. The AuNSs/GO hybrids have been assembled by means of electrostatic interactions created by a deposited layer of Poly (diallyldimethyl ammonium chloride) (PDDA) on GO nanosheets and then exploited for bilirubin detection. The resulting SERS substrate combines the EM effect originated by hotspots generated from AuNSs and the ability of GO nanosheets to adsorb bilirubin molecules by means of strong electrostatic and π-π interactions as shown by kinetics measurements. In addition, the SERS performances are improved by the superquenching of fluorescence by both GO nanosheets and GNSs. The SERS substrate showed an LOD as low as 0.436 µM for free bilirubin in blood serum, thus holding considerable properties for clinical translation in accurate diagnosis of jaundice and its related diseases. In Table 2, the performance of hybrid compounds containing metal nanoparticles with different morphologies is reported.

Table 2. Performances of the SERS platform based on different shaped nanoparticles.

NANORODS						
System	Molecule Used to Calculate LOD	Limit of Detection (LOD)	Enhancement Factor (EF)	Molecule Used to Calculate EF	Equation Used to Calculate EF	Reference
Graphene oxide/Au nanorods hybrids (GO-AuNRs)	Crystal violet (CV), neutral red (NR), blue (TB), ponceau S (PS)	Not calculated	Not calculated			[66]
Silver nanorods/reduced graphene oxide nanosheets hybrids (AgNR/rGO)	Rhodamine 6G (Rh6G) / Iodine ion	0.2 nmol/L– 0.004 µmol/L	Not calculated			[67]
Au/Ag core–shell nanorods and reduced graphene oxide hybrid structure (Au@AgNRs/rGO)	Thiram	5.12×10^{-3} µM	$(5.0 \pm 0.2) \times 10^{8}$	Rhodamine-6G (R6G))	$(I_{SERS}/I_{bulk}) \times (N_{bulk}/N_{SERS})$	[68]

Table 2. *Cont.*

Au NANOTRIANGLES						
System	**Molecule used to calculate LOD**	**Limit of Detection (LOD)**	**Enhancement Factor (EF)**	**Molecule used to calculate EF**	**Equation used to calculate EF**	**Reference**
Reduced graphene oxide/silver nanotriangles hybrid structures (rGO/AgNT)	dopamine (DA)	1.2 µmol/L	Not calculated			[65]
Ag-nanoplates/ graphene hybrids (Ag-NP@GH)	thiram / Methyl parathion (MP)	40 Nm–600 nM	4.7×10^8	Rhodamine 6G (R6G)	$(I_{SERS} \times C_0)/(I_0 \times C_{SERS})$	[66]
Au NANOSTARS						
System	**Molecule used to calculate LOD**	**Limit of Detection (LOD)**	**Enhancement Factor (EF)**	**Molecule used to calculate EF**	**Equation used to calculate EF**	**Reference**
Graphene oxide/Au nanostars hybrid structure (GO/AuNSs)	Crystal violet (CV)	10^{-11} M	1.7×10^5	Crystal violet (CV)	$(I_{SERS}/N_{SERS})/(I_{Nor}/N_{Nor})$	[69]
Popcorn-shaped gold nanoparticles and graphene oxide hybrid structures	Methicillin-resistant Staphylococcus aureus (MRSA)	10 CFU/mL	3.8×10^{11}	Rh6G	$(I_{SERS}/I_{bulk}) \times (M_{bulk}/M_{ads})$	[70]
Graphene oxide/Au nanostars hybrid structure (GO/AuNSs)	Bilirubin	0.436 µM	$2.43 \times I_{SERS}/I_{bulk}$	4-nitrothiophenol (4-NTP)	$(I_{SERS}/N_{SERS}) \times (N_{bulk}/I_{bulk})$	[71]

The effect of nanoparticle morphology has been reviewed in this section, but it is noteworthy to mention that in other works, structures formed by plasmonic metal nanohole arrays hybridized with graphene have been exploited with promising results in SERS analysis [74,75].

6. Engineered 2D Nanomaterials

Tailoring the anisotropy in nanoparticles morphology amplifies the SERS signal by means of EM field intensification near the nanoparticles; nevertheless, the SERS enhancement also relies on the number of the hotspots created by the structure used as a substrate, so that alternative approaches optimize the hybrid composite structure in such a way as to create the greatest possible number of hotspots by engineering the bidimensional material according to different strategies. Some strategies, for example, exploited the intrinsic properties of the bidimensional material to amplify the SERS efficacy. Recently, in the work of Su [76], a SERS substrate composed of 1T-MoS$_2$ nanosheets decorated with silver nanocubes (1T-MoS$_2$/AgNCs) and assembled on filter paper has been developed and used for thiram (TRM) and thiabendazole (TBZ) residues in apple fruit detection. The two different phases of MoS$_2$, the 1T (metallic, trigonal) and 2H (semiconducting, hexagonal), have been used to create the hybrids with Ag nanocubes (1T-MoS$_2$/AgNCs and 2H-MoS$_2$/AgNCs) and tested in SERS analysis with the model molecules rhodamine 6G (R6G) in order to evaluate the effect of the MoS$_2$ phase on the SERS performances. The best performances obtained for the composites containing the metallic 1T phase have been explained with a superior electron transfer from Ag to 1T-MoS$_2$ compared to the 2H-MoS$_2$ phase, depending on the absence of band gap, the lower binding energies of 1T-MoS$_2$ compared to 2H-MoS$_2$ and the more abundant density of state (DOS) near the Fermi level. In the work of Koratkar [77], a SERS substrate composed of monolayer MoS_2 decorated with AuNPs has been engineered to obtain higher SERS analysis performances. By means of low-power focused laser cutting, artificial edges have been sculpted in monolayer MoS_2, on which AuNPs, when deposited by drop-casting, tend to predominantly accumulate. The huge density of AuNPs along these artificial edges concentrates the plasmonic effects in this region, so that hotspots are generated exclusively along these artificial edges. Calculations of first-principles density functional theory (DFT) suggested that AuNPs are strongly coupled to the artificial edges through dangling bonds that are widespread along the unpassivated edges cut by the laser. Moreover, according to DFT calculations, as a result of AuNP binding, there is an enriched availability of conduction channels around the Fermi level so that artificial edges decorated with AuNPs displayed a higher electrical conductivity. The dense assemblage of AuNPs and the increased electrical conductivity generate along the artificial edges regions of mobile charge oscillating in phase with the laser light that drastically enhanced the magnetic field and the SERS response. Using Raman mapping it has been possible to localize the hotspots along the MoS_2 edges cut by the laser. Inspired by the intrinsic properties of boron nitride (BN) nanosheets, Li et al. [78] have developed a SERS substrate composed of faceted Au nanoparticles synthesized over BN nanosheets by a simple sputtering and annealing method. The stronger resistance to oxidation renders more advantageous the use of BN nanosheets as reusable SERS substrates as they support the heating at high temperatures in air necessary to remove the analyte molecules for reusing. Furthermore, different from graphene which introduces intrinsic Raman band of high intensity in SERS spectra [79], BN nanosheets only display a Raman G band [80] of a low intensity that are barely enhanced by AuNPs, so that interferences are not created and only Raman signals from analytes are shown. The performances in SERS analysis have been tested using rhodamine 6G (R6G) as the probe molecule and silicon oxide (SiO2/Si), atomically thin BN and bulk hBN substrates decorated with AuNPs obtaining the greater enhancement signal for structures comprising BN nanosheets. Even though AuNPs are able to adsorb a certain quantity of R6G molecules, as a result of π-π interactions the BN surface can adsorb a drastically higher number of molecules that are localized in the hotspots between AuNPs, thus enhancing the SERS signal. As expected on the basis of the stronger resistance to oxidation of BN, the experimental results confirmed the ability of hybrid composites to be reused as SERS substrates, being able to sustain multiple thermal regeneration cycles. The performances in SERS analysis for hybrid compounds containing engineered 2D nanomaterials are reported in Table 3.

Table 3. Performances of the SERS platform based on hybrid compounds containing engineered 2D nanomaterials.

ENGINEERED 2D NANOMATERIAL						
System	Molecule Used to Calculate LOD	Limit of Detection (LOD)	Enhancement Factor (EF)	Molecule Used to Calculate EF	Equation Used to Calculate EF	Reference
Ag nanocube-decorated 1T-MoS_2 nanosheet composites (1T-MoS_2/AgNCs)	Thiram (TRM) Thiabendazole (TBZ)	0.62 Nm–50 Nm	1.78×10^7	Rhodamine 6G (R6G)	$(I_{SERS} \times C_{Raman})/(I_{Raman} \times C_{SERS})$	[76]
Gold nanoparticle-decorated MoS_2 nanosheets (n-MoS_2@AuNP)	R = Rhodamine B (RhB)	10^{-10} M	$\sim10^4$	R = Rhodamine B (RhB)	Peak intensity ratio of the SERSactive regions and the flake surface	[77]
Gold nanoparticles decorated boron nitride (BN) nanosheets (Au/BN)	Rhodamine 6G (R6G)	5.12×10^{-3} µM	$(5.0 \pm 0.2) \times 10^8$	Rhodamine 6G (R6G)	$(I_{SERS}/I_{bulk}) \times (N_{bulk}/N_{SERS})$	[78]

7. Three-Dimensional Structures

Despite the huge steps forward, the development of the SERS platforms is still mainly limited to plasmonic metal nanoparticles hybridized with bidimensional nanomaterial, but three-dimensional (3D) nanostructures start to be used to create hybrid materials. In his work [81], Yin prepared three-dimensional MoS_2 nanohybrids according to the microwave irradiation hydrothermal synthesis strategy (3D MoS_2-NS@Au-NPs) and the system created has been compared with bidimensional MoS_2/Au nanoparticle hybrids and tested as a SERS platform for melamine in milk detection. From comparison, it can be seen that the SERS activity of 3D MoS_2-NS@Au-NPs structures is improved by almost 56.4-fold in EF compared to 2D MoS_2-NST@Au-NPs hybrid structures. The amplification of EF has been ascribed to the larger surface area for adsorbing probe molecules and the higher number of hot spots generated to benefit the SERS performances supplied by the three-dimensional structure. Once optimized, the effectiveness of the generated hot spots by tuning the size and the density of the Au nanoparticles of the composite, the optimized structures have been tested for melamine quantitative detection in milk, finding a LOD of 1 ppb, a value lower than the maximum level of melamine as 2.5 ppm in food prescribed by the U.S. Food and Drug Administration. In the same year, Wang et al. [82] fabricated hierarchical MoS_2-microspheres (MoS_2-MS) decorated with "cauliflower-like" AuNP arrays (CF-AuNPs), by means of a new synthetic route, which have been successively investigated and tested for SERS analysis. According to the obtained results, it is possible to tailor the average size of CF-AuNPs@MoS_2-MS nanocomposites by tailoring the molar ratio between MoS_2-MS and $HAuCl_4$ and, once optimized to achieve the best performances in SERS analysis, they have been tested for molecular detection. In addition to R6G and methylene blue (MB) molecule sensing, that showed an LOD, respectively, of 10^{-14} M and 10^{15} M, the composites have been tested for the detection of various metabolites in human early morning urine with promising results. Furthermore, the CF-AuNPs@MoS_2-MS nanocomposites have been inserted in cellulose acetate membrane (CAM) to fabricate flexible wafer-scale flexible SERS substrates. From the results obtained by exploiting three-dimensional nanostructures hybridized with metal plasmonic nanoparticles as substrates in SERS analysis, it has been shown that the main advantage of these composites resides in their huge surface area. In fact, due to their large surface area the composites can adsorb a higher number of probe molecules compared to composites containing bidimensional materials and, in addition, are able to generate hotspots distributed in the space that are more efficient in enhancing the SERS signal by means of electrochemical mechanism. The effect of surface area in SERS analysis efficacy has been explored by Singh [83]. By means of facile hydrothermal method, a series of MoS_2 nanoflowers with a surface area ranging from 5 m^2/g to 20 m^2/g has been synthesized and tested in SERS analysis using the R6G as a probe molecule. It has been found a linear dependency of SERS signals originating from different substrates with the surface area, thus correlating the surface area of the composites with the intensity of the Raman signal. In Table 4, the performances of the hybrid compounds with a three-dimensional structure are reported.

Table 4. Performances of the SERS platform based on hybrid compounds with a three-dimensional structure.

THREE DIMENSIONAL STRUCTURES						
System	Molecule Used to Calculate LOD	Limit of Detection (LOD)	Enhancement Factor (EF)	Molecule Used to Calculate EF	Equation Used to Calculate EF	Reference
Gold nanoparticle-decorated three-dimensional (3D) MoS_2 nanospheres ((3D MoS_2-NS@Au-NPs)	Melamine	1 ppb	7.9×10^7	4-mercaptophenol (4-MPH)	$(I_{SERS}/N_{ads})/(I_{bulk}/N_{bulk})$	[81]
Hierarchical MoS_2-microspheres decorated with "cauliflower-like" AuNP arrays (CF-AuNPs@MoS_2-MS)	Rhodamine 6G (R6G)	10^{-14}–10^{-15}	Not calculated			[82]
	Methylene blue (MB)					
Gold nanoparticle-decorated boron nitride (BN) nanosheets (Au/BN)	Rhodamine 6G (R6G)	5.12×10^{-3} µM	$(5.0 \pm 0.2) \times 10^8$	Rhodamine 6G (R6G)	$(I_{SERS}/I_{bulk}) \times (N_{bulk}/N_{SERS})$	[78]

8. Nanospacers

Other approaches introduced bidimensional nanomaterial as a nanospacer between layers of plasmonic structures in order to create dense three-dimensional hotspots that support the striking SERS enhancement. In the early work of Zhu [84], the interactions between light and the sub-nanometer gap were systematically investigated. With a simple fabrication technique, a structure composed of graphene sandwiched between two layers of vertically stacked Au NPs has been developed and investigated as a SERS substrate. Performing numerical simulations based on the Finite element method (FEM) to investigate the effect of graphene in the sandwich structure, it has been found that the electric field is strongly amplified in the gap defined by the graphene film between two vertically stacked layers of AuNPs, leading to electric field enhancement of up to 88 times, much higher compared to that of 14 times in the horizontal gaps between Au nanoparticles without graphene. In addition, by changing the number of graphene films it is possible to control the nanogap between the vertical AuNPs layers, and, as consequence, the SERS enhancement. To investigate the effect of the gap induced by a different number of graphene layers, composites of two vertical layers of AuNPs containing a various number of graphene film have been created and used as a substrate in SERS analysis, obtaining the best performance for structures containing graphene monolayer, thus deducing that the coupling between the layers of plasmonic structures decays exponentially with their distance. Using the composite containing the graphene monolayer (4 nm Au/1LG/4 nm Au) as a SERS substrate, an LOD of 10^{-9} M for rhodamine B (RhB) has been found, hence showing a higher sensitivity compared to 4 nm Au/4 nm Au films, that showed an LOD of 10^{-7} for RhB molecules when used as substrates. The same results have been obtained for R6G molecules, with composites containing graphene resulting in a higher sensitivity. The use of 4 nm Au/1LG/4 nm Au in practical applications has been tested on by means of Sudan III and methylene blue detection, finding an LOD of 0,1 nM, thus showing a potential use in areas of food safety, medical diagnostics, biological imaging and environmental pollutant detection. Similar results showing the extraordinary performances in SERS enhancement obtained by exploiting graphene as a nanospacer have been obtained in the work of Man [85]. In this work, Au nanoparticles, (AuNPs), silver nanoparticles (AgNPs) and graphene have been combined in order to form a sandwiched structure, AgNPs/graphene@AuNPs, to be exploited in order to achieve unique performances in SERS analysis. By means of rhodamine R6G and crystal violet (CV) molecules, the composite has been experimentally tested in SERS analysis generating a huge amplification in Raman signal intensity. The excellent signal SERS enhancement obtained has been explained as the combination of chemical mechanism (CM) and electromagnetic mechanism (EM). The graphene film induced the CM, thus enhancing the SERS activity, but can also act as a nanospacer able to control the hot spots' size by changing the number of graphene layers. In fact, the electromagnetic enhancement was the result of three-dimensional hotspots generated by lateral nanogaps (AuNPs-AuNPs, AgNPs-AgNPs) and vertical nanogaps (AgNP-AuNPs), tunable by graphene layers. To investigate the effect of the graphene film, different composites containing a different number of graphene layers have been explored using R6G molecules as probes, finding that the graphene bilayer offers the best performances. A single layer of graphene induced the plasmon tunneling phenomenon due to the short distance (<0.5 nm) between Ag and Au nanoparticles which reduces the plasmonic coupling effect weakening the Raman signal, whereas for higher numbers of graphene layers, the nanogap also increases so that the electromagnetic enhancement is reduced because the enhanced local electric field will exponentially decay with distance. This composite system AgNPs/graphene@AuNPs has been tested for Malachite green (MG) detection in sea water finding an LOD of 10^{-11} M, thus demonstrating a potential ability in practical applications. Motivated by the few studies exploiting WS_2 bidimensional nanomaterial in SERS analysis and on the presumption that, due to its structure, WS_2 could promote both chemical enhancement (CM) by means of charge transfer between substrate and probe molecules, and electromagnetic enhancement (EM) by means of the strong coupling between WS_2 and metallic nanostructures through surface plasmon excitation, thus enhancing the SERS signal, Jiang in his work [86] exploited bidimensional WS_2 nanomaterial as a nanospacer in hybrid nanostructures. A remarkable SERS platform based on AuNPs/WS_2@AuNPs nanohybrids has been designed and developed in a multi-step process. Firstly, annealing an Au film deposited onto a SiO_2 substrate, a layer of Au NPs has been created. Successively, by means of a thermal decomposition process, a bilayer WS_2 film has been grown onto the AuNPs surface, and finally, a second layer of AuNPs was deposited onto the WS_2 film by means of a further annealing, thus obtaining the AuNPs/WS_2@AuNPs composites. Introducing the bilayer WS_2 film as a nanospacer between the two layers of plasmonic structures, a highly enhanced local electromagnetic field has been generated. Dense 3D hotspots occurring through this hybrid plasmonic nanostructures are responsible for the greatly enhanced SERS performances. Using rhodamine R6G as a probe molecule to test the performance in SERS analysis, the AuNPs/WS_2@AuNPs nanohybrids showed an excellent sensitivity with the minimum detectable concentration of 10^{-11} M. In addition, the AuNPs/WS_2@AuNPs nanohybrids showed extremely satisfying performances in detecting other probe molecules such as crystal violet (CV) molecules. The results obtained in the presented works illustrate the role of bidimensional nanomaterial used as a nanospacer between the layer of plasmonic metal nanostructures in the enhancement of the SERS signal. When used as a nanospacer, the bidimensional nanomaterial is able to create three-dimensional hotspots generated by the combination of lateral nanogaps (gaps inside plasmonic layer) and vertical nanogaps (gaps between plasmonic nanolayers). The great benefit of using a nanospacer is the possibility to finely tune the vertical gap by changing the number of bidimensional nanomaterial layers and taking into account that an exponential decay controls the coupling between two plasmonic layers, so that increasing the nanogap with a higher number of 2D nanomaterial layers reduces the SERS enhancement, but also for too small nanogaps, the Raman signal is weakened by the plasmon tunneling phenomenon. The performances of hybrid structures containing nanospacers in SERS detection are listed in Table 5.

Table 5. The performances of the SERS platform based on hybrid compounds engineered with nanospacers are listed.

NANOSPACERS						
System	Molecule Used to Calculate LOD	Limit of Detection (LOD)	Enhancement Factor (EF)	Molecule Used to Calculate EF	Equation Used to Calculate EF	Reference
Graphene sandwiched between two layers of vertically stacked Au NPs (Au NP/graphene/ Au NP)	Sudan III Methylene blue	0.1 nM	1.6×10^8–2.5×10^8	Rhodamine B (RhB) Rhodamine 6G (R6G)	$(I_{SERS}/I_{bulk})/$ (N_{bulk}/N_{SERS})	[84]
Graphene nanosheet sandwiched between a layer of AuNPs and AgNPs (AgNPs/graphene@AuNPs)	Malachite green (MG) in deionized (DI) water Malachite green (MG) in sea water	10^{-11} M–10^{-8} M	Not calculated			[85]
WS$_2$ nanosheets sandwiched between two Au nanoparticle layers (AuNPs/WS$_2$@AuNPs)	Rhodamine 6G (R6G)	10^{-11} M	Not calculated			[86]

9. Bidimensional Nanomaterials Used to Veil AuNP Arrays

In the majority of the hybrid composites that have been developed, the bidimensional nanomaterial acts as a support for the plasmonic metal structures; however, this configuration only provides a limited number of contact points between 2D nanomaterials and metal nanostructures: to attain a large enhancement of the SERS signal, an efficient contact between the metal framework and the bidimensional nanomaterial is necessary [87]. On this basis, different strategies wrapped the 2D nanomaterial around plasmonic metal nanoparticles in such a way that the number of contact points is increased, thus optimizing the Raman enhancement. A SERS substrate has been created veiling an array of silver nanocubes (AgNCs) with a graphene oxide (GO) film by means of a simple GO deposition process based on GO self-assembly on the metal surface [88]. According to the finite element method (FEM) calculations, the maximum intensity is 70% reduced when a 7 nm-thick GO layer is added on the metal structure surface and a more spread E-field distribution along the cluster edges and at the interface between particles is generated after supporting a 7 nm-thick GO layer on AgNCs, in contrast with well-localized hot spots observed for bare cubes. In conformity with FEM calculations, a reduction in SERS efficacy is expected, but the experimental results pointed out a superior SERS activity including more resolved peaks with higher signal intensity and larger reproducibility. The excellent SERS activity has been explained with a chemical enhancement deriving from a combination of π-π interactions and charge transfer from the oxygen-rich functional groups of GO to the probe molecules and in addition, with the GO ability to catch different compounds that therefore are accumulated on its surface thus intensifying the SERS signal. Comparing the performances, it has been found that wrapping plasmonic nanostructures enables a greater enhancement of the SERS signal than supporting on the bidimensional material. Intrigued by the great potential of Au nanoparticles hybridized with graphene nanosheets as SERS substrates, Cerruti [89] produced Au nanostars wrapped by graphene oxide (GO) nanosheets which further improved the SERS platform. Previously synthesized AuNSs have been functionalized with the positively charged Cysteamine, that create electrostatic interactions with negatively charged GO, so that GO-wrapped AuNSs (Au NSt@nGO) have been produced and tested in SERS analysis. As a result of AuNSs being wrapped with GO, the Raman signal of nGO by 5.3-fold compared to samples in which nGO is in contact with the nanostars but does not wrap them, whereas there is a higher enhancement for wrapped AuNSs, thus confirming the efficiency of wrapping to improve the SERS signal. SERS signals of typical Raman reporter such as rhodamine B (RhB), crystal violet (CV) and R6G sandwiched between AuNSs and GO nanosheets were higher compared to the signal obtained when the molecules are adsorbed on the nanostar surface. Together with an increase in SERS efficiency, wrapping AuNSs with GO results in a greater physiological stability, depending on a prevented RhB desorption in physiological conditions. A more detailed interpretation of SERS efficacy generated by plasmonic nanoparticle wrapping with a bidimensional material is given in the work of Chen [90]. Inspired by the strong surface adsorption of airborne hydrocarbon and aromatic molecules of thin boron nitride (BN) nanosheets, a SERS platform has been created placing an atomically thin BN nanosheet over an Au nanoparticle array produced via physical processes. Using R6G as probe molecules and BN nanosheets with different thicknesses, it has been found that Raman signals were most prominent for the lower thickness of BN, but reduced when the layer thickness increased and the stronger Raman signals were attributed to hotspots. Thinner BN nanosheets, due to their greater flexibility are able to better conform to the underlying AuNPs so that the analyte molecules were closer to the plasmonic hotspots. For increasing thickness, the BN nanosheets were much less deformed so that analyte molecules were more distant from the plasmon-induced EM field which decays exponentially with the distance. In addition, more R6G molecules were attracted depending on the strong adsorption ability of BN nanosheets towards aromatic molecules by means of π-π interactions. Conformational changes explain the stronger adsorption capability, with BN nanosheet polarity not contributing to such effect. Wrapping plasmonic metal nanostructures with graphene nanosheets has revealed great potential, thus expanding applications of hybrid composites from SERS detection to SERS bioimaging. Graphene oxide (GO)-wrapped Au nanorods (GO@GNRs) have been created by Wu [91]. Assessing the cytotoxicity showed a greatly enhanced biocompatibility of GO@GNRs, provided by the encapsulation enabling a reduced contact with the surrounding environment, thus decreasing the amount of residual CTAB that induces cytotoxicity. The SERS activity in the near infrared (NIR) has been investigated using six dye molecules as probes showing extremely intense SERS signals and highly enhanced activities of NIR SERS multiple effects, such as Au nanorods LSPR, the charge transfer between graphene nanosheets and probe molecules and the enrichment of dye molecules on the GO sheets. The enhanced NIR SERS activity and the improved biocompatibility enable a successful application of GO@GNRs as a robust nanoplatform for ultrafast NIR SERS bioimaging. On these bases, exploiting bidimensional materials to veil arrays of plasmonic nanostructures produces a greater SERS activity compared to that obtained when 2D material is used as a support. Due to its flexibility, the bidimensional nanomaterial is able to bring analyte molecules into the proximity of the hotspots originated from plasmonic structures thus enhancing the SERS efficacy. Moreover, using 2D nanomaterials as veiling medium offers the additional advantage of protecting the arrays of plasmonic nanostructures, thus increasing the stability of the SERS substrate and broadening the composite range of applications. Wrapping plasmonic nanostructures imparts a biocompatibility to the system, thus that can also be exploited in the biomedical field.

10. Future Perspectives

In order to give a better understanding of its enormous potentiality, the future perspectives of the SERS technique will be provided considering a range of directions in which the research in this field is focused. Since its development, SERS has affirmed a great potential as a powerful technique to detect simple and more complex molecules, in contact or adjacent to a plasmonic substrate that generally consists of a metal surface, and recently, also of hybrid materials. Such enormous potential, combined with advances in the development of associated instrumentation has produced an outbreak of research, moving forward many different applications ranging from materials and environmental science through biology and medicine. Some clinical implementations of SERS that could find an application in the near future are: 1) the detection of tumor margins during surgery [92], as it has been demonstrated that tags were sufficient to improve tracking of tumor margins employing a portable Raman microscope instead of a benchtop instrument; 2) in optical fiber-guided imaging procedures such as endoscopy, colonoscopy, or others used to detect and to visualize superficial diseased tissues within the body [93]; and 3) in liquid biopsy, a term widely comprising the identification of disease biomarkers in blood or other bodily fluids [94]. Liquid biopsy holds great potential to simplify disease detection and monitoring and make it less painful for the patient. In particular, the monitoring of the disease progresses and the response to therapy would ideally be allowed on a daily basis, thus avoiding the dependence on imaging approaches that could not detect changes in the size of the tumor. The SERS analysis has also been revealed as a valid tool to detect inorganic and highly toxic organic pollutants as well as to monitor bacterial contaminant with a detection threshold in the parts for a billion range. In food analytics, detection limits in quality control and nutrient quantification down to the nanomolar range have been reached. The difficulties in analyzing the surface residues, an issue of significance in a variety of areas including health and safety, homeland security, forensics, etc., paved the way to the development of flexible SERS substrates, still a young research area within the progress of the SERS technique. The greater part of recent flexible substrates designed for point of care analysis can be classified according to two categories: sticky "SERS tapes" and adsorptive "SERS swabs". Typically, SERS tapes are adhesive and flexible plastic films, bearing plasmonic particles on the surface, that can be pressed and peeled from the sample surface to extract the molecules for in situ analysis [95], whereas SERS swabs can be exploited to collect chemical compounds by dabbing the surface of the sample [96]. A main obstacle in the present applications of SERS is the difficulty in analyzing complex real samples, containing a large variety of chemical species and micro/macro-contaminants in addition to the target analyte molecules. Often, these impurities are present in much higher concentrations than the analyte, thus they can interfere with the analysis resulting in significantly reduced sensitivity and reproducibility. The method developed to address this issue still requires sophisticated equipment for the pre-treatment of the samples so that the analyses are restricted to a laboratory setting and must be performed by trained professionals. In this scenario, the difficulties in analyzing SERS spectra from complex samples for disease diagnosis and food analytics pave the way to the implementation of Artificial Intelligence [97]: the acquired SERS spectra of such samples could be spectrally unmixed so that the concentration profiles for better quantification could be estimated. The algorithm offers the additional advantages to characterize the disease while evaluating the adulterants and toxins in food processing as well. Thus, while SERS has experienced enormous progress, a broader range of use is limited by high running costs, as well as their inefficacy in point-of-care analysis. In view of the tremendous progress in the implementation of the SERS technique as an analytical tool, a great challenge is a rapid evolution in commercial products such as compact setups, tailored sensing platforms or efficient imaging methods that are able to compete or complete goods in current use in a wide range of technologies. In the next steps, SERS must be developed such that its processes of application will be simplified and its use of routine will be enabled by non-specialists. In light of the commercialization, adequacy for automated mass production and stability during storage must be examined.

11. Conclusions

Between the plethora of approaches focused on improving the performances and making SERS analysis a routine technique to enhance the quality of life of people, in recent years the development of sensing platforms based on structures composed of plasmonic metal nanoparticles hybridized with bidimensional nanoparticles have attracted great interest due to their outstanding properties. In this review, the different approaches used to enhance the outcomes of hybrid compounds have been discussed. Initially, the simplest structures based on spherical metal nanoparticles hybridized with graphene and its derivates (the most common bidimensional nanomaterial) have been considered, and then it has been considered how the structures have been evolved by changing the particle shape or engineering the bidimensional material to obtain the highest values in the enhancement factor (EF). Due to a greater number of analyte molecules adsorbed and the number of hotspots formed, three-dimensional (3D) structures of 2D nanomaterials and plasmonic nanoparticles showed the best results, so that, most likely, the future steps will be focused on the development of such compounds. The future perspectives of the technique have been discussed from a broader point of view considering the various directions in which the technique is progressing and not only from a point of view of the hybrid compounds. In this scenario, multifunctional substrates combining several of the characteristics considered will be developed in order to render SERS a daily technique for better living conditions.

Author Contributions: Conceptualization, C.S., A.F. and E.C.B.A.A.; methodology, C.S.; writing—original draft preparation, C.S., A.F. and E.C.B.A.A.; writing—review and editing, C.S., A.F. and E.C.B.A.A.; supervision, A.F., E.C.B.A.A. and M.V.; funding acquisition, A.F., E.C.B.A.A. and M.V. All authors have read and agreed to the published version of the manuscript.

Funding: This research was supported by EU funds through the FEDER European Regional Development Fund project LISBOA-01-0145-FEDER-031311, by Portuguese national funds provided by FCT—Fundação para a Ciência e Tecnologia, through grant SFRH/BD/09347/2021 and PTDC/QUI-QIN/29778/2017 project and Instituto Politécnico de Lisboa (IPL/2021/WASTE4CAT_ISEL and IPL/2021/MuMiAS-2D_ISEL).

Institutional Review Board Statement: Not applicable.

Informed Consent Statement: Not applicable.

Data Availability Statement: Not applicable.

Conflicts of Interest: The authors declare no conflict of interest.

References

1. Li, Y.; Yang, C.; Guo, X. Single-Molecule Electrical Detection: A Promising Route toward the Fundamental Limits of Chemistry and Life Science. *Accounts Chem. Res.* **2019**, *53*, 159–169. [CrossRef] [PubMed]
2. Chen, X.; Zhou, C.; Guo, X. Ultrasensitive Detection and Binding Mechanism of Cocaine in an Aptamer-based Single-molecule Device. *Chin. J. Chem.* **2019**, *37*, 897–902. [CrossRef]
3. Fang, S.; Hu, Y.H. Open the door to the atomic world by single-molecule atomic force microscopy. *Matter* **2021**, *4*, 1189–1223. [CrossRef]
4. Akkilic, N.; Geschwindner, S.; Höök, F. Single-molecule biosensors: Recent advances and applications. *Biosens. Bioelectron.* **2020**, *151*, 111944. [CrossRef] [PubMed]
5. Langer, J.; Aberasturi, D.J. Present and Future of Surface-Enhanced Raman Scattering. *ACS Nano* **2020**, *14*, 28–117. [CrossRef] [PubMed]
6. Li, C.; Huang, Y.; Li, X.; Zhang, Y.; Chen, Q.; Ye, Z.; Alqarni, Z.; Bell, S.E.J.; Xu, Y. Towards practical and sustainable SERS: A review of recent developments in the construction of multifunctional enhancing substrates. *J. Mater. Chem. C* **2021**, *9*, 11517–11552. [CrossRef]
7. Xia, M. 2D Materials-Coated Plasmonic Structures for SERS Applications. *Coatings* **2018**, *8*, 137. [CrossRef]
8. Fleischmann, M.; Hendra, P.J.; McQuillan, A.J. Raman spectra of pyridine adsorbed at a silver electrode. *Chem. Phys. Lett.* **1974**, *26*, 163–166. [CrossRef]
9. D'Elia, V.; Rubio-Retama, J.; Ortega-Ojeda, F.E.; García-Ruiz, C.; Montalvo, G. Gold nanorods as SERS substrate for the ultra trace de-tection of cocaine in non-pretreated oral fluid samples. *Colloids Surf. A Physicochem. Eng. Asp.* **2018**, *557*, 43–50. [CrossRef]
10. Qin, Y.; Wu, Y.; Wang, B.; Wang, J.; Yao, W. Facile synthesis of Ag@Au core-satellite nanowires for highly sensitive SERS detection for tropane alkaloids. *J. Alloy. Compd.* **2021**, *884*, 161053. [CrossRef]
11. Wang, K.; Sun, D.-W.; Pu, H.; Wei, Q. Polymer multilayers enabled stable and flexible Au@Ag nanoparticle array for nondestructive SERS detection of pesticide residues. *Talanta* **2021**, *223*, 121782. [CrossRef] [PubMed]
12. Lee, P.C.; Meisel, D. Adsorption and sur-face-enhanced Raman of dyes on silver and gold sols. *J. Phys. Chem.* **1982**, *86*, 3391–3395. [CrossRef]
13. Eustis, S.; El-Sayed, M.A. Why gold nanoparticles are more precious than pretty gold: Noble metal surface plasmon resonance and its enhancement of the radiative and nonradiative properties of nanocrystals of different shapes. *Chem. Soc. Rev.* **2005**, *35*, 209–217. [CrossRef] [PubMed]
14. Jain, P.K.; Lee, K.S.; El-Sayed, I.H.; El-Sayed, M.A. Calculated Absorption and Scattering Properties of Gold Nanoparticles of Different Size, Shape, and Composition: Applications in Biological Imaging and Biomedicine. *J. Phys. Chem. B* **2006**, *110*, 7238–7248. [CrossRef] [PubMed]
15. Kumar, C.S. *Raman Spectroscopy for Nanomaterials Characterization*; Springer: Berlin/Heidelberg, Germany, 2012.
16. Anderson, D.J.; Moskovits, M. A SERS-Active System Based on Silver Nanoparticles Tethered to a Deposited Silver Film. *J. Phys. Chem. B* **2006**, *110*, 13722–13727. [CrossRef] [PubMed]
17. Stiles, P.L.; Dieringer, J.A.; Shah, N.C.; Duyne, R.P.V. Surface-Enhanced Raman Spectroscopy. *Annu. Rev. Anal. Chem.* **2008**, *1*, 601–626. [CrossRef]

18. Guillot, N.; de la Chapelle, M.L. The electromagnetic effect in surface enhanced Raman scattering: Enhancement optimization using precisely controlled nanostructures. *J. Quant. Spectrosc. Radiat. Transf.* **2012**, *113*, 2321–2333. [CrossRef]

19. Pilot, R.; Signorini, R.; Durante, C.; Orian, L.; Bhamidipati, M.; Fabris, L. A Review on Surface-Enhanced Raman Scattering. *Biosensors* **2019**, *9*, 57. [CrossRef]

20. Schlücker, S. Surface-Enhanced Raman Spectroscopy: Concepts and Chemical Applications. *Angew. Chem. Int. Ed.* **2014**, *53*, 4756–4795. [CrossRef]

21. Li, W.; Camargo, P.; Lu, X.; Xia, Y. Dimers of Silver Nanospheres: Facile Synthesis and Their Use as Hot Spots for Surface-Enhanced Raman Scattering. *Nano Lett.* **2008**, *9*, 485–490. [CrossRef]

22. Hao, E.; Schatz, G.C. Electromagnetic fields around silver nanoparticles and dimers. *J. Chem. Phys.* **2004**, *120*, 357–366. [CrossRef] [PubMed]

23. Álvarez-Puebla, R.A.; Contreras-Cáceres, R.; Pastoriza-Santos, I.; Pérez-Juste, J.; Liz-Marzán, L.M. Au@pNIPAM Colloids as Molecular Traps for Surface-Enhanced, Spectroscopic, Ultra-Sensitive Analysis. *Angew. Chem.* **2008**, *121*, 144–149. [CrossRef]

24. Guerrini, L.; Garcia-Ramos, J.V.; Domingo, C.; Sanchez-Cortes, S. Functionalization of Ag Nanoparticles with Dithiocarbamate Calix[4]arene As an Effective Supramolecular Host for the Surface-Enhanced Raman Scattering Detection of Polycyclic Aromatic Hydrocarbons. *Langmuir* **2006**, *22*, 10924–10926. [CrossRef] [PubMed]

25. Li, X.; Zhu, J.; Wei, B. Hybrid nanostructures of metal/two-dimensional nanomaterials for plasmon-enhanced applications. *Chem. Soc. Rev.* **2016**, *45*, 3145–3187. [CrossRef] [PubMed]

26. Manzeli, S.; Ovchinnikov, D.; Pasquier, D.; Yazyev, O.V.; Kis, A. 2D transition metal dichalcogenides. *Nat. Rev. Mater.* **2017**, *2*, 17033. [CrossRef]

27. Yu, X.; Cai, H.; Zhang, W.; Li, X.; Pan, N.; Luo, Y.; Wang, X.; Hou, J.G. Tuning Chemical Enhancement of SERS by Controlling the Chemical Reduction of Graphene Oxide Nanosheets. *ACS Nano* **2011**, *5*, 952–958. [CrossRef]

28. Xu, H.; Xie, L.; Zhang, H.; Zhang, J. Effect of Graphene Fermi Level on the Raman Scattering Intensity of Molecules on Graphene. *ACS Nano* **2011**, *5*, 5338–5344. [CrossRef]

29. Xie, L.; Ling, X.; Fang, Y.; Zhang, J.; Liu, Z. Graphene as a Substrate To Suppress Fluorescence in Resonance Raman Spectroscopy. *J. Am. Chem. Soc.* **2009**, *131*, 9890–9891. [CrossRef]

30. Swathi, R.S.; Sebastian, K.L. Resonance energy transfer from a dye molecule to graphene. *J. Chem. Phys.* **2008**, *129*, 054703. [CrossRef]

31. Ling, X.; Fang, W.; Lee, Y.H.; Araujo, P.T.; Zhang, X.; Rodriguez-Nieva, J.F.; Lin, Y.; Zhang, Y.; Kong, J.; Dresselhaus, M.S. Raman Enhancement Effect on Two-Dimensional Layered Materials: Graphene, h-BN and MoS2. *Nano Lett.* **2014**, *14*, 3033–3040. [CrossRef]

32. Novoselov, K.S.; Geim, A.K.; Morozov, S.V.; Jiang, D.; Katsnelson, M.I.; Grigorieva, I.V.; Dubonos, S.V.; Firsov, A.A. Two-dimensional gas of massless Dirac fermions in graphene. *Nature* **2005**, *438*, 197–200. [CrossRef] [PubMed]

33. Han, W.-Q.; Wu, L.; Zhu, Y.; Watanabe, K.; Taniguchi, T. Structure of chemically derived mono- and few-atomic-layer boron nitride sheets. *Appl. Phys. Lett.* **2008**, *93*, 223103. [CrossRef]

34. Splendiani, A.; Sun, L.; Zhang, Y.; Li, T.; Kim, J.; Chim, C.-Y.; Galli, G.; Wang, F. Emerging Photoluminescence in Monolayer MoS2. *Nano Lett.* **2010**, *10*, 1271–1275. [CrossRef] [PubMed]

35. Ren, W.; Fang, Y.; Wang, E. A Binary Functional Substrate for Enrichment and Ultrasensitive SERS Spectroscopic Detection of Folic Acid Using Graphene Oxide/Ag Nanoparticle Hybrids. *ACS Nano* **2011**, *5*, 6425–6433. [CrossRef]

36. Liu, M.; Chen, W. Graphene nanosheets-supported Ag nanoparticles for ultrasensitive detection of TNT by surface-enhanced Raman spectroscopy. *Biosens. Bioelectron.* **2013**, *46*, 68–73. [CrossRef] [PubMed]

37. Zhang, P.; Huang, Y.; Lu, X.; Zhang, S.; Li, J.; Wei, G.; Su, Z. One-Step Synthesis of Large-Scale Graphene Film Doped with Gold Nanoparticles at Liquid–Air Interface for Electrochemistry and Raman Detection Applications. *Langmuir* **2014**, *30*, 8980–8989. [CrossRef]

38. Su, S.; Zhang, C.; Yuwen, L.; Chao, J.; Zuo, X.; Liu, X.; Song, C.; Fan, C.; Wang, L. Creating SERS Hot Spots on MoS2 Nanosheets with in Situ Grown Gold Nanoparticles. *ACS Appl. Mater. Interfaces* **2014**, *6*, 18735–18741. [CrossRef]

39. Shorie, M.; Kumar, V.; Kaur, H.; Singh, K.; Tomer, V.K.; Sabherwal, P. Plasmonic DNA hotspots made from tungsten disulfide nanosheets and gold na-noparticles for ultrasensitive aptamer-based SERS detection of myoglobin. *Microchim. Acta* **2018**, *185*, 158. [CrossRef]

40. Kelly, K.L.; Coronado, E.; Zhao, L.L.; Schatz, G.C. The Optical Properties of Metal Nanoparticles: The Influence of Size, Shape, and Dielectric Environment. *J. Phys. Chem. B* **2003**, *107*, 668–677. [CrossRef]

41. Link, S.; Mohamed, M.B.; El-Sayed, M.A. Simulation of the Optical Absorption Spectra of Gold Nanorods as a Function of Their Aspect Ratio and the Effect of the Me-dium Dielectric Constant. *J. Phys. Chem. B* **1999**, *103*, 3073–3077. [CrossRef]

42. Ni, W.; Kou, X.; Yang, Z.; Wang, J. Tailoring Longitudinal Surface Plasmon Wavelengths, Scattering and Absorption Cross Sections of Gold Nanorods. *ACS Nano* **2008**, *2*, 677–686. [CrossRef] [PubMed]

43. Chu, M.-W.; Myroshnychenko, V.; Chen, C.H.; Deng, J.-P.; Mou, C.-Y.; de Abajo, F.J.G. Probing Bright and Dark Surface-Plasmon Modes in Individual and Coupled Noble Metal Nanoparticles Using an Electron Beam. *Nano Lett.* **2008**, *9*, 399–404. [CrossRef] [PubMed]

44. Guiton, B.S.; Iberi, V.; Li, S.; Leonard, D.N.; Parish, C.M.; Kotula, P.G.; Varela, M.; Schatz, G.C.; Pennycook, S.J.; Camden, J.P. Correlated Optical Measurements and Plasmon Mapping of Silver Nanorods. *Nano Lett.* **2011**, *11*, 3482–3488. [CrossRef] [PubMed]
45. Smitha, S.; Gopchandran, K.; Smijesh, N.; Philip, R. Size-dependent optical properties of Au nanorods. *Prog. Nat. Sci.* **2013**, *23*, 36–43. [CrossRef]
46. Pardehkhorram, R.; Bonaccorsi, S.; Zhu, H.; Gonçales, V.; Wu, Y.; Liu, J.; Lee, N.A.; Tilley, R.D.; Gooding, J.J. Intrinsic and well-defined second generation hot spots in gold nanobipyramids versus gold nanorods. *Chem. Commun.* **2019**, *55*, 7707–7710. [CrossRef] [PubMed]
47. Yougbaré, S.; Chou, H.-L.; Yang, C.-H.; Krisnawati, D.I. Facet-dependent gold nanocrystals for effective photothermal killing of bacteria. *J. Hazard. Mater.* **2021**, *407*, 124617. [CrossRef]
48. LI, N.; HAN, S.; ZHANG, C.; LIN, S.; SHA, X.-Y.; HASI, W. Detection of Chlortetracycline Hydrochloride in Milk with a Solid SERS Substrate Based on Self-assembled Gold Nanobipyramids. *Anal. Sci.* **2020**, *36*, 935–940. [CrossRef]
49. Nelayah, J.; Kociak, M.; Stéphan, O.; de Abajo, F.J.G.; Tencé, M.; Henrard, L.; Taverna, D.; Pastoriza-Santos, I.; Liz-Marzán, L.M.; Colliex, C. Mapping surface plasmons on a single metallic nanoparticle. *Nat. Phys.* **2007**, *3*, 348–353. [CrossRef]
50. Brioude, A.; Pileni, M.P. Silver Nanodisks: Optical Properties Study Using the Discrete Dipole Approximation Method. *J. Phys. Chem. B* **2005**, *109*, 23371–23377. [CrossRef]
51. Tan, T.; Tian, C.; Ren, Z.; Yang, J.; Chen, Y.; Sun, L.; Li, Z.; Wu, A.; Yin, J.; Fu, H. LSPR-dependent SERS performance of silver nanoplates with highly stable and broad tunable LSPRs prepared through an improved seed-mediated strategy. *Phys. Chem. Chem. Phys.* **2013**, *15*, 21034–21042. [CrossRef]
52. Hao, F.; Nehl, C.L.; Hafner, J.H.; Nordlander, P. Plasmon Resonances of a Gold Nanostar. *Nano Lett.* **2007**, *7*, 729–732. [CrossRef] [PubMed]
53. Nehl, C.L.; Liao, H.; Hafner, J.H. Optical Properties of Star-Shaped Gold Nanoparticles. *Nano Lett.* **2006**, *6*, 683–688. [CrossRef] [PubMed]
54. Kumar, P.S.; Pastoriza-Santos, I.; Rodriguez-Gonzalez, B.; De Abajo, F.J.G.; Liz-Marzán, L.M. High-yield synthesis and optical response of gold nanostars. *Nanotechnology* **2007**, *19*, 015606. [CrossRef] [PubMed]
55. Hao, E.; Bailey, R.C.; Schatz, G.C.; Hupp, J.T.; Li, S. Synthesis and Optical Properties of "Branched" Gold Nanocrystals. *Nano Lett.* **2004**, *4*, 327–330. [CrossRef]
56. Alvarez-Puebla, R.A.; Liz-Marzán, L.M. SERS-Based Diagnosis and Biodetection. *Small* **2010**, *6*, 604–610. [CrossRef]
57. Khoury, C.G.; Vo-Dinh, T. Gold Nanostars for Surface-Enhanced Raman Scattering: Synthesis, Characterization and Optimization. *J. Phys. Chem. C* **2008**, *112*, 18849–18859. [CrossRef]
58. Mers, S.V.S.; Umadevi, S.; Ganesh, V. Controlled Growth of Gold Nanostars: Effect of Spike Length on SERS Signal Enhancement. *ChemPhysChem* **2017**, *18*, 1358–1369. [CrossRef]
59. Hong, S.; Li, X. Optimal Size of Gold Nanoparticles for Surface-Enhanced Raman Spectroscopy under Different Conditions. *J. Nanomater.* **2013**, *2013*, 790323. [CrossRef]
60. Kundu, S. A new route for the formation of Au nanowires and application of shape-selective Au nanoparticles in SERS studies. *J. Mater. Chem. C* **2013**, *1*, 831–842. [CrossRef]
61. Li, M.; Cushing, S.K.; Zhang, J.; Lankford, J.; Aguilar, Z.P.; Ma, D.; Wu, N. Shape-dependent surface-enhanced Raman scattering in gold–Raman-probe–silica sandwiched nanoparticles for biocompatible applications. *Nanotechnology* **2012**, *23*, 115501. [CrossRef]
62. Shvalya, V.; Filipič, G.; Zavašnik, J.; Abdulhalim, I.; Cvelbar, U. Surface-enhanced Raman spectroscopy for chemical and biological sensing using nanoplas-monics: The relevance of interparticle spacing and surface morphology. *Appl. Phys. Rev.* **2020**, *7*, 031307. [CrossRef]
63. Mejía-Salazar, J.R.; Oliveira, O.N., Jr. Plasmonic bio-sensing: Focus review. *Chem. Rev.* **2018**, *118*, 10617–10625. [CrossRef] [PubMed]
64. Soler, M.; Lechuga, L.M. Principles, technologies, and applications of plasmonic biosensors. *J. Appl. Phys.* **2021**, *129*, 111102. [CrossRef]
65. Abdulhalim, I. Coupling configurations between extended surface electromagnetic waves and localized surface plasmons for ultrahigh field enhancement. *Nanophotonics* **2018**, *7*, 1891–1916. [CrossRef]
66. Hu, C.; Rong, J.; Cui, J.; Yang, Y.; Yang, L.; Wang, Y.; Liu, Y. Fabrication of a graphene oxide–gold nanorod hybrid material by electrostatic self-assembly for surface-enhanced Raman scattering. *Carbon* **2013**, *51*, 255–264. [CrossRef]
67. Liu, Q.; Zhang, X.; Wen, G.; Luo, Y.; Liang, A.; Jiang, Z. A Sensitive Silver Nanorod/Reduced Graphene Oxide SERS Analytical Platform and Its Application to Quantitative Analysis of Iodide in Solution. *Plasmonics* **2014**, *10*, 285–295. [CrossRef]
68. Yu, J.; Ma, Y.; Yang, C.; Zhang, H.; Liu, L.; Su, J.; Gao, Y. SERS-active composite based on rGO and Au/Ag core-shell nanorods for analytical applications. *Sens. Actuators B Chem.* **2018**, *254*, 182–188. [CrossRef]
69. Luo, Y.; Ma, L.; Zhang, X.; Liang, A.; Jiang, Z. SERS Detection of Dopamine Using Label-Free Acridine Red as Molecular Probe in Reduced Graphene Oxide/Silver Nanotriangle Sol Substrate. *Nanoscale Res. Lett.* **2015**, *10*, 230. [CrossRef]
70. Wang, X.; Zhu, C.; Hu, X.; Xu, Q.; Zhao, H.; Meng, G.; Lei, Y. Highly sensitive surface-enhanced Raman scattering detection of organic pesticides based on Ag-nanoplate decorated graphene-sheets. *Appl. Surf. Sci.* **2019**, *486*, 405–410. [CrossRef]
71. Krishnan, S.K.; Godoy, Y.C. Deep Eutectic Sol-vent-Assisted Synthesis of Au Nanostars Supported on Graphene Oxide as an Efficient Substrate for SERS-Based Molecular Sensing. *ACS Omega* **2020**, *5*, 1384–1393. [CrossRef]

72. Fan, Z.; Kanchanapally, R.; Ray, P.C. Hybrid Graphene Oxide Based Ultrasensitive SERS Probe for Label-Free Biosensing. *J. Phys. Chem. Lett.* **2013**, *4*, 3813–3818. [CrossRef]

73. Pan, X.; Li, L.; Lin, H.; Tan, J.; Wang, H.; Liao, M.; Chen, C.; Shan, B.; Chen, Y.; Li, M. A graphene oxide-gold nanostar hybrid based-paper bio-sensor for label-free SERS detection of serum bilirubin for diagnosis of jaundice. *Biosens. Bioelectron.* **2019**, *145*, 111713. [CrossRef] [PubMed]

74. Zhu, X.; Shi, L.; Schmidt, M.S.; Boisen, A.; Hansen, O.; Zi, J.; Xiao, S.; Mortensen, N.A. Enhanced Light–Matter Interactions in Graphene-Covered Gold Nanovoid Arrays. *Nano-Lett.* **2013**, *13*, 4690–4696. [CrossRef] [PubMed]

75. Zhao, Y.; Chu, B.; Zhang, L.; Zhao, F.; Yan, J.; Li, X.; Liu, Q.; Lu, Y. Constructing sensitive SERS substrate with a sandwich structure separated by single layer graphene. *Sensors Actuators B Chem.* **2018**, *263*, 634–642. [CrossRef]

76. Tegegne, W.A.; Su, W.-N.; Tsai, M.-C.; Beyene, A.-B.; Hwang, B.-J. Ag nanocubes decorated 1T-MoS2 nanosheets SERS substrate for reliable and ultrasensitive detection of pesticides. *Appl. Mater. Today* **2020**, *21*, 100871. [CrossRef]

77. Rani, R.; Yoshimura, A.; Das, S.; Sahoo, M.R.; Kundu, A.; Sahu, K.K.; Meunier, V.; Nayak, S.K.; Koratkar, N.; Hazra, K.S. Sculpting Artificial Edges in Monolayer MoS2 for Controlled Formation of Surface-Enhanced Raman Hotspots. *ACS Nano* **2020**, *14*, 6258–6268. [CrossRef]

78. Cai, Q.; Li, L.H.; Yu, Y.; Liu, Y.; Huang, S.; Chen, Y.; Watanabe, K.; Taniguchi, T. Boron nitride nanosheets as improved and reusable sub-strates for gold nanoparticles enabled surface enhanced Raman spectroscopy. *Phys. Chem. Chem. Phys.* **2015**, *17*, 7761–7766. [CrossRef]

79. Xu, C.; Wang, X. Fabrication of Flexible Metal-Nanoparticle Films Using Graphene Oxide Sheets as Substrates. *Small* **2009**, *5*, 2212–2217. [CrossRef]

80. Gorbachev, R.V.; Riaz, I.; Nair, R.R.; Jalil, R.; Britnell, L.; Belle, B.D.; Hill, E.W.; Novoselov, K.S.; Watanabe, K.; Taniguchi, T.; et al. Hunting for Monolayer Boron Nitride: Optical and Raman Signatures. *Small* **2011**, *7*, 465–468. [CrossRef]

81. Liang, X.; Zhang, X.-J.; You, T.-T.; Yang, N.; Wang, G.-S.; Yin, P.-G. Three-dimensional MoS2-NS@Au-NPs hybrids as SERS sensor for quantitative and ultra-sensitive detection of melamine in milk. *J. Raman Spectrosc.* **2018**, *49*, 245–255. [CrossRef]

82. Qiu, H.; Wang, M.; Li, L.; Li, J.; Yang, Z.; Cao, M. Hierarchical MoS2-microspheres decorated with 3D AuNPs arrays for high-efficiency SERS sensing. *Sensors Actuators B Chem.* **2017**, *255*, 1407–1414. [CrossRef]

83. Singh, J.; Kumar, R.S.; Soni, R.K. Synthesis of 3D-MoS2 nanoflowers with tunable surface area for the application in photocatalysis and SERS based sensing. *J. Alloy. Compd.* **2020**, *849*, 156502. [CrossRef]

84. Zhao, Y.; Li, X.; Du, Y.; Chen, G.; Qu, Y.; Jiang, J.; Zhu, Y. Strong light–matter interactions in sub-nanometer gaps defined by monolayer graphene: Toward highly sensitive SERS substrates. *Nanoscale* **2014**, *6*, 11112–11120. [CrossRef] [PubMed]

85. Zhang, C.; Li, C.; Yu, J.; Jiang, S.; Xu, S.; Yang, C.; Liu, Y.J.; Gao, X.; Liu, A.; Man, B. SERS activated platform with three-dimensional hot spots and tunable nanometer gap. *Sensors Actuators B Chem.* **2018**, *258*, 163–171. [CrossRef]

86. Lu, Z.; Si, H.; Li, Z.; Yu, J.; Liu, Y.; Feng, D.; Zhang, C.; Yang, W.; Man, B.; Jiang, S. Sensitive, reproducible, and stable 3D plasmonic hybrids with bilayer WS2 as nanospacer for SERS analysis. *Opt. Express* **2018**, *26*, 21626–21641. [CrossRef]

87. Xu, W.; Xiao, J.; Chen, Y.; Chen, Y.; Ling, X.; Zhang, J. Graphene-Veiled Gold Substrate for Surface-Enhanced Raman Spectroscopy. *Adv. Mater.* **2013**, *25*, 928–933. [CrossRef]

88. Banchelli, M.; Tiribilli, B.; De Angelis, M.; Pini, R.; Caminati, G.; Matteini, P. Controlled Veiling of Silver Nanocubes with Graphene Oxide for Improved Surface-Enhanced Raman Scattering Detection. *ACS Appl. Mater. Interfaces* **2016**, *8*, 2628–2634. [CrossRef]

89. Jalani, G.; Cerruti, M. Nano graphene oxide-wrapped gold nanostars as ultrasensitive and stable SERS nanoprobes. *Nanoscale* **2015**, *7*, 9990–9997. [CrossRef]

90. Cai, Q.; Mateti, S.; Watanabe, K.; Taniguchi, T.; Huang, S.; Chen, Y.; Li, L.H. Boron Nitride Nanosheet-Veiled Gold Nanoparticles for Sur-face-Enhanced Raman Scattering. *ACS Appl. Mater. Interfaces* **2016**, *8*, 15630–15636. [CrossRef]

91. Qiu, X.; You, X.; Chen, X.; Chen, H.; Dhinakar, A.; Liu, S.; Guo, Z.; Wu, J.; Liu, Z. Development of graphene oxide-wrapped gold nanorods as robust nanoplatform for ultrafast near-infrared SERS bioimaging. *Int. J. Nanomed.* **2017**, *12*, 4349–4360. [CrossRef]

92. Karabeber, H.; Huang, R.; Iacono, P.; Samii, J.M.; Pitter, K.; Holland, E.C.; Kircher, M.F. Guiding Brain Tumor Resection Using Surface-Enhanced Raman Scattering Na-noparticles and a Hand-Held Raman Scanner. *ACS Nano* **2014**, *8*, 9755–9766. [CrossRef]

93. Wang, Y.W.; Kang, S.; Khan, A.; Bao, P.Q.; Liu, J.T. In vivo multiplexed molecular imaging of esophageal cancer via spectral endoscopy of topically applied SERS nanoparticles. *Biomed. Opt. Express* **2015**, *6*, 3714–3723. [CrossRef] [PubMed]

94. Zhang, Y.; Mi, X.; Tan, X.; Xiang, R. Recent Progress on Liquid Biopsy Analysis using Surface-Enhanced Raman Spectroscopy. *Theranostics* **2019**, *9*, 491–525. [CrossRef] [PubMed]

95. Chen, J.; Huang, Y.; Kannan, P.; Zhang, L.; Lin, Z.; Zhang, J.; Chen, T.; Guo, L. Flexible and Adhesive Surface Enhance Raman Scattering Active Tape for Rapid Detection of Pesticide Residues in Fruits and Vegetables. *Anal. Chem.* **2016**, *88*, 2149–2155. [CrossRef] [PubMed]

96. Ma, Y.; Wang, Y.; Luo, Y.; Duan, H.; Li, D.; Xu, H.; Fodjo, E.K. Rapid and sensitive on-site detection of pesticide residues in fruits and vegetables using screen-printed paper-based SERS swabs. *Anal. Methods* **2018**, *10*, 4655–4664. [CrossRef]

97. Perumal, J.; Wang, Y.; Attia, A.B.E.; Dinish, U.S.; Olivo, M. Towards a point-of-care SERS sensor for biomedical and agri-food analysis applications: A review of recent advancements. *Nanoscale* **2021**, *13*, 553–580. [CrossRef] [PubMed]

biosensors

MDPI

Article

A Novel Enzyme-Based SPR Strategy for Detection of the Antimicrobial Agent Chlorophene

Gabriela Elizabeth Quintanilla-Villanueva [1,2], Donato Luna-Moreno [3,*], Edgar Allan Blanco-Gámez [1,2], José Manuel Rodríguez-Delgado [4], Juan Francisco Villarreal-Chiu [1,2] and Melissa Marlene Rodríguez-Delgado [1,2,*]

[1] Laboratorio de Biotecnología, Facultad de Ciencias Químicas, Universidad Autónoma de Nuevo León, Av. Universidad S/N Ciudad Universitaria, San Nicolás de los Garza C.P. 66455, Nuevo León, Mexico; gabriela.quintanillavl@uanl.edu.mx (G.E.Q.-V.); edgar.blancogmz@uanl.edu.mx (E.A.B.-G.); juan.villarrealch@uanl.edu.mx (J.F.V.-C.)

[2] Centro de Investigación en Biotecnología y Nanotecnología (CIByN), Facultad de Ciencias Químicas, Universidad Autónoma de Nuevo León. Parque de Investigación e Innovación Tecnológica, Km. 10 Autopista al Aeropuerto Internacional Mariano Escobedo, Apodaca C.P. 66629, Nuevo León, Mexico

[3] Centro de Investigaciones en Óptica AC, Div. de Fotónica, Loma del Bosque 115, Col. Lomas del Campestre, León C.P. 37150, Guanajuato, Mexico

[4] Tecnológico de Monterrey, School of Engineering and Sciences, Av. Eugenio Garza Sada Sur No. 2501, Col. Tecnológico, Monterrey C.P. 64849, Nuevo León, Mexico; jmrd@tec.mx

[*] Correspondence: dluna@cio.mx (D.L.-M.); melissa.rodriguezdl@uanl.edu.mx (M.M.R.-D.); Tel.: +52-47-7441-4200 (ext. 200) (D.L.-M.); +52-81-8329-4000 (ext. 6361) (M.M.R.-D.)

Abstract: Chlorophene is an important antimicrobial agent present in disinfectant products which has been related to health and environmental effects, and its detection has been limited to chromatographic techniques. Thus, there is a lack of research that attempts to develop new analytical tools, such as biosensors, that address the detection of this emerging pollutant. Therefore, a new biosensor for the direct detection of chlorophene in real water is presented, based on surface plasmon resonance (SPR) and using a laccase enzyme as a recognition element. The biosensor chip was obtained by covalent immobilization of the laccase on a gold-coated surface through carbodiimide esters. The analytical parameters accomplished resulted in a limit of detection and quantification of 0.33 mg/L and 1.10 mg/L, respectively, fulfilling the concentrations that have already been detected in environmental samples. During the natural river's measurements, no significant matrix effects were observed, obtaining a recovery percentage of 109.21% ± 7.08, which suggested that the method was suitable for the fast and straightforward analysis of this contaminant. Finally, the SPR measurements were validated with an HPLC method, which demonstrated no significant difference in terms of precision and accuracy, leading to the conclusion that the biosensor reflects its potential as an alternative analytical tool for the monitoring of chlorophene in aquatic environments.

Keywords: SPR biosensor; enzyme; laccase; chlorophene; emerging pollutant; water sample

Citation: Quintanilla-Villanueva, G.E.; Luna-Moreno, D.; Blanco-Gámez, E.A.; Rodríguez-Delgado, J.M.; Villarreal-Chiu, J.F.; Rodríguez-Delgado, M.M. A Novel Enzyme-Based SPR Strategy for Detection of the Antimicrobial Agent Chlorophene. *Biosensors* **2021**, *11*, 43. https://doi.org/10.3390/bios11020043

Received: 14 January 2021
Accepted: 6 February 2021
Published: 9 February 2021

1. Introduction

Emerging pollutants are persistent chemicals in the environment, classified as pharmaceutical compounds or their metabolites (human and veterinary drugs). These include personal care products (e.g., disinfectants, fragrances, insect repellents, cosmetics and sunscreens) and endocrine disrupting compounds (e.g., bisphenol A, triclosan and pesticides) [1]. In particular, halogenated phenolic compounds comprise the vast majority of the active ingredients employed in the manufacture of personal care products [2]. In this sense, chlorophene (4-chloro-2-(phenylmethyl)phenol) is an antimicrobial agent widely applied in disinfectants for cleaning activities and for farming, industrial and household environments [3,4], as well as preservatives in cosmetics and wood [5]. According to the Environmental Protection Agency (EPA), chlorophene has been included in the list of priority toxic pollutants [6,7]. It has been related to mutagenic effects in mammals [5], fertility alterations and kidney damage through prolonged exposure [4]. The occurrence of chlorophene (CP) has been reported in water [7] and soil [8]. For example, concentrations up to 0.13 mg/L of CP were detected in a backwater stream in Kerala (India) [7]. Meanwhile, 50 mg/L was quantified in activated sludge sewage, and 10 μg/L was quantified in treatment plant effluent [9].

These micropollutants enter the environment through anthropogenic pathways [7]. A trace amount results in ecological risks, such as biomagnification along the food chain due to accumulation in organisms by hydrophobic properties [3]. Such is the case of CP's occurrence in male bream bile from the Dommel river (7 μg/mL) [10]. The presence of this emerging pollutant has been commonly detected by high performance liquid chromatography mass spectrometry (HPLC-MS) [4] and gas chromatography mass spectrometry (GC-MS) [3], powerful analytical methods for detecting and quantifying trace amounts of compounds. For instance, Rayaroth et al. (2015) [7] identified the presence of chlorophene in a backwater stream using the liquid chromatography quadrupole time of flight MS (LC-QTOF-MS), with a C18 column set at 35 °C and a gradient elution of acetonitrile:formic acid in water (0.1%). Meanwhile, Chen et al. (2018) [11] established the quantification of CP using an HPLC instrument with a UV absorbance detector, an SB-C18 column and a mixture of formic acid as a mobile phase and methanol. On the other hand, the use of the GC-MS technique has also been reported, employing a 5% phenyl methyl siloxane capillary column, splitless injection at 250 °C, an oven temperature from 70 to 280 °C (10 °C/min) and helium as the carrier gas [3]. It is worth highlighting that in prior chromatography—mass spectrometry analysis, sample pretreatment needed to be performed, commonly a purified process by solid-phase extraction (SPE) using cartridges [11] or solvent extraction followed by a concentration step [3]. Consequently, the time-consuming sample preparation and lab environments' restrictions remain significant drawbacks that limit chromatographic techniques. Thus, there is an increasing interest in developing new analytical tools that provide fast, sensitive, and in situ measurements, such as biosensor systems. In this sense, the surface plasmon resonance (SPR) technique has had significant relevance in the environmental field. For example, it was employed in the detection of endocrine disruptors (estrogen [12] and bisphenol A [13]), organophosphate pesticides like chlorpyrifos [14] and industrial pollutants such as polychlorinated biphenyls [15]. Nevertheless, no attempt to use biosensors to detect CP has been explored.

On the other hand, diverse treatment processes have been applied to remove CP, such as MnO_2 oxidation, persulfate treatment and ozonation [11]. However, the use of laccase enzymes in removing chlorophene and dichlorophene is worth noticing [16]. Laccases are phenoloxidases produced in extracellular form by a diverse variety of organisms, from higher plants to fungi [17,18] and bacteria [19]. These enzymes catalyze the oxidation of organic compounds by the concomitant reduction of oxygen [20]. In particular, the removal of CP by laccase catalysis was demonstrated by Shi et al. (2016), suggesting a direct polymerization as the principal mechanism for elimination [16].

Therefore, this work establishes the immobilization of laccase enzymes for their use as a receptor in the detection of chlorophene using an SPR technique. The use of enzymes as recognition elements is very uncommon in SPR techniques [21,22], considering that most applications rely on antigen—antibody interactions [23], aptamer recognition [24,25] and nucleic acid hybridization [26]. In particular, studies of laccase as a bioreceptor in SPR are very scarce [27]. The proposed enzyme-based SPR biosensor's analytical parameters, such as the limit of detection, sensitivity and working range, were studied. Finally, fortified real water samples were analyzed by the SPR biosensor, and the results obtained were compared in terms of accuracy and precision against a well-known HPLC method.

2. Materials and Methods

2.1. Reagents

All the chemical compounds employed in the enzyme immobilization and the biosensing process, such as 16-mercaptohexadecanoic acid (MHDA), 11-mercaptoundecanol

(MUD), ethanolamine hydrochloride, N-hydroxysuccinimide (NHS) and 1-ethyl-3-(3-dimethylamino-propyl) carbodiimide hydrochloride (EDC)) were purchased from Sigma-Aldrich (St. Louis, MO, USA). The laccase enzymes (*Rhus vernicifera*) and salts employed in buffers were purchased from Sigma-Aldrich (St. Louis, MO, USA).

HPLC analysis was performed by employing a Zorbax ODS C18, 25 cm × 4.6 mm, 5 µm column, which was purchased from SUPELCO Analytical (St. Louis, MO, USA) with an HPLC system model YL9100 (Younglin Instrument Co., Ltd., Gyeonggi-do, Korea). The HPLC-grade acetonitrile and water were purchased from Merck (Darmstadt, Germany).

The standards of chlorophene (analytical grade, Sigma-Aldrich, Mexico City, Mexico) were prepared as stock solutions (10 mg/mL) in ethanol:water (90:10, %v/v) 96 and completed at 10 mL with ultrapure water. From these stock solutions, working solutions were prepared by serial dilution in a water:phosphate-buffered saline solution with a pH of 7.3 (90:10, %v/v) in a concentration range from 0 to 10 mg/L.

Real water samples were obtained from a river and filtered with Whatman grade 40 filter as the only pretreatment. Then, one level of fortification was prepared by spiking the river samples with chlorophene at a concentration of 3 mg/L, followed by its analysis by HPLC and the SPR technique.

2.2. Sample Collection and Characterization

Water samples were collected from a river located in León city, Guanajuato-México (21°09′54.0″ N, 101°43′30.6″ W). Sample collection was performed following the Mexican standards established in NOM-230-SSA1-2002 [28]. Briefly, at the sampling site, water samples were collected in polyethylene bottles (pre-rinsed with distillate water). The temperature and pH were measured in the area with a multiparameter probe (WTW Multi 350i). Analyses of the sulfate, total alkalinity and acidity were performed according to the Mexican standards NMX-AA-036-SCFI-2001 [29], as well as the hardness following the methods of NMX-AA-072-SCFI-2001 [30] and the chlorides according to NMX-AA-073-SCFI-2001 [31]. The total organic carbon (COT) and heavy metals were measured by a Shimadzu analyzer by catalytic oxidation of combustion (TOC-L, Shimadzu, Kyoto, Japan) and atomic absorption (Thermo Jarrell Ash Scan1, Franklin, MA, USA), respectively, under NMX-AA-051-SCFI-2016 [32].

2.3. Cr/Au Thin Film Deposition

Homogeneous thin films of Cr/Au were deposited on thin glass substrates by electron gun evaporation, following the method described by Luna-Moreno [33]. Briefly, the chromium layer was evaporated up to a 3 nm thickness. Then, a gold film of 50 nm was deposited by thermal evaporation at a rate of 5 Å/s and 8×10^{-6} mbar. The thickness of the thin films was evaluated by employing a quartz crystal microbalance thickness monitor (Leybold Inficon XTC/2 Depositions controllers).

2.4. SPR Instrumentation

The SPR setup was a homemade platform described previously by Sánchez-Alvarez et al. (2018) [34], based on a Kretschmann configuration and comprising two stacked rotation plates, configured for synchronized movement according to a θ-2θ system by a stepper motor. The measuring cell in the SPR system consisted of a sandwich configuration integrated by a Teflon cell, a gold thin film chip and a hemicylindrical BK7 glass prism. The substrate's glass surface was optically coupled to the prism using an oil matching index (n = 1.51). Meanwhile, the chip's gold-coated surface was facing against the flow of the Teflon cell, which had an inlet and outlet that allowed the solutions to come in contact with the gold through its inner channel (Figure 1). Our design allowed for adjusting to customized measurement chips (different size and thickness), depending on the desired application compared to commercial cells.

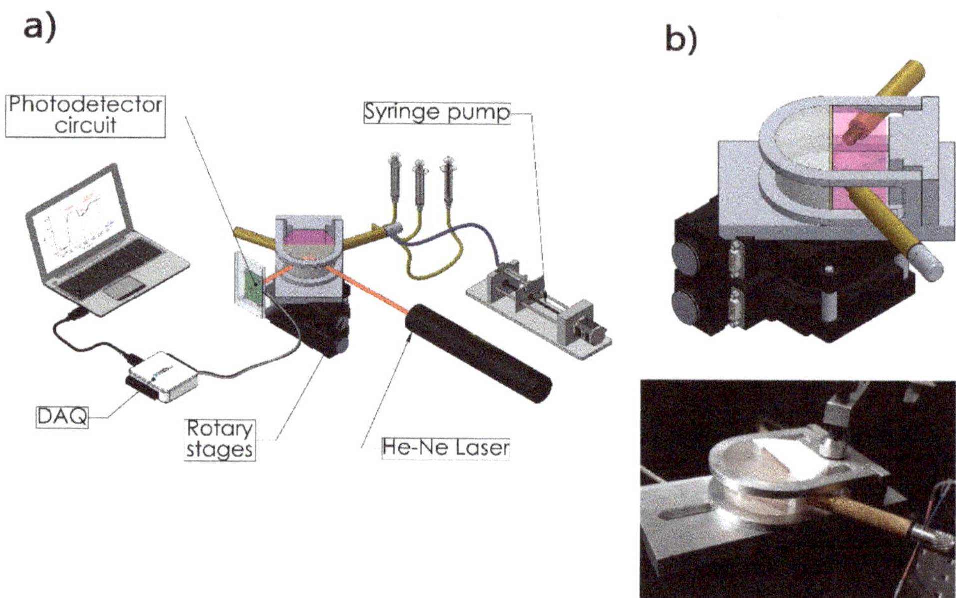

Figure 1. (**a**) Surface plasmon resonance (SPR) setup and (**b**) scheme of the measuring cell.

The chemical solutions continuously flowed through the measuring cell via a syringe pump (Legato 100) at a rate of 30 µL min^{-1}. Furthermore, a photodetector (Hamamatsu, model S1226-8Bk) was used to capture the reflected light of a He-Ne laser (Uniphase mod. 1101P) that passed through the prism.

2.5. Enzymatic Activity

The laccase enzymatic activity was measured through the spectrophotometric UV-Vis assay established by Zhang et al. (2018) [35]. For the assay, 200 µL of the enzyme was added to a reaction mixture (2 mL) containing 10 mM of 2,2′-azino-bis(3-ethylbenzothiazoline-6-sulfonic acid) (ABTS) in a 0.1 M sodium acetate buffer with a pH of 4.5. The reaction occurred at room temperature, and absorbance changes were recorded at 420 nm in a UV-Vis spectrophotometer (Cary 50, Varian Inc., Palo Alto, CA, USA).

Enzyme activity was expressed as a function of the amount of enzyme necessary to produce 1 µM of product per minute (U) and was calculated by the following equation:

$$Activity = \frac{\left[\left(\frac{\Delta Abs}{min}\right) \times V_t\right]}{\varepsilon \times 10^4 \times 1 \times Vm}$$

where ΔAbs is the change in absorbance, V_t is the total volume of the cell, ε is the molar extinction coefficient of ABTS (36,000 M^{-1}cm^{-1}) at 420 nm and Vm is the volume of the laccase sample [35].

2.6. Enzyme Bioreceptor: Chip Functionalization and Laccase Immobilization

Before the immobilization process, a functionalization treatment on the gold substrate (50 nm chips) was performed. Briefly, the gold chips were cleaned by consecutive immersion in acetone and ethanol (30 s in each solvent) and then dried with air. Then, the clean chips were immersed for 12 h at room temperature in a solution of alkanethiols MHDA:MUD (250 µM in ethanol) [36]. The sulfur group from the alkanethiols bound to the gold. Meanwhile, the free carboxylic group on the other end of the chain served as the binding site for the further immobilization of the enzyme.

Once the chip's surface was functionalized with the alkanethiols, the carboxylic groups were activated using the EDC/NHS crosslinkers [37]. A solution of EDC/NHS (EDC 0.2 M/NHS 0.05 M) in an MES buffer (100 mM, 500 mM NaCl, pH 5.0) was flowed on the gold surface, allowing the formation of carbodiimide esters. Then, a solution of 200 U mg-1 of laccase was injected. By creating an amide bond between the amino acids of the enzyme and the activated carboxylic terminal group, the laccase's attachment occurred, ending the immobilization process. Finally, the remaining active esters were deactivated with a solution of ethanolamine (1 M, pH 8.5), preventing unspecific bindings. Once the immobilization process concluded, a washing step was performed, flowing over the sensor surface a phosphate buffer solution (PBS) to remove non-bonded molecules.

2.7. SPR Measurements: Chlorophene Detection

Once the chip was mounted on the SPR setup, the working angle was established at the slope's midpoint, formed in the SPR curve (approaching the critical angle). At this point, greater sensitivity to changes in light intensity, caused by the interaction of the receptor with the analyte, was achieved.

The analysis of chlorophene (CP) was performed by a direct enzyme–substrate assay, where the immobilized laccase enzymes catalyzed the oxidation reaction of CP in the sample. The obtained signals (enzyme–substrate binding) were directly proportional to the concentration of the analyte in the samples, since a shift in the conformation of the enzyme occurred as a result of CP binding in the active site of the laccase, observed as a change in the refractive index measured by the photodetector [38]. The PBS buffer was set as a running solution during the measurement process, and the samples containing CP (0–10 mg/mL) were flowed at 30 μL/min over the sensor surface. The sensor surface was then washed with a PBS buffer injection to remove weakly bound CP from the biofunctionalized chip. Finally, a regeneration solution (NaOH 10 mM) was injected for 20 s to release the bioreceptor and prepare it for a new measurement cycle. All measurements were performed in triplicate. The obtained average SPR signals were plotted as a CP concentration function in the sample. The calibration curve generated was employed to establish the analytical parameters of the biosensor. The sensitivity of the method corresponded to the slope of the curve. The detection limit was evaluated as three times the standard deviation of the baseline, while the limit of quantitation was 10 times the standard deviation. The recovery and reproducibility of the analytical procedure were established using spiked real samples. The determination of the recovery and precision of the SPR was also performed on natural samples, evaluating possible matrix effects.

2.8. HPLC Measurements

The HPLC analysis (YL9300, Thermofisher-USA) was performed according to the method previously described [4]. Briefly, a solution of acetonitrile:water (85:15) was employed as a mobile phase at a flow rate of 1 mL/min, using a Zorbax ODS C18, 25 cm × 4.6 mm (5 μm particle size) column and a UV detector at 290 nm [7]. The calibration standards of the CP stock were prepared in ethanol from a stock solution of 100 mg/L (working range from 0–10 mg/L). The recovery and reproducibility of the analytical procedure were established using spiked real samples.

3. Results and Discussion

3.1. Chip Functionalization and Laccase Immobilization

The immobilization process of the enzymes was initiated with the functionalization of a gold-coated chip. After 12 h of incubation, the binding alkanethiols were activated through the EDC/NHS cross-linkers, forming an amide bond that attached to the laccase enzymes. Figure 2a shows the critical angle displacement of the SPR curve of the alkanethiol-coated chip (blue line) in comparison with the SPR curve after laccase immobilization occurred (red line), showing a shift of 3.2 degrees (from 69.2° to 72.4°) in the resonance angle.

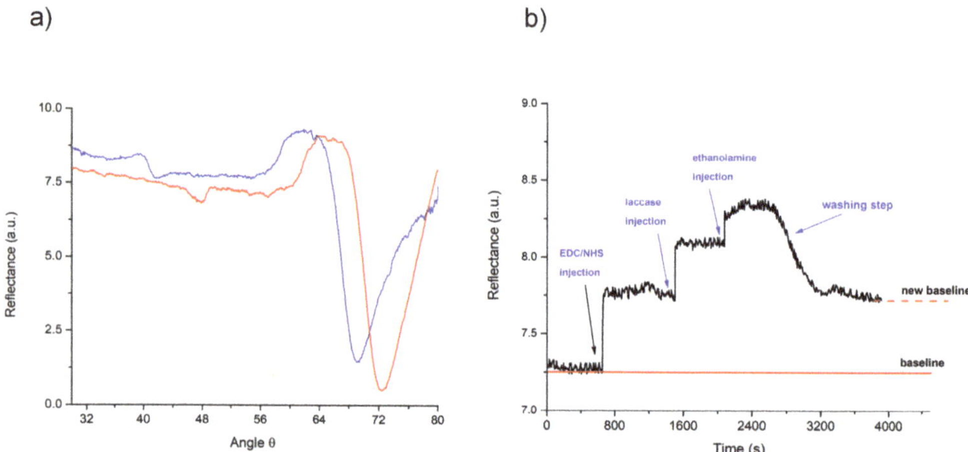

Figure 2. (a) Reflectance spectra obtained by the angular sweep of a sensor chip with alkanethiols (blue line) and after immobilization of the laccase (red line). (b) SPR sensorgram of laccase immobilization.

The shift was attributed to the increase of the mass density due to the bound enzymes on the surface. In this sense, several studies have quantified the immobilization yield on a surface through a conversion factor of 1 ng/mm^2 of biomolecules or protein. The conversion factor was related to a change of 0.1° in the SPR angle (1000 refractive units) [39–41]. Therefore, the angle displacement obtained in this study would represent a density of 32 ng/mm^2 of enzyme onto the gold-coated chip.

The immobilization process was also monitored in real time at a fixed angle of 66.8° (highest sensitivity from the SPR curve slope), obtained by the angular sweep of the immobilized SPR chip (Figure 2a). The sensorgram obtained from the immobilization process in-flow is observed in Figure 2b. The increase of the SPR baseline signal was notable after the subsequent addition of the EDC/NHS crosslinkers, the laccase enzyme and the ethanolamine. However, after the washing step, the signal decreased, suggesting the removal of those molecules weakly bonded on the surface. Once the washing process concluded, it was noticeable that the SPR signal was higher than the initial baseline (prior immobilization), inferring the successful linkage of the molecules that remained on the surface. These results agree with the ones obtained by the angular swept measurements (Figure 2a).

3.2. SPR Measurements: Chlorophene Detection

A direct enzyme–substrate assay was performed to detect chlorophene ranging from 0–10 mg/mL, using an SPR gold-coated chip immobilized with laccase enzymes. In this biosensor, the immobilized laccase catalyzed the oxidation reaction of CP in the sample, observed as a change in the refractive index (SPR signal) due to a shift in the conformation of the enzyme as a result of the concentration of analyte binding to its active site [38]. The active site of the laccase enzymes comprised four copper atoms (type I, type II and two type III copper atoms) [42]. The enzyme's catalytic mechanism involved the substrate oxidation in the type I copper site, followed by an internal electron transfer from the reduced type I atoms to the type II and type III trinuclear cluster, where the reduction of dioxygen to water occurred [42].

According to Enguita et al. (2004) [43], apolar groups in chemical structures are attached to a hydrophobic binding site in laccase, which is located in proximity to the type I Cu site of the enzyme through the His497 residue (one of the type I copper ligands). In this sense, the phenyl group of the chlorophene molecule might present a close approach toward the aromatic ring of the His497 residue in laccase, favoring the electron transfer from CP (oxidation process) to the type I copper and subsequent internal transfer to the trinuclear cluster [43]. Jabbari et al. (2017) [27] reported this electron transfer mechanism during the study to detect the bromocriptine drug by the SPR technique using laccase from Bacillus sp. HR03 [27]. The SPR signal value at the plateau (saturation in the binding event) obtained from the sensorgrams was plotted as a function of the CP concentrations to generate a calibration curve (Figure 3).

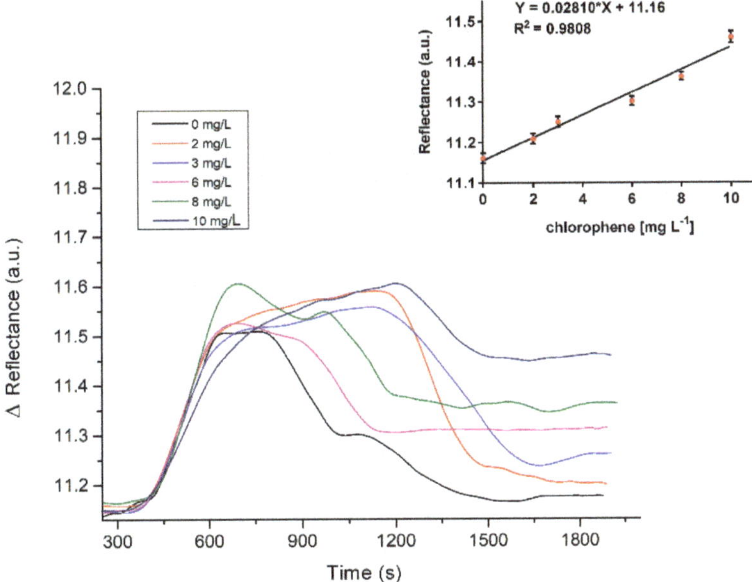

Figure 3. SPR sensorgrams for chlorophene (CP) detection at different concentrations and calibration curves in PBS ($n = 3$).

The analytical parameters obtained in this study are summarized in Table 1. The results obtained meet the concentration detected in a backwater stream in Kerala (India) [7] and in activated sludge sewage [9] by HPLC-MS and GC-MS. Additionally, the immobilized enzyme withstood 35 regeneration cycles by using 0.1 M NaOH before any significant loss of recognition capacity was observed. Currently, there is a lack of biosensors that address the detection of CP, since the analytical tools have been mainly limited to chromatographic techniques. Thus, this research represents the first approach to the fabrication of a robust SPR platform for the routine monitoring of chlorophene in water.

Table 1. Analytical parameters of SPR-based biosensors for chlorophene detection.

Bioreceptor	LOD (mg mL⁻¹)	LOQ (mg mL⁻¹)	Sensitivity (Reflectance/mg mL⁻¹)	Dynamic Range (mg mL⁻¹)
Laccase enzyme	0.33 ± 0.01	1.1 ± 0.01	0.0281 ± 0.0001	0–10

3.3. Evaluation of SPR Performance with River Water: Study of Matrix Effects

Certain components in the real sample matrix could lead to false positives or unspecific bindings that interfere with bioreceptor recognition [44]. Thus, this is a critical issue to determine the performance of a method during the analysis of real samples. Possible matrix effects due to river water composition need to be evaluated. Thus, a natural water sample was injected as a control. Apparently, no significant SPR signal was shown in the preliminary sensitivity assay result with the river water, suggesting that no significant enzyme–substrate reaction occurred. However, it is essential to perform a selectivity assay in the presence of related compounds (interferences) to confirm the method's feasibility for the selective detection of chlorophene. Then, spiked river samples with 3 mg/mL of CP (triplicate) were measured, obtaining nearly identical SPR responses among them under these conditions (see Figure 4).

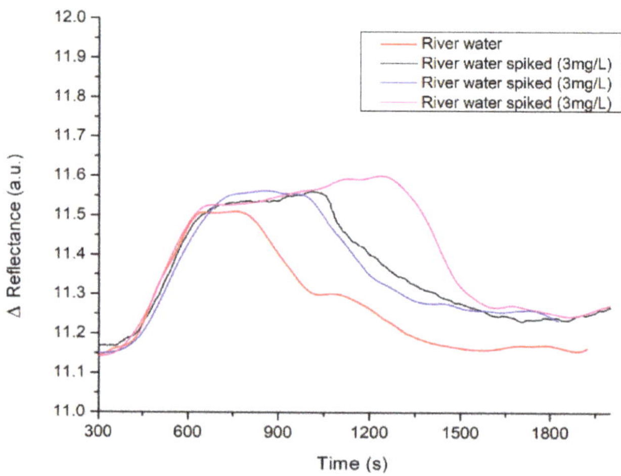

Figure 4. Evaluation of nonspecific signals due to matrix effects from the river water.

The characterization of the river samples can be observed in Table 2, where it is worth noticing that the concentration of organic matter did not represent a considerable effect on the recovery percentage analysis, showing a recovery percentage of 109.21% under the method conditions (see Table 3). No significant differences were found between the theoretical concentration and the one obtained experimentally ($p = 0.05$, $n = 3$).

Table 2. Components and significant ion concentrations (mg/L) in the river water samples.

pH	Total Hardness	Calcium-Based Hardness	Magnesium-Based Hardness	Chloride	Sulfates	Total Organic Carbon	Inorganic Carbon	As	Cu	Pb
8.05 ± 0.07	220.7 ± 6.4	151.3 ± 10.3	69.5 ± 3.9	10.99 ± 0.49	32.11 ± 1.85	23.66	60.75	<0.003	<0.1	<0.2

Table 3. Determination of spiked river samples by the SPR and HPLC methods ($n = 3$).

Fortification Level (mg mL^{-1})	SPR Method		HPLC Method	
3	Mean (mg mL^{-1}) 3.28 ± 0.27	Recovery (%) 109.21 ± 7.08	Mean (mg mL^{-1}) 3.04 ± 0.01	Recovery (%) 101.33 ± 3.55

On the other hand, the laccase enzymes were stable under the concentration of dissolved chlorides and the alkaline pH, since those are conditions that tend to affect the enzymatic activity of these enzymes [45,46].

3.4. Comparison of SPR Protocol with the HPLC Method

Samples spiked in a range from 0 to 30 mg/L were analyzed by HPLC. The linear regression analysis provided a correlation of 0.9995, a limit of detection (LOD) of 0.07 mg/L and a limit of quantification (LOQ) of 0.22 mg/L, with an operating range of 0–30 mg/L.

Then, the river samples spiked with chlorophene at concentrations of 3 mg/L were analyzed. A Student's *t*-test with 95% confidence was evaluated, and no significant discrepancies were found when comparing the SPR and HPLC methods, indicating excellent agreement between the techniques (Table 3).

4. Conclusions

This work established the use of a homemade SPR biosensor based on using laccase enzymes as a bioreceptor for the real-time detection of the hazardous antimicrobial chlorophene in real waters. To the best of our knowledge, this study is the first attempt to develop a biosensor to detect chlorophene. The analytical parameters accomplished by the SPR biosensor fulfilled the concentrations that have already been detected in natural water samples. The biosensor method resulted in a limit of detection and quantification of 0.33 mg/L and 1.10 mg/L, respectively. Although no apparent matrix effects were detected in the analytical response of the SPR measurements of the river samples, it is essential to perform a selectivity assay to confirm the method's feasibility in the selective detection of chlorophene. Furthermore, the method's reliability could be improved by analyzing certified samples to complement fortified natural water results. Finally, the comparison of SPR measurements with an HPLC conventional method demonstrated no significant difference in precision and accuracy. These results show a considerable advantage, due to its lack of a pretreatment process, which is required by traditional techniques, suggesting a suitable and straightforward analysis of this contaminant in natural water.

Author Contributions: Conceptualization, M.M.R.-D. and D.L.-M.; methodology, G.E.Q.-V., D.L.-M., J.M.R.-D. and E.A.B.-G.; validation, G.E.Q.-V. and M.M.R.-D.; formal analysis, G.E.Q.-V., J.F.V.-C. and M.M.R.-D.; investigation, G.E.Q.-V., D.L.-M., J.M.R.-D. and E.A.B.-G.; writing—original draft preparation, G.E.Q.-V. and M.M.R.-D.; writing—review and editing, D.L.-M., J.F.V.-C. and M.M.R.-D. All authors have read and agreed to the published version of the manuscript.

Funding: This research was funded by UANL's Programa de Apoyo a la Investigación Científica y Tecnológica (PAICYT), grant number CE866-19.

Acknowledgments: The authors would like to thank the Consejo Nacional de Ciencia y Tecnología (Conacyt) for Gabriela Quintanilla-Villanueva scholarship #740156.

Conflicts of Interest: The authors declare no conflict of interest, personal, financial, or otherwise, with the manuscript's material.

References

1. Félix-Cañedo, T.E.; Durán-Álvarez, J.C.; Jiménez-Cisneros, B. The occurrence and distribution of a group of organic micropollutants in Mexico City's water sources. *Sci. Total Environ.* **2013**, *454*, 109–118. [CrossRef] [PubMed]
2. Daughton, C.G.; Terne, A. Pharmaceuticals and Personal Care Products in the Environment: Agents of Subtle Change? *Environ. Health Perspect.* **1999**, *107* (Suppl. S6), 907–938. [CrossRef]
3. Zhang, H.; Huang, C.H. Oxidative transformation of triclosan and chlorophene by manganese oxides. *Environ. Sci. Technol.* **2003**, *37*, 2421–2430. [CrossRef] [PubMed]
4. ECHA. *Chlorophene Product-type 2 (Disinfectants and Algaecides not Intended for Direct Application to Humans or Animals): Assesment Report*; ECHA: Helsinki, Finland, 2017.
5. Yamarik, T. Safety assessment of dichlorophene and chlorophene. *Int. J. Toxicol.* **2004**, *23*, 1–27.
6. US EPA. *Summary Document Registration Review: Initial Docket, Chemical Safety and Pollution Prevention for o-benzyl-p-chlorophenol (7510P)*; United States Environmental Protection Agency: Washington, DC, USA, 2011. Available online: http://www.regulations.gov/ (accessed on 18 March 2020).
7. Rayaroth, M.; Nejumal, K.; Subha, S.; Usha, A.; Charuvila, A. Identification of chlorophene in a backwater stream in Kerala (India) and its sonochemical degradation studies. *CLEAN Soil Air Water* **2015**, *43*, 1338–1343. [CrossRef]
8. Bolobajev, J.; Bilgin, N.; Öncü, M.; Viisimaa, M.; Trapido, M.; Goi, A.; Balcıoğlu, I. Column experiment on activation aids and biosurfactant application to the persulphate treatment of chlorophene-contaminated soil. *Environ. Technol.* **2015**, *36*, 348–357. [CrossRef]
9. Sirés, I.; Garrido, J.A.; Rodriguez, R.M.; Brillas, E.; Oturan, N.; Oturan, M.A. Catalytic behavior of the Fe^{3+}/Fe^{2+} system in the electro-Fenton degradation of the antimicrobial chlorophene. *Appl. Catal. B* **2007**, *72*, 382–394. [CrossRef]

10. Houtman, M.; van Oostveen, C.J.; Brouwer, A.M.; Lamoree, A.; Legler, J.H. Identification of estrogenic compounds in fish bile using bioassay-directed fractionation. *Environ. Sci. Technol.* **2004**, *38*, 6415–6423. [CrossRef] [PubMed]

11. Chen, J.; Wu, N.; Xu, X.; Qu, R.; Li, C.; Pan, X.; Wei, Z.; Wang, Z. Fe(VI)-Mediated Single-Electron Coupling Processes for the Removal of Chlorophene: A Combined Experimental and Computational Study. *Environ. Sci. Technol.* **2018**, *52*, 12592–12601. [CrossRef]

12. Habauzit, D.; Armengaud, J.; Roig, B.; Chopineau, J. Determination of estrogen presence in water by SPR using estrogen receptor dimerization. *Anal. Bioanal. Chem.* **2008**, *390*, 873–883. [CrossRef]

13. Marchesini, G.R.; Meulenberg, E.; Haasnoot, W.; Irth, H. Biosensor immunoassays for the detection of bisphenol A. *Anal. Chim. Acta* **2005**, *528*, 37–45. [CrossRef]

14. Mauriz, E.; Calle, A.; Lechuga, L.M.; Quintana, J.; Montoya, A.; Manclús, J.J. Real-time detection of chlorpyrifos at part per trillion levels in ground, surface and drinking water samples by a portable surface plasmon resonance immunosensor. *Anal. Chim. Acta* **2006**, *561*, 40–47. [CrossRef]

15. Hong, S.; Kang, T.; Oh, S.; Moon, J.; Choi, I.; Choi, K.; Yi, J. Label-free sensitive optical detection of polychlorinated biphenyl (PCB) in an aqueous solution based on surface plasmon resonance measurements. *Sens. Actuators B Chem.* **2008**, *134*, 300–306. [CrossRef]

16. Shi, H.; Peng, J.; Li, J.; Mao, L.; Wang, Z.; Gao, S. Laccase-catalyzed removal of the antimicrobials chlorophene and dichlorophen from water: Reaction kinetics, pathway and toxicity evaluation. *J. Hazard. Mater.* **2016**, *317*, 81–89. [CrossRef]

17. Thurston, C.F. The structure and function of fungal laccases. *Microbiology* **1994**, *140*, 19–26. [CrossRef]

18. Bollag, J.M. Decontaminating soil with enzymes. *Environ. Sci. Technol.* **1992**, *26*, 1876–1881. [CrossRef]

19. Sondhi, S.; Sharma, P.; George, N.; Chauhan, P.; Puri, N.; Gupta, N. An extracellular thermo-alkali-stable laccase from Bacillus tequilensis SN4, with a potential to biobleach softwood pulp. *3 Biotech* **2015**, *5*, 175–185. [CrossRef]

20. Moss, P. *Enzyme Nomenclature*; Academic Press: San Diego, CA, USA, 1992.

21. Myszka, D.G. Analysis of small-molecule interactions using Biacore S51 technology. *Anal. Biochem.* **2004**, *329*, 316–323. [CrossRef] [PubMed]

22. Gestwicki, J.E.; Hsieh, H.V.; Pitner, J.B. Using receptor conformational change to detect low molecular weight analytes by surface plasmon resonance. *Anal. Chem.* **2001**, *73*, 5732–5737. [CrossRef] [PubMed]

23. Daems, D.; Lu, J.; Delport, F.; Mariën, N.; Orbie, L.; Aernouts, B.; Adriaens, I.; Huybrechts, T.; Saeys, W.; Spasic, D.; et al. Competitive inhibition assay for the detection of progesterone in dairy milk using a fiber optic SPR biosensor. *Anal. Chim. Acta* **2017**, *950*, 1–6. [CrossRef]

24. Lou, Z.; Han, H.; Mao, D.; Jiang, Y.; Song, J. Qualitative and quantitative detection of PrPSc based on the controlled release property of magnetic microspheres using surface plasmon resonance (SPR). *Nanomaterials* **2018**, *8*, 107. [CrossRef]

25. Yuan, C.; Lou, Z.; Wang, W.; Yang, L.; Li, Y. Synthesis of Fe3C@ C from Pyrolysis of Fe3O4-Lignin clusters and its application for quick and sensitive detection of PrPSc through a sandwich SPR detection assay. *Int. J. Mol. Sci.* **2019**, *20*, 741. [CrossRef]

26. Su, X.; Wu, Y.-J.; Knoll, W. Comparison of surface plasmon resonance spectroscopy and quartz crystal microbalance techniques for studying DNA assembly and hybridization. *Biosens. Bioelectron.* **2005**, *21*, 719–726. [CrossRef] [PubMed]

27. Jabbari, S.; Dabirmanesh, B.; Arab, S.S.; Amanlou, M.; Daneshjou, S.; Gholami, S.; Khajeh, K. A novel enzyme based SPR-biosensor to detect bromocriptine as an ergoline derivative drug. *Sens. Actuators B Chem.* **2017**, *240*, 519–527. [CrossRef]

28. SSA. NOM-230-SSA1-2002, Enviromental Health. Water for Human Use and Consumption, Sanitary Requirements Must Be Met in Public and Private Supply Systems during Water Management. Sanitary Procedures for Sampling. 2002. Available online: http://www.salud.gob.mx/unidades/cdi/nom/230ssa102.html (accessed on 6 April 2020).

29. SE. NMX-AA-036 SCFI-2001: Determination of Acidity and Alkalinity in Natural Water, Wastewater and Treated Wastewater-Test Method, Diario Oficial de la Federacion 2001. Available online: https://www.gob.mx/cms/uploads/attachment/file/166776/NMX-AA-036-SCFI-2001.pdf (accessed on 16 March 2020).

30. CONAGUA. NMX-AA-072-SCFI-2001: Water Analysis-Determination of Total Hardness in Natural Water, Wastewater and Treated Wastewater-Test Method. 2001. Available online: http://lasa.ciga.unam.mx/monitoreo/images/biblioteca/45NMX-AA-072-SCFI-2001_Dureza.pdf (accessed on 6 April 2020).

31. SCFI. NMX-AA-073-SCFI-2001: Water Analysis- Determination of Total Chloride in Natural Water, Wastewater and Treated Wastewater-Test Method. 2001. Available online: http://www.aniq.org.mx/pqta/pdf/NMX-AA-quimicosgpo2.pdf (accessed on 6 April 2020).

32. DOF. NMX-AA-051-SCFI-2016: Water Analysis -Measurement of Metals by Atomic Absorption in Natural Water, Drinking Water, Wastewater and Treated Wastewater-Test Method. 2016. Available online: http://www.dof.gob.mx/nota_detalle.php?codigo=5464459&fecha=07/12/2016 (accessed on 6 April 2020).

33. Luna-Moreno, D.; Sánchez-álvarez, A.; Islas-Flores, I.; Canto-Canche, B.; Carrillo-Pech, M.; Villarreal-Chiu, J.; Rodríguez-Delgado, M. Early detection of the fungal banana black sigatoka pathogen Pseudocercospora fijiensis by an SPR immunosensor method. *Sensors (Basel)* **2019**, *19*, 465. [CrossRef]

34. Sánchez-Alvarez, A.; Luna-Moreno, D.; Hernández-Morales, J.A.; Zaragoza-Zambrano, J.O.; Castillo-Guerrero, G.H. Control of Stepper Motor Rotary Stages applied to optical sensing technique using LabView. *Opt. Int. J. Light Electron. Opt.* **2018**, *164*, 65–71. [CrossRef]

35. Zhang, Z.; Liu, J.; Fan, J.; Wang, Z.; Li, L. Detection of catechol using an electrochemical biosensor based on engineered Escherichia coli cells that surface-display laccase. *Anal. Chim. Acta* **2018**, *1009*, 65–72. [CrossRef]
36. Leonardo, S.; Toldrà, A.; Rambla-Alegre, M.; Fernández-Tejedor, M.; Andree, K.; Ferreres, L.; Campbell, K.; Elliott, C.; O'Sullivan, C.; Pazos, Y.; et al. Self-assembled monolayer-based immunoassays for okadaic acid detection in seawater as monitoring tools. *Mar. Environ. Res.* **2017**, *133*, 6–14. [CrossRef] [PubMed]
37. Soler, M.; Estevez, M.-C.; Alvarez, M.; Otte, M.A.; Sepulveda, B.; Lechuga, L.M. Direct detection of protein biomarkers in human fluids using site-specific antibody immobilization strategies. *Sensors* **2014**, *14*, 2239–2258. [CrossRef] [PubMed]
38. Sota, H.; Hasegawa, Y.; Iwakura, M. Detection of conformational changes in an immobilized protein using surface plasmon resonance. *Anal. Chem.* **1998**, *70*, 2019–2024. [CrossRef]
39. Yang, C.Y.; Brooks, E.; Li, Y.; Denny, P.; Ho, C.M.; Qi, F.; Shi, W.; Wolinsky, L.; Wu, B.; Wong, D.T.; et al. Detection of picomolar levels of interleukin-8 in human saliva by SPR. *Lab. Chip* **2005**, *5*, 1017–1023. [CrossRef]
40. Song, H.Y.; Zhou, X.; Hobley, J.; Su, X. Comparative study of random and oriented antibody immobilization as measured by dual polarization interferometry and surface plasmon resonance spectroscopy. *Langmuir* **2011**, *28*, 997–1004. [CrossRef]
41. Vashist, S.K.; Dixit, C.K.; MacCraith, B.D.; O'Kennedy, R. Effect of antibody immobilization strategies on the analytical performance of a surface plasmon resonance-based immunoassay. *Analyst* **2011**, *136*, 4431–4436. [CrossRef] [PubMed]
42. Madhavi, V.; Lele, S.S. Laccase: Properties and applications. *Bioresources* **2009**, *4*, 1694–1717.
43. Enguita, F.J.; Marcal, D.; Martins, L.O.; Grenha, R.; Henriques, A.O.; Lindley, P.F.; Carrondo, M.a. Substrate and Dioxygen Binding to the Endospore Coat Laccase from Bacillus subtilis. *J. Biol. Chem.* **2004**, *279*, 23472–23476. [CrossRef] [PubMed]
44. Estevez, M.-C.; Belenguer, J.; Gomez-Montes, S.; Miralles, J.; Escuela, A.M.; Montoya, A.; Lechuga, L.M. Indirect competitive immunoassay for the detection of fungicide Thiabendazole in whole orange samples by Surface Plasmon Resonance. *Analyst* **2012**, *137*, 5659–5665. [CrossRef]
45. Pardo, I.; Rodríguez-Escribano, D.; Aza, P.; de Salas, F.; Martínez, A.; Camarero, S. A highly stable laccase obtained by swapping the second cupredoxin domain. *Sci. Rep.* **2018**, *8*, 1–10. [CrossRef]
46. Muthukumarasamy, N.P.; Jackson, B.; Joseph-Raj, A.; Sevanan, M. Production of Extracellular Laccase from Bacillus subtilis MTCC 2414 Using Agroresidues as a Potential Substrate. *Biochem. Res. Int.* **2015**, *2015*, 1–9. [CrossRef]

biosensors

MDPI

Article

Optimization of High-Density Fe-Au Nano-Arrays for Surface-Enhanced Raman Spectroscopy of Biological Samples

Giovanni Marinaro [1,*], **Maria Laura Coluccio** [2] **and Francesco Gentile** [2]

[1] Institute of Process Engineering and Environmental Technology, Technische Universität Dresden, 01069 Dresden, Germany

[2] Nanotechnology Research Center, Department of Experimental and Clinical Medicine, University of Magna Graecia, 88100 Catanzaro, Italy; coluccio@unicz.it (M.L.C.); francesco.gentile@unicz.it (F.G.)

* Correspondence: giovanni.marinaro@kaust.edu.sa

Abstract: The method of realizing nanostructures using porous alumina templates has attracted interest due to the precise geometry and cheap cost of nanofabrication. In this work, nanoporous alumina membranes were utilized to realize a forest of nanowires, providing a bottom-up nanofabrication method suitable for surface-enhanced Raman spectroscopy (SERS). Gold and iron were electroplated through the straight channels of the membrane. The resulting nanowires are, indeed, made of an active element for plasmonic resonance and SERS as the hexagonal distribution of the nanowires and the extreme high density of the nanowires allows to excite the plasmon and detect the Raman signal. The method to reduce the distance between pores and, consequently, the distance of the nanowires after electrodeposition is optimized here. Indeed, it has been predicted that the light intensity enhancement factor is up to 10^{12} when the gap is small than 10 nm. Measurements of Raman signal of thiol groups drying on the gold nanowires show that the performance of the device is improved. As the thiol group can be linked to proteins, the device has the potential of a biosensor for the detection of a few biomolecules. To assess the performance of the device and demonstrate its ability to analyze biological solutions, we used it as SERS substrates to examine solutions of IgG in low abundance ranges. The results of the test indicate that the sensor can convincingly detect biomolecules in physiologically relevant ranges.

Keywords: plasmonic nanowires; molecular sensing; surface-enhanced Raman spectroscopy; porous alumina

Citation: Marinaro, G.; Coluccio, M.L.; Gentile, F. Optimization of High-Density Fe-Au Nano-Arrays for Surface-Enhanced Raman Spectroscopy of Biological Samples. *Biosensors* **2021**, *11*, 181. https://doi.org/10.3390/bios11060181

Received: 3 May 2021
Accepted: 2 June 2021
Published: 5 June 2021

Publisher's Note: MDPI stays neutral with regard to jurisdictional claims in published maps and institutional affiliations.

1. Introduction

Over the past years, surface-enhanced Raman spectroscopy (SERS) has become a powerful tool allowing non-destructive, highly sensitive studies of molecules, chemicals or biological samples [1–3]. Further improvements of this technique could spur considerable progress in areas such as single-molecule sensing, early cancer detection and in situ analyte detection in microfluidics.

SERS takes advantage of highly packed sensitive nanostructures elements positioned in areas of few squared micrometers. Similar devices, fabricated thanks to recent advances in nanotechnology, enable the ultrasensitive, label-free detection of analytes. This detection can be enhanced, in turn, through integration with microfluidics that allows tight control over the volumes, flows and velocities of the biological liquids under examination. Microfluidics-assisted SERS has found applications in several fields, including biomedical engineering, proteomics, life science, and cellomics [4,5]. Notably, the combination of nanoscale devices and the manipulation of nano-liquids has enabled, among other things, the separation and identification of complex mixtures in very low abundance ranges [6,7].

While the technique achieves high sensitivity and ultra-low detection limit, nevertheless, the characteristics of sensitivity, precision, and selectivity have to be improved—this would make SERS devices suitable for the detection of biomarkers in complex solutions and biological fluids.

SERS devices amplify the Raman signal. Raman spectroscopy is a method that provides structural, chemical and conformational information about biomolecules such as proteins and DNA. In Raman spectroscopy, visible light and infrared radiation interact with a molecule, producing, as a result, a spectrum that describes the energy associated to the vibrational states of that molecule. Raman spectroscopy has a number of features, such as requiring minimal sample preparation and being label-free, non-destructive and non-invasive. The principal limitation of this spectroscopy is the extremely low Raman scattering cross-section (typically about 10^{-24} to 10^{-27} photons per events per molecule [8]) which is insufficient to characterize many biological systems, especially in low concentration ranges. Therefore, it is of primary importance to provide means to increase the signal intensity [9]. Among others, an efficient method for enhancing the Raman signal is to use plasmonic surfaces. Similar surfaces, typically made of a metal nanomaterial, manipulate and enhance the local electromagnetic (EM) field of several orders of magnitude. The EM field enhancement is the result of the collective, resonant oscillation of the electrons on the nano-metal's surface [10]. The EM field increment has, as a consequence, the *surface enhancement* of the *Raman spectroscopy* signal, from which the acronym SERS is derived. The SERS effect provides access to otherwise unattainable information of biological systems, drugs, diluted analytes and biomolecules that are not detectable with conventional techniques of analysis [3].

The theoretical upper bound for SERS enhancement is 10^{12} [11]. Remarkably, since the SERS efficiency shows a very high sensitivity on the *geometry* of the substrate, in recent years, a variety of techniques have been developed to fabricate nanoscale structures with maximum resolution, maximum precision, and minimal tolerances [12–17]. Moreover, the design and fabrication methods for efficient SERS substrates should allow reliability and reproducibility over sufficient large areas to provide, at the same time, enhancement of the EM field, stability over time, and the ability to resist mechanical and environmental vibrations and noise. This leaves a lot of room for improvements in the design and fabrication of the final devices. Due to the topological requirements, the fabrication of SERS substrates involves nanotechnology techniques. Some representative examples are given in the following: Kattumenu et al. exploited nanorod-decorated nanowires to observe the Raman enhancement of thiolic molecules [18]. A super-hydrophobic surface made of micropillars was used to concentrate and detect a few molecules dissolved in a droplet [2]. Optical properties of a hexagonal array of metal nanopillars for plasmonic applications were investigated by Zhang et al. [19]. Menvod et al. functionalized graphene nanosheets with cationic poly (diallyldimethylammonium) (PDDA) and citrate-capped gold nanoparticles (AuNPs) for SERS bio-detection application [20]. Zhang et al. fabricated large-scale Au nanodisk arrays on Si substrate via x-ray interference for the detection of Rhodamine 6G as low as 10^{-5} mM [21]. Gentile et Al. dispersed silver nanoparticles into the pores of a superhydrophobic surfaces to guarantee superior SERS capabilities [1].

The present approach regards the use of plasmonic devices with large surface area and high-density hotspots whose increased detection efficacy is due to the strong plasmonic coupling of the nanowires. The sensitive device area is in the range of centimeters. While the single plasmonic elements can be made from 30 to 300 nm, their coupling distance can be adjusted between 3 nm and 20 nm. All these properties in the same device enable sensitive analysis of biological solutions, statistical significance, reliability, and repeatability. Moreover, the proposed technology is cheap and can be used in future translational biological medicine studies.

In this paper, the method has been used to detect Benzenedithiol molecules which were chemisorbed on the gold nanowire surface and, in another case, immunoglobuline IgG. The performance and the results of the biosensors will be presented in the next sections.

2. Materials and Methods

2.1. Nanowires Fabrication

In order to efficiently and reproducibly fabricate NW-based substrates, the electrochemical deposition of NWs into nanoporous alumina templates was utilized (Figure 1). For a detailed description of the porous alumina fabrication, the protocol in a previous paper [8] was considered. The process involves two steps of anodization of a one-inch aluminum disk which create a thin layer of aluminum oxide with a highly ordered nanopore distribution. The first anodization step yields an alumina layer with poorly organized pores at the top, but high regularity at the bottom.

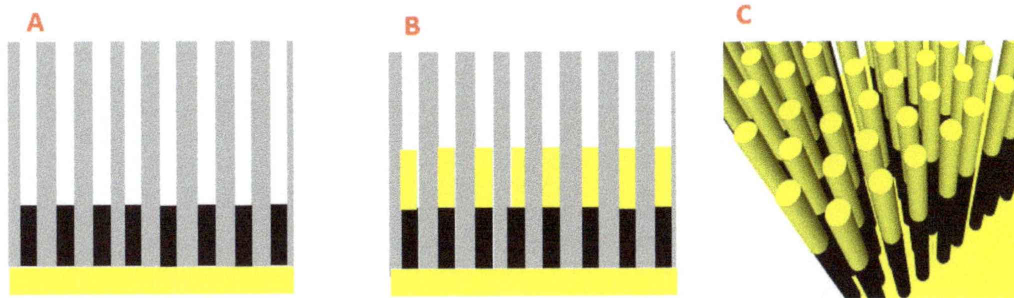

Figure 1. Sketch of Fe-Au electrodeposition through the pores of porous alumina: iron (**A**), addition of gold (**B**), and 3D illustration (**C**).

The pore distribution and size homogeneity are shown in Figure 2. The pores of alumina templates have a diameter of about 60 nm and are distributed in a hexagonal lattice with a constant distance of 40 nm. The electrolyte for the iron segment growth was realized with 6 g of iron sulfate heptahydrate ($FeSO_4 \cdot 7H_2O$, Sigma-Aldrich), 1 g of boric acid (H_3BO_3, Sigma-Aldrich) and 1 g of ascorbic acid ($C_6H_8O_6$, Sigma-Aldrich). Boric acid was added to improve the purity of iron while ascorbic acid was added to adjust the pH to 3.

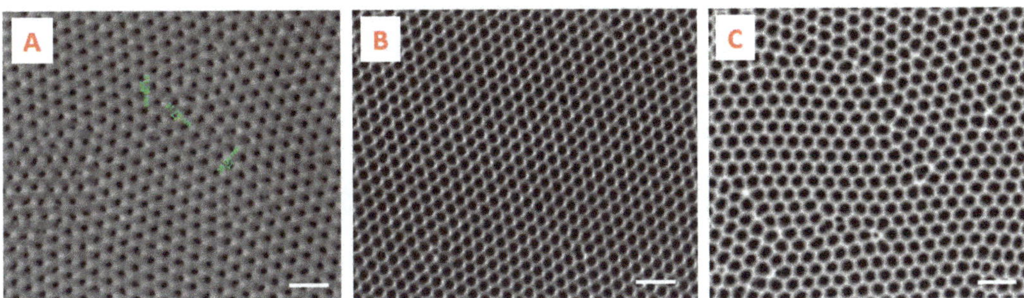

Figure 2. SEM micrographs. Porous alumina membranes with a distance between pores of 60 nm (**A**), 85 nm (**B**) and 95 nm (**C**). The white scale bar is 200 nm.

The electrolyte for the gold segment growth was prepared with 0.1 potassium dicyanoaurate ($KAu(CN)_2$, Sigma-Aldrich, St. Louis, MO, USA) and 4 g of boric acid (H_3BO_3, Sigma-Aldrich). Boric acid was added in order to adjust the pH and work in safe acidic conditions. In order to realize three different sizes of the pores, and consequently, three different aspect ratios of nanowires, the membranes were dipped in 5% H_3PO_4 in water (w/w) for different lengths of time. Porous alumina was then etched away in an acidic chrome solution (CrO_3/H_3PO_4 in water) at 40 °C overnight (Figure S1) in order to obtain free-standing nanowires.

Removal of the alumina reveals highly ordered features on the surface of the Al substrate, which facilitates the growth of ordered pores upon the second anodization step. The pores obtained with the second anodization are straight channels arranged in a hexagonal pattern. The diameter of pores and the distance between them can be tuned by controlling the parameters of anodization, while their length depends on the anodization duration. One of the main challenges in using porous membranes for SERS is to reduce the gap between the active elements of the nanostructure because, based on numerical predictions, the light intensity shows the best enhancement factor (up to 10^{12}) when the gap is smaller than 10 nm.

A way to address the problem to a solution is to gently dip porous alumina in a diluted phosphoric acid solution in order to widen the pores and, consequently, reduce the inter-distance to a few 10s of nanometers. Simulations have already predicted that nanostructures obtained by this method are suitable for SERS devices [8].

After removing the backlayer substrate of alumina with an etching solution, an electric contact was made using a PECVD by depositing a thin layer of 20 nm of ITO (Indium Tin Oxide) on the surface of the porous alumina membrane. The electrolyte for the iron segment growth was realized with 6 g of iron sulfate heptahydrate ($FeSO_4 \cdot 7H2O$, Sigma-Aldrich), 1 g of boric acid (H_3BO_3, Sigma-Aldrich) and 1 g of ascorbic acid ($C_6H_8O_6$, Sigma-Aldrich). Boric acid was added to improve the purity of the iron while ascorbic acid was added to adjust the pH to 3. The electrolyte for the gold segment growth was prepared with 0.1 potassium dicyanoaurate ($KAu(CN)_2$, Sigma-Aldrich) and 4 g of boric acid (H_3BO_3, Sigma-Aldrich). Boric acid was added in order to adjust the pH and work in safe acidic conditions.

2.2. SEM Analysis

SEM analysis of the plasmonic nanowires was conducted with a Zeiss GeminiSEM 500 at Dresden Center for Nanoanalysis (DCN), TU Dresden. The samples were already conductive, so there was no need to sputter gold. The samples were fixed on a stub with a long pin and then mounted on a carousel 9×9 mm sample holder. In order to fix the samples, a small amount of silver paint was applied between the edge of the aluminum disc and the stub. A further copper lever was screwed in order to secure the sample on the stub. Several images of metal nanowires were acquired in high vacuum mode at 5 kV, a magnification of 300,000 and a working distance of about 3 mm with an InLens Detector (ZEISS) for secondary electrons. In order to reduce the noise, a frame integration (N = 14) was performed. With this setup, every frame was scanned and averaged 14 times.

2.3. Light Transmission Measurements

A Nikon Eclipse Ni with an integrated Thor spectrometer (a ray diagram is shown in Figure 3) was used to measure the wavelength band of white light in transmission.

To do so, a variation in the fabrication step was realized: an ITO nanofilm was sputtered to the back of the alumina membrane as a conductive layer for the electrodeposition instead of sputtering gold. ITO is a material that is conductive and transparent at the same time. This guaranteed the electrodeposition of nanowires through the pore channels and the ability to perform optical light measurements.

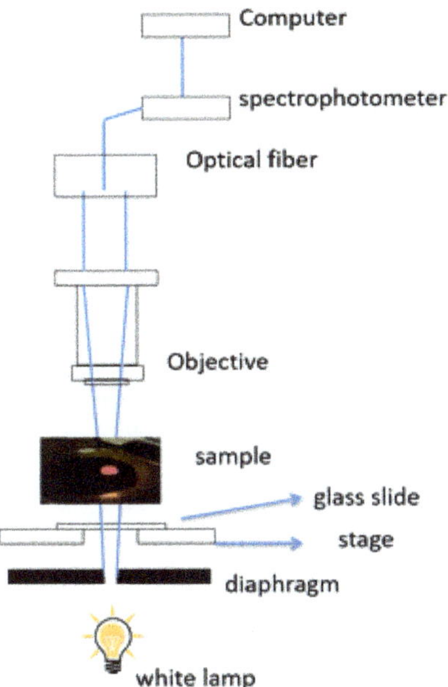

Figure 3. Ray diagram of light transmission setup. The onset shows the plasmonic device illuminated from the backside.

2.4. Raman Spectroscopy

A Renishaw InVia Raman spectrometer with a 1200 line/mm grating for the SERS measurements was used for the measurement of the Raman signals. The samples were excited by 830 nm laser line in backscattering configuration through 100× objective (NA = 0.9) using the respective edge filters to stop the laser lines. The scattering was collected in the range of 200 to 2000 cm^{-1}. The spectra were analyzed with WiRE 3. Benzenedithiol 1,4 molecules were deposited on the substrates by immersion in solution and subsequent rinsing in MilliQ water. A droplet of about 10^{-1} mM of benzedithiol 1,4 in ethanol was gently deposited on the surface of the device with a pipette and allowed to dry for about one hour in order to stabilize the disulphuric links of thiols with the gold nanowires. SERS spectrum of Benzenedithiol 1,4 was excited with a laser power of 2 mW. Further investigations were carried out with an HR 800 Raman spectrometer with a micro-Raman spectral acquisition images. The samples were excited with a 795 nm laser line through a 100× objective (NA = 0.9). The spectra were exported as text files and analyzed with an in-house script written and run in Matlab (R2017b MathWorks).

2.5. SERS Analysis of IgG

A drop of immunoglobulin (IgG) at 0.44 mg/mL was positioned on the substrate and left to dry. The Raman spectra of IgG adsorbed on the substrate were collected by an InVia Raman spectrometer with a 1200 line/mm grating, equipped with a 100× optical microscope objective. Samples were excited by an 830 nm laser line, setting the laser power at 1.6 mW. The Raman signals were recorded on maps of different sizes in a spectral range of 800 to 1800 cm^{-1} and an integration time of 2 s for each point. The map spectra were analyzed with WiRE 3 and elaborated with Wolfram Mathematica (The Wolfram Centre, Oxford, United Kingdom), analyzing each map on the basis of characteristic peak intensities (1250, 1330 and 1450 cm^{-1}). Individual Raman spectra were baseline corrected using a polynomial passing through at least eight points uniformly distributed in the spectral range. Then, spectra were rescaled in the intensity direction using min–max normalization, whereby the minimum value of a spectrum intensity is transformed into a 0, the maximum value is transformed into a 1, and every other value is transformed into a decimal between 0 and 1 [22].

3. Results—SERS Device

3.1. Characterization of Nanowires

The SERS device, realized according to the description in the previous session, was characterized both with SEM during different steps of fabrication and with a spectrometer connected to an optical microscope, as described in the materials and methods session. Figure 2A shows the alumina membrane after the two steps of anodization. Figure 2B shows the template after 100 min of the widening process, which allows 15 nm of pore distance; Figure 2C shows the template after 120 min of the widening process, with a homogeneous pore distance of 5 nm. An ITO nanofilm was sputtered to the back of the alumina membrane as a conductive layer for the electrodeposition. Porous templates were then used to grow metal nanowires through the pores after electrodeposition. The electrodeposition Fe-Au nanowires form a composite material together with the alumina template near the bottom (Figure 4A), and after the removal of the alumina, free-standing Au-Fe nanowires were finally obtained(Figure 4B,C).

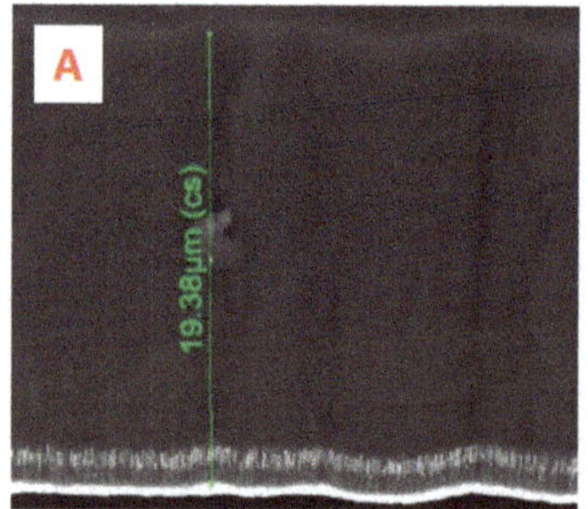
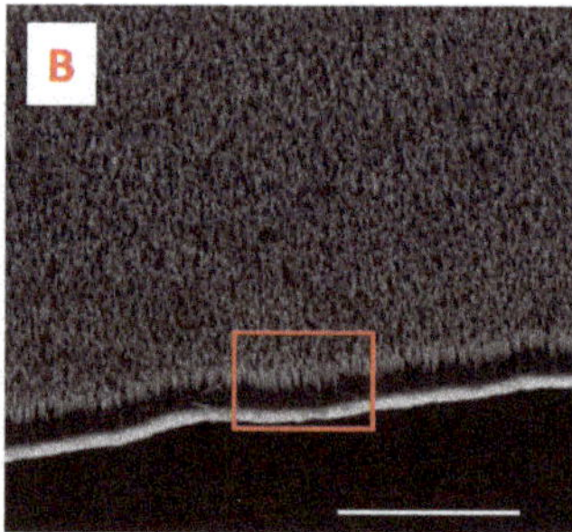
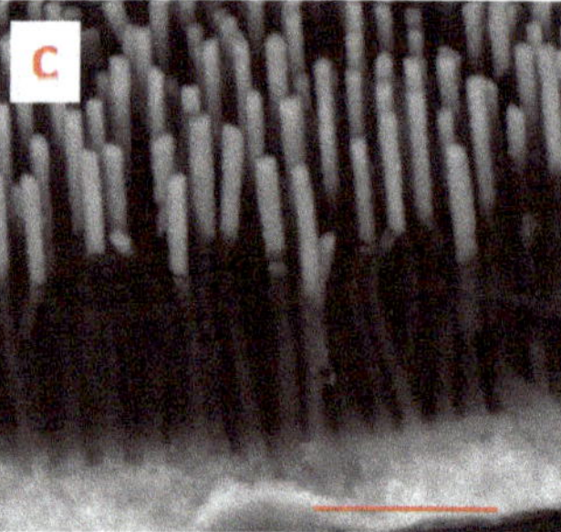

Figure 4. SEM micrographs. (**A**). Porous alumina with electroplated nanowires at the bottom near the metal contact. The thickness of the membrane is roughly 20 μm, as indicated from the green line. (**B**). Nanowires after alumina etch(scale bar 5 μm). (**C**). Magnified region marked as red square in Figure 3B; the scale bar is 500 nm.

Since three different typologies of porous alumina templates based on pore diameter were realized, an increasing plating time was applied for small, medium and large pores in order to fill the different hollow volumes of the nanoporous material and ensure about the same height of nanowires. Figure 5A shows the top of the gold nanorods obtained with a porous alumina template without the widening process. Figure 5B,C shows Au-Fe nanowires with a distance, respectively, of 15 and 5 nm.

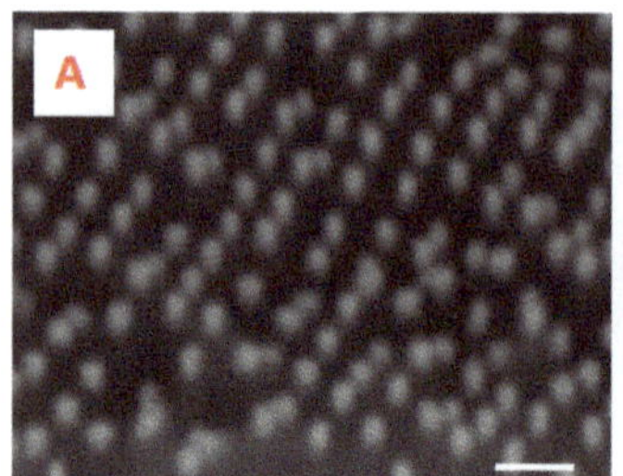
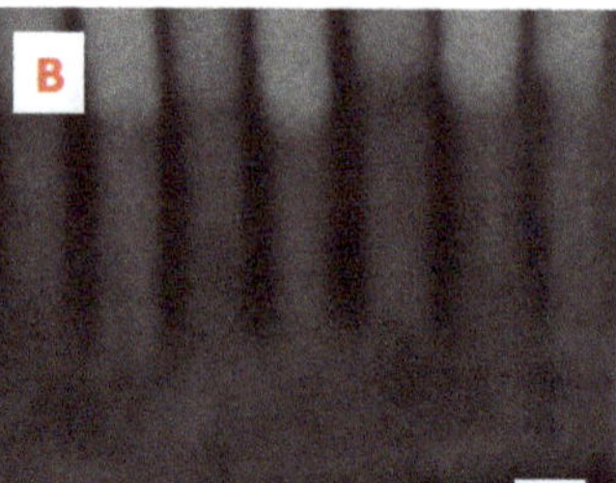

Figure 5. SEM micrographs. Fe-Au nanowires with distance between wires 40 nm (**A**), scale bar = 200 nm, 15 nm (**B**), scale bar = 90 nm and 5 nm (**C**), scale bar = 180 nm.

Transmission spectra of the nanowires was acquired and then convoluted by using a Savitzky–Golay filter. The peaks at 747 nm, 820 nm and 810 nm were transmitted with higher intensity by the nanowires with smaller distance (Figure S2), suggesting that the plasmonic resonance of the gold nanowires is higher when they are excited with an infrared laser line. For this reason, Raman spectroscopy was conducted with a laser line of 833 nm.

3.2. Detection of Benzenedithiol and IgG Solution Dried on the Biosensor

Free-standing Fe-Au nanowires were fabricated using the electrochemical method after two consecutive steps of electrodeposition. The SERS spectrum of Benzenedithiol 1,4 is shown in Figure 6. The performance of the device with 5 nm gap nanowires, as shown in Figure 6, is much better than the other cases when the distance is larger. The estimation of the signal intensity is of the order of 10^4 with respect to the thicker nanowire device. The signal of Benzenedithiol for the small gap spacing was measured at different points by mapping a region of the biosensor. The signal in Figure 6 (blue curve) is extracted from one point of the map. We repeated the spectra acquisition several times at that point and did not observe relevant changes. We mapped the other surfaces (with higher gaps between nanowires) and detected a poor signal.

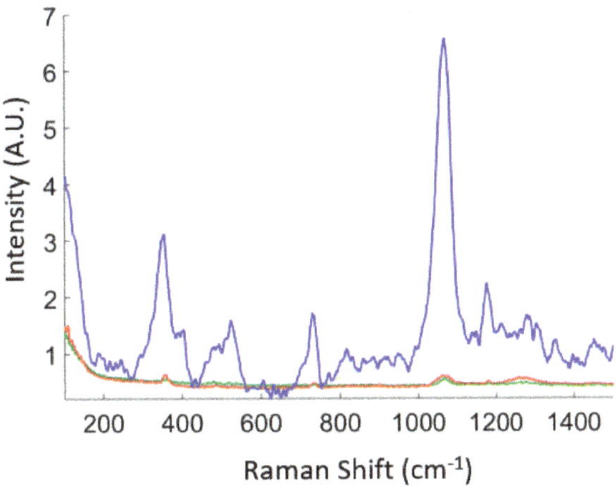

Figure 6. Raman spectra of Benzenedithiol 1,4 chemiosorbed on Au nanowires with large distance (green line), nanowires with middle distance (red), and nanowires with small distance between nanowires (blue line).

As predicted by simulations in a previous work, the SERS signal of Benzenedithiol 1,4 shows a higher intensity due to the fact that the dipoles of gold nanorods on the head of nanowires, realized in this work, have, in the infrared region, a better electric-field enhancement, which was estimated to have a factor of 10^4. In Figure S4 of the Supporting Information, we report the SERS signal coming from Benzenedithiol (BDT) measured by the nanowire sensor device with three different configurations (large, middle, small gap) compared to the Raman spectrum of BDT acquired over a flat non-SERS substrate (flat silicon surface). In the image, all spectra are individually normalized to the maximum peak in the spectral range. Remarkably, the SERS and Raman signal are significantly different, proving the selective adsorption and enhanced Raman activity of BDT over the SERS nanowires device, compared to the spontaneous Raman scattering of the molecule in standard conditions.

To further assess the performance of the device, we used the biosensor to examine a solution of antibodies in low abundance ranges. Upon casting a sample drop containing IgG with an initial concentration of 0.44 mg/mL on the sensor surface, we waited until the complete evaporation of the solvent and analyzed the residual using the Raman setup described in the methods of the paper. The Raman maps reported in Figure 7 were collected over a region of the sensor device partially coated with the sample drop. The maps in Figure 7a report the normalized Raman spectrum intensity measured at 1250 cm^{-1}, 1330 cm^{-1} and 1450 cm^{-1}, respectively. Notably, the signal distribution in the maps matches with very high accuracy with the originating layout of the sample drop on the device (Figure 7b). The signal is high within the contact area of the sample with the sensor surface, then it sharply falls to nearly zero, moving away from the biological sample towards the free sensor surface. The very high correspondence between the expected spatial distribution of the biological sample with the Raman maps indicates that the device and the entire method is effective in performing the analysis of biological solutions. The spatial frequencies of 1250 cm^{-1}, 1330 cm^{-1} and 1450 cm^{-1} that we have used as a reference in the analysis are central lines where the IgG vibrational activity is preferentially expressed. They represent the fingerprint of IgG.

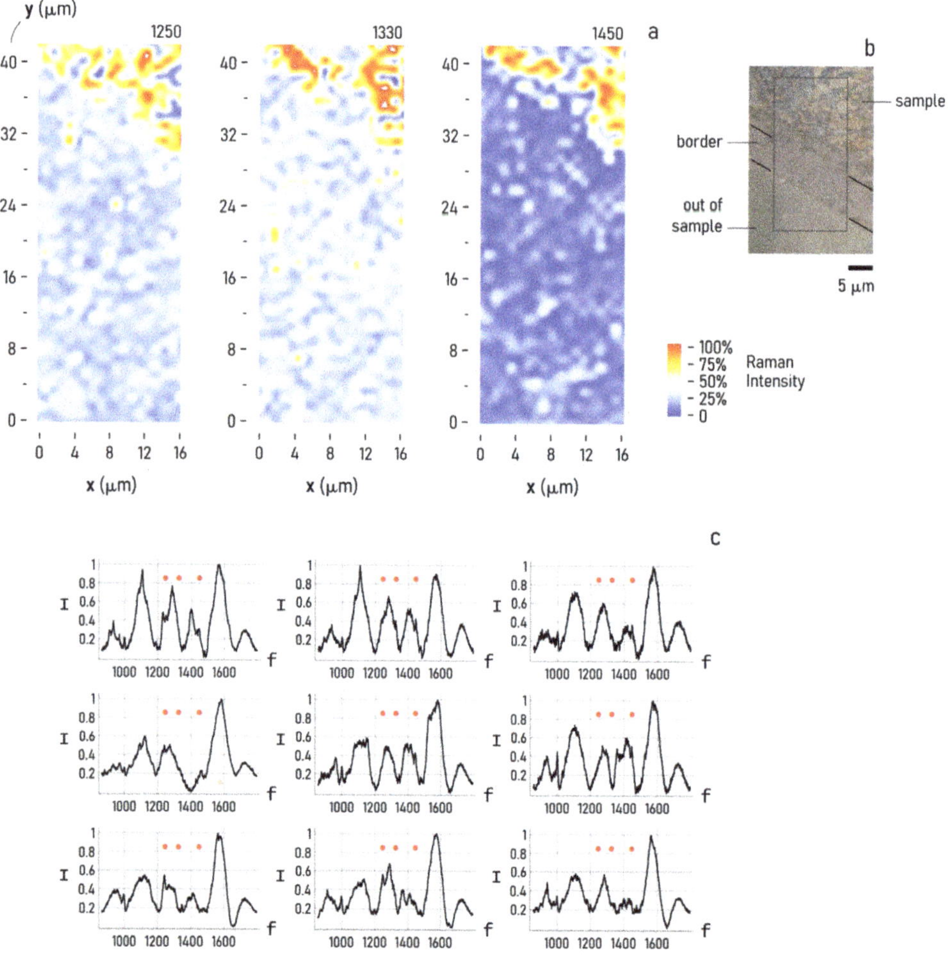

Figure 7. SERS maps of IgG positioned on the sensor device, acquired at 1250, 1330 and 1450 cm^{-1}, respectively (**a**). Optical image of the sample drop after evaporation on the device, and the region of the sample surface interrogated through SERS analysis (**b**). Collection of nine spectra randomly sampled from the SERS maps, where the characteristic peaks of IgG have been highlighted (**c**).

IgG molecules are characterized by a significant percentage (47%) of β-sheet conformation and only 7% of α-helix [23]. The β-sheet secondary structure is identified by the amide I broad band at 1673 cm^{-1} and by the amide III region with a slight band at around 1243 cm^{-1}. In the amide III region, the band at 1336 cm^{-1} is also observable, evidencing the α-helix secondary structure portion [24]. The CH2 deformation (ρCH2) band is observed at around 1450 cm^{-1}, associated with the protein structures. Other minor peaks are related to amino-acidic residues (e.g., phenylalanine at 1004 cm^{-1}) and to the backbone skeletal vC–C vibration bands around 1030 to 1170 cm^{-1} [25].

In Figure 7c, we report a number of Raman spectra extracted from the full-field Raman maps described above. Each of those spectra, randomly sampled from the maps, exhibit characteristic peaks at no less than one of the following frequencies: 1250 cm^{-1}, 1330 cm^{-1} and 1450 cm^{-1}. Remarkably, above roughly 1500 cm^{-1}, the Raman spectra in the grid seem to convey no or little information about the biological sample, perhaps because environmental or instrumental noise obscure the sample emission. Below the 1500 cm^{-1} limit, the frequency content of the signal is consistent with the biological sample being IgGs.

The Raman spectra that we have reported in Figure 7c are representative examples of a larger set of data, all having the property of showing a very high Raman signal at 1250 cm^{-1}, 1330 cm^{-1}, 1450 cm^{-1}, or a combination of these three frequencies. This is illustrated from the Raman maps reported in Figure 7a, where the signal calculated in correspondence with those reference values shows minimal irregularity. To demonstrate repeatability more convincingly, we report in Figure S3 of the Supporting Information section the complete set of Raman spectra acquired over the active area of the sensor device.

4. Discussions

The SERS signal coming from molecules in close proximity to a plasmonic nanomaterial shows a very high sensitivity to the distance of the molecules to the surface and the molecule orientation, which are parameters that are not or are minimally influenced by the operator during the measurement. As a result, for its nature, the SERS analysis of a compound and corresponding Raman spectra show poor reproducibility and repeatability. This limits the use of the technique as a quantitative method of analysis of biological systems. Nevertheless, while not perfectly identical, the Raman spectra that we have shown in Figure 7c all show a pattern similarity. Centered either at 1250 cm^{-1}, 1330 cm^{-1} or 1450 cm^{-1}, the spectra have a peak that is characteristic of IgG, as reflected by the maps reported in part of the same figure, where the Raman intensity calculated for those peaks shows very high uniformity over the sample sensing area.

To understand the clinical implications of the results, it is useful to put the IgG concentration of 0.44 mg/mL that we have used in this work in context. The typical values of IgG normally found in adults fall in the 7 to 15 mg/mL interval, called the reference range [26], which are between 15 and 30 times higher than our study's sample concentration. Thus, even without testing the device with ultra-low concentrated solutions, the results of the work indicate that this biosensor is suitable to detect IgG fluctuations downward or upward relative to the reference range. Upward oscillations (high levels of IgG) may be indicative of pathological states including chronic infection, such as HIV, multiple myeloma, chronic hepatitis, and multiple sclerosis. On the other hand, downward oscillations (low levels of IgG) occur, such as, for example, in macroglobulinemia; in some types of leukemia; and in nephrotic syndrome, a type of kidney damage. Further to this end, in a more sophisticated evolution of the device that will be developed over time, the gold sensor surface will be functionalized with antibodies [27] or aptamers [28] for the selective capture of biomarkers. This sensing device will achieve the recognition of antigens in complex mixtures in very low abundance ranges, combining the characteristics of low cost, high resolution, high sensitivity, and selectivity.

5. Conclusions

In the present work, a method of SERS device nanofabrication using alumina template was optimized for molecular sensing. To do so, alumina templates were fabricated according to two anodization steps. The SERS device, realized according to the description in the previous sessions, was characterized geometrically and optically with SEM during different steps of fabrication and with a spectrometer. Porous templates were used to grow metal nanowires through the pores after electrodeposition. A series of SERS measurements at 830 nm using a forest of nanowires, organized in a hexagonal lattice, with a gap of only 5 nm to detect the Raman signal of chemisorbed benzenedithiol 1,4 shows a large increase in the local intensity with respect to the forest of nanowires with larger distance. As reported in a previous paper [8], the enhancement factor of the Raman signal for this device is of the order of 10^4. As the thiol group can be linked to many biomolecules of interest such as proteins, the device has the potential to be used as a biosensor for the detection of a few biomolecules in different fields such as microfluidics, proteomics and optoelectronics. Further analysis of solutions of biomedical interest—i.e., IgG—in low abundance ranges confirm the ability of the device to detect biomolecules with potential applications in the treatment and diagnosis of diseases.

Supplementary Materials: The following are available online at https://www.mdpi.com/article/10.3390/bios11060181/s1, Figure S1: Etching process of porous alumina. The samples are upside down and floating on chrome solution which is kept at 40 °C. Figure S2: Etching process of porous alumina. The samples are upside down and floating on chrome solution which is kept at 40 °C. Figure S3: Complete set of Raman spectra acquired over the active area of the sensor device. Figure S4: SERS signal coming from Benzenedithiol (BDT) measured by the nanowires sensor device with three different configurations (big, middle, small gap) compared to the Raman spectrum of BDT acquired over a flat non-SERS substrate (flat Silicon surface). In the image, all spectra are individually normalized to the maximum peak in the spectral range.

Author Contributions: G.M. fabricated the SERS devices, acquired the SEM images, prepared the solution of BDT, performed the measurements of BDT sensing using Raman spectroscopy and wrote the original draft, M.L.C. performed the measurements of IgG sensing using Raman spectroscopy and contributed to the analysis of the results. F.G. analyzed and discussed the results. G.M. and F.G. wrote, reviewed and edited the manuscript. All authors have read and agreed to the published version of the manuscript.

Funding: This research received no external funding.

Institutional Review Board Statement: Not applicable.

Informed Consent Statement: Not applicable.

Data Availability Statement: The data that support the findings of this study are available from the corresponding authors upon reasonable request.

Acknowledgments: G.M. would like to thank the Dresden Center for Nanoanalysis of the Technische University Dresden for support provided for the SEM analysis.

Conflicts of Interest: The authors declare no conflict of interest.

References

1. Gentile, F.; Coluccio, M.L.; Accardo, A.; Marinaro, G.; Rondanina, E.; Santoriello, S.; Marras, S.; Das, G.; Tirinato, L.; Perozziello, G.; et al. Tailored Ag nanoparticles/nanoporous superhydrophobic surfaces hybrid devices for the detection of single molecule. *Microelectron. Eng.* **2012**, *97*, 349–352. [CrossRef]
2. De Angelis, F.; Gentile, F.; Mecarini, F.; Das, G.; Moretti, M.; Candeloro, P.; Coluccio, M.L.; Cojoc, G.; Accardo, A.; Liberale, C.; et al. Breaking the diffusion limit with super-hydrophobic delivery of molecules to plasmonic nanofocusing SERS structures. *Nat. Photonics* **2011**, *5*, 682–687. [CrossRef]
3. Di Fabrizio, E.; Schlücker, S.; Wenger, J.; Regmi, R.; Rigneault, H.; Calafiore, G.; West, M.; Cabrini, S.; Fleischer, M.; Van Hulst, N.F.; et al. Roadmap on biosensing and photonics with advanced nano-optical methods. *J. Opt.* **2016**, *18*, 063003. [CrossRef]
4. Marinaro, G.; Riekel, C.; Gentile, F. Size-exclusion particle separation driven by micro-flows in a quasi-spherical droplet: Modelling and experimental results. *Micromachines* **2021**, *12*, 185. [CrossRef] [PubMed]

5. Laurell, T.; Petersson, F.; Nilsson, A. Chip integrated strategies for acoustic separation and manipulation of cells and particles. *Chem. Soc. Rev.* **2007**, *36*, 492–506. [CrossRef]

6. Shen, Y.; Tolic, N.; Masselon, C.; Pasa-Tolic, L.; Camp, D.G. Nanoscale proteomics. *Anal. Bioanal. Chem.* **2004**, *378*, 1037–1045. [CrossRef]

7. Shen, Y.; Zhao, R.; Berger, S.J.; Anderson, G.A.; Rodriguez, N.; Smith, R.D. High-efficiency nanoscale liquid chromatography coupled on-line with mass spectrometry using nanoelectrospray ionization for proteomics. *Anal. Chem.* **2002**, *74*, 4235–4249. [CrossRef]

8. Marinaro, G.; Das, G.; Giugni, A.; Allione, M.; Torre, B.; Candeloro, P.; Kosel, J.; Di Fabrizio, E. Plasmonic nanowires for wide wavelength range molecular sensing. *Materials* **2018**, *11*, 827. [CrossRef]

9. Mikoliunaite, L.; Rodriguez, R.D.; Sheremet, E.; Kolchuzhin, V.; Mehner, J.; Ramanavicius, A.; Zahn, D.R. The substrate matters in the Raman spectroscopy analysis of cells. *Sci. Rep.* **2015**, *5*, 13150. [CrossRef]

10. Stiles, P.L.; Dieringer, J.A.; Shah, N.C.; Van Duyne, R.P. Surface-enhanced Raman spectroscopy. *Annu. Rev. Anal. Chem.* **2008**, *1*, 601–626. [CrossRef]

11. Camden, J.P.; Dieringer, J.A.; Wang, Y.; Masiello, D.J.; Marks, L.D.; Schatz, G.C.; Van Duyne, R.P. Probing the structure of single-molecule surface-enhanced Raman scattering hot spots. *J. Am. Chem. Soc.* **2008**, *130*, 12616–12617. [CrossRef]

12. Moreno Garcia, J.; Khan, M.A.; Ivanov, Y.P.; Lopatin, S.; Holguin Lerma, J.A.; Marinaro, G.; Ooi, B.S.; Idriss, H.; Kosel, J. Growth of ordered iron oxide nanowires for photo-electrochemical water oxidation. *ACS Appl. Energy Mater.* **2019**, *2*, 8473–8480. [CrossRef]

13. Marinaro, G.; Accardo, A.; De Angelis, F.; Dane, T.; Weinhausen, B.; Burghammer, M.; Riekel, C. A superhydrophobic chip based on SU-8 photoresist pillars suspended on a silicon nitride membrane. *Lab. Chip* **2014**, *14*, 3705–3709. [CrossRef] [PubMed]

14. Marinaro, G.; Accardo, A.; De Angelis, F.; Dane, T.; Weinhausen, B.; Burghammer, M.; Riekel, C. Networks of neuroblastoma cells on porous silicon substrates reveal a small world topology. *Integr. Biol.* **2015**, *7*, 184–197. [CrossRef] [PubMed]

15. Chirumamilla, M.; Das, G.; Toma, A.; Gopalakrishnan, A.; Zaccaria, R.P.; Liberale, C.; De Angelis, F.; Di Fabrizio, E. Optimization and characterization of Au cuboid nanostructures as a SERS device for sensing applications. *Microelectron. Eng.* **2012**, *97*, 189–192. [CrossRef]

16. Das, G.; Patra, N.; Gopalakrishnan, A.; Zaccaria, R.P.; Toma, A.; Thorat, S.; Di Fabrizio, E.; Diaspro, A.; Salerno, M. Fabrication of large-area ordered and reproducible nanostructures for SERS biosensor application. *Analyst* **2012**, *137*, 1785–1792. [CrossRef]

17. Nisticò, R.; Rivolo, F.; Novara, C.; Giorgis, F. New branched flower-like Ag nanostructures for SERS analysis. *Colloids Surf. Physicochem. Eng. Asp.* **2019**, *578*, 123600. [CrossRef]

18. Kattumenu, R.; Lee, C.H.; Tian, L.; McConney, M.E.; Singamaneni, S. Nanorod decorated nanowires as highly efficient SERS-active hybrids. *J. Mater. Chem.* **2011**, *21*, 15218–15223. [CrossRef]

19. Zhang, Y.; Ashall, B.; Doyle, G.; Zerulla, D.; Lee, G.U. Highly ordered fe–au heterostructured nanorod arrays and their exceptional near-infrared plasmonic signature. *Langmuir* **2012**, *28*, 17101–17107. [CrossRef]

20. Mevold, A.H.; Hsu, W.W.; Hardiansyah, A.; Huang, L.Y.; Yang, M.C.; Liu, T.Y.; Chan, T.Y.; Wang, K.S.; Su, Y.A.; Jeng, R.J.; et al. Fabrication of gold nanoparticles/graphene-PDDA nanohybrids for bio-detection by SERS nanotechnology. *Nanoscale Res. Lett.* **2015**, *10*, 397. [CrossRef]

21. Zhang, P.; Yang, S.; Wang, L.; Zhao, J.; Zhu, Z.; Liu, B.; Zhong, J.; Sun, X. Large-scale uniform Au nanodisk arrays fabricated via x-ray interference lithography for reproducible and sensitive SERS substrate. *Nanotechnology* **2014**, *25*, 245301. [CrossRef]

22. Di Mascolo, D.; Coclite, A.; Gentile, F.; Francardi, M. Quantitative micro-Raman analysis of micro-particles in drug delivery. *Nanoscale Adv.* **2019**, *1*, 1541–1552. [CrossRef] [PubMed]

23. Marquart, M.; Deisenhofer, J.; Huber, R.; Palm, W. Crystallographic refinement and atomic models of the intact immunoglobulin molecule Kol and its antigen-binding fragment at 3.0 A and 1.0 A resolution. *J Mol. Biol.* **1980**, *141*, 369–391. [CrossRef]

24. Cai, S.; Singh, B.R. A distinct utility of the amide III infrared band for secondary structure estimation of aqueous protein solutions using partial least squares methods. *Biochemistry* **2004**, *43*, 2541–2549. [CrossRef] [PubMed]

25. Kengne-Momo, R.P.; Daniel, P.; Lagarde, F.; Jeyachandran, Y.L.; Pilard, J.F.; Durand-Thouand, M.J.; Thouand, G. Protein interactions investigated by the Raman spectroscopy for biosensor applications. *Int. J. Spectrosc.* **2012**, *2012*, 462901. [CrossRef]

26. Fischbach, F.T.; Dunning, M.B. *A Manual of Laboratory and Diagnostic Tests*; Wolters Kluwer Health/Lippincott Williams & Wilkins: Philadelphia, PA, USA, 2009.

27. Della Ventura, B.; Gelzo, M.; Battista, E.; Alabastri, A.; Schirato, A.; Castaldo, G.; Corso, G.; Gentile, F.; Velotta, R. Biosensor for point-of-care analysis of immunoglobulins in urine by metal enhanced fluorescence from gold nanoparticles. *ACS Appl. Mater. Interfaces* **2019**, *11*, 3753–3762. [CrossRef] [PubMed]

28. Minopoli, A.; Della Ventura, B.; Lenyk, B.; Gentile, F.; Tanner, J.A.; Offenhäusser, A.; Mayer, D.; Velotta, R. RUltrasensitive antibody-aptamer plasmonic biosensor for malaria biomarker detection in whole blood. *Nat. Commun.* **2020**, *11*, 6134. [CrossRef]

Article

Plasmonic Interferometers as TREM2 Sensors for Alzheimer's Disease

Dingdong Li [1], Rachel Odessey [1], Dongfang Li [1] and Domenico Pacifici [1,2,*]

[1] School of Engineering, Brown University, 184 Hope St, Providence, RI 02912, USA;
Dingdong_Li@alumni.brown.edu (D.L.); Rachel_Odessey@brown.edu (R.O.);
Dongfang_Li@alumni.brown.com (D.L.)

[2] Department of Physics, Brown University, 182 Hope St, Providence, RI 02912, USA

* Correspondence: Domenico_Pacifici@brown.edu

Abstract: We report an effective surface immobilization protocol for capture of Triggering Receptor Expressed on Myeloid Cells 2 (TREM2), a receptor whose elevated concentration in cerebrospinal fluid has recently been associated with Alzheimer's disease (AD). We employ the proposed surface functionalization scheme to design, fabricate, and assess a biochemical sensing platform based on plasmonic interferometry that is able to detect physiological concentrations of TREM2 in solution. These findings open up opportunities for label-free biosensing of TREM2 in its soluble form in various bodily fluids as an early indicator of the onset of clinical dementia in AD. We also show that plasmonic interferometry can be a powerful tool to monitor and optimize surface immobilization schemes, which could be applied to develop other relevant antibody tests.

Keywords: TREM2 sensors; Alzheimer's disease; plasmonic interferometry; optical biosensor; surface functionalization

1. Introduction

Alzheimer's disease (AD) is a chronic neurodegenerative disorder that affects more than five million Americans and approximately 50 million people worldwide [1]. AD causes loss of memory followed by loss of ability to think and communicate and, finally, loss of life [2]. As it progresses, AD has devastating effects on the ability of subjects to carry out the events of their day-to-day lives and can create significant mental and emotional distress for loved ones whose identities and relationships to the patient are forgotten. AD is one of the costliest disorders to society, costing over a quarter of a trillion dollars in 2017 alone in the United States [3]. Despite the many incentives and the correspondingly tremendous efforts of biopharmaceutical researchers, no disease-modifying therapy is yet available for AD and the drug candidates put forward to treat or prevent the onset of AD symptoms continue to fail in clinical trials [4].

AD is traditionally diagnosed by monitoring subjects' behavioral changes because it is challenging to diagnose more definitively without an invasive examination of the brain. This inexact method, which has a misdiagnosis rate of up to 45%, contributes to the untenably high attrition rate of drug candidates by creating an incomplete understanding of disease etiology and, as such, a lack of robust and valid biomarkers on the causal path of the disease [5]. Such biomarkers are essential to effective patient care and, specifically, the efficient development of drug treatments because they enable early and more accurate (i) diagnosis and stratification of patients during trial enrollment, (ii) measurement of target engagement and modulation, and (iii) testing of the therapeutic hypothesis in clinical trials that are already extremely long and costly [6].

1.1. TREM2 as Biomarker for Early-Onset Detection of AD

Most cases of AD have a complex, highly polygenic architecture [7]. A number of causal genes for AD have been identified in recent years, many of which play important

Citation: Li, D.; Odessey, R.; Li, D.; Pacifici, D. Plasmonic Interferometers as TREM2 Sensors for Alzheimer's Disease. *Biosensors* **2021**, *11*, 217. https://doi.org/10.3390/bios11070217

Received: 1 June 2021
Accepted: 26 June 2021
Published: 1 July 2021

Publisher's Note: MDPI stays neutral with regard to jurisdictional claims in published maps and institutional affiliations.

roles in myeloid cells such as microglia, immune cells in the brain [8–10]. One such gene is Triggering Receptor Expressed on Myeloid Cells 2 (TREM2) [11–13], which senses brain tissue damage due to aging or neurodegeneration by triggering a microglial response aimed at scavenging and clearing brain tissue debris [14–17]; genetic variants which impair this function also increase the risk of AD more than threefold [18,19]. Recent studies have shown that TREM2 was abnormally elevated 5 years before the expected onset of symptoms in AD patients [13,16,17]. These findings suggest that TREM2 is in the causal path to disease and among the strongest genetic risk factors for AD [20]. Moreover, TREM2 could be used as an effective biomarker for early stage detection of AD [12] and other neurodegenarative diseases [21].

TREM2 may be found in its soluble form in cerebrospinal fluid (CSF) and other bodily fluids, such as saliva [22]. CSF concentrations of soluble TREM2 have been shown to be higher in AD cases than in controls, they correlate with markers of neurodegeneration, and may be used to quantify glial activation in AD [23]. Recently, soluble CSF TREM2 has also been proposed as a surrogate immune biomarker of neuronal injury in Parkinson's disease [24]. Because of its extremely low concentration in CSF ($\sim$1–5 ng/mL [13,16,17]), TREM2 is conventionally detected and quantified by enzyme-linked immunosorbent assay (ELISA) [25], a well-established and highly sensitive plate-based assay which uses a multilayered format with a labeled secondary antibody. ELISA, while widely in use and highly sensitive, involves a multi-step incubation protocol that usually calls for 2 to 8 h of preparation time for each step. This, together with its stringent washing and blocking protocols and fluorescent labeling steps, makes ELISA a time-consuming assay that is hardly integrable into point-of-care, portable, or multiplexed biosensing platforms.

1.2. Plasmonic Interferometry for Sensing Applications

Several optics- and nanostructure-based alternatives to conventional biosensing methods have been developed, including sensing based on magnetic nanoparticles [26], carbon nanotubes and other nanostructures [27], quantum dots [28], and surface plasmon polaritons (SPPs) in gold, especially gold nanoparticles [29]. SPPs are collective oscillations of electrons that may occur when light interacts with the interface between a dielectric and a metal. One promising use of SPPs for biosensing is in plasmonic interferometry, which has been demonstrated to host biosensors with extremely high sensitivity and selectivity [30–44]. In this method, SPPs are generated and propagate within a micrometer-scale optical interferometer such as the one in Figure 1a,b: light is incident on a groove-slit-groove (GSG) geometry patterned onto a metal film; optical scattering at the subwavelength-width grooves couples a fraction of the incident light field into SPPs that travel across the surface of the metal, toward the slit. SPPs accrue a propagative phase that depends on the physical distance traveled (that is, the interferometer arm length) and on the SPP refractive index, which is given by [45]:

$$\tilde{n}_{\text{SPP}}(\lambda) = n_{\text{SPP}}(\lambda) + i\kappa_{\text{SPP}}(\lambda) = \sqrt{\frac{\epsilon_{\text{m}}(\lambda)\epsilon_{\text{d}}(\lambda)}{\epsilon_{\text{m}}(\lambda) + \epsilon_{\text{d}}(\lambda)}}, \tag{1}$$

where n_{SPP} and κ_{SPP} are the real and imaginary parts of $\tilde{n}_{\text{SPP}}$, respectively, and ϵ_{m} and ϵ_{d} are the complex dielectric functions of the corresponding metal and dielectric material. A small change in ϵ_{d} (which may be due, for example, to the presence of molecules near the surface) can produce a significant change in the optical path length traveled by an SPP across the surface of the metal [30]. When the counter propagating SPPs arrive at the subwavelength-width slit, they interfere with each other and with the optical beam incident at the slit location before coupling back into free space through the slit. This interference process modulates the light intensity that is transmitted through the slit and detected in the far-field [30,44,46,47]. Since SPPs are highly confined near the metal-dielectric interface, their propagative phase is highly sensitive to the refractive index at the surface. Therefore, the presence of an analyte adsorbed to the surface can be detected even in submonolayer concentrations by observing changes in the transmission spectra determined by plasmonic interferometry [30,39,42,43,46]. The sensitivity of a device based on plasmonic interferometry can be enhanced by simply increasing the interferometer arm length, and the signal to noise ratio can be improved by simply changing the geometrical parameters (such as slit/groove width, depth, and length) and the incident wavelength [30,42,44,46].

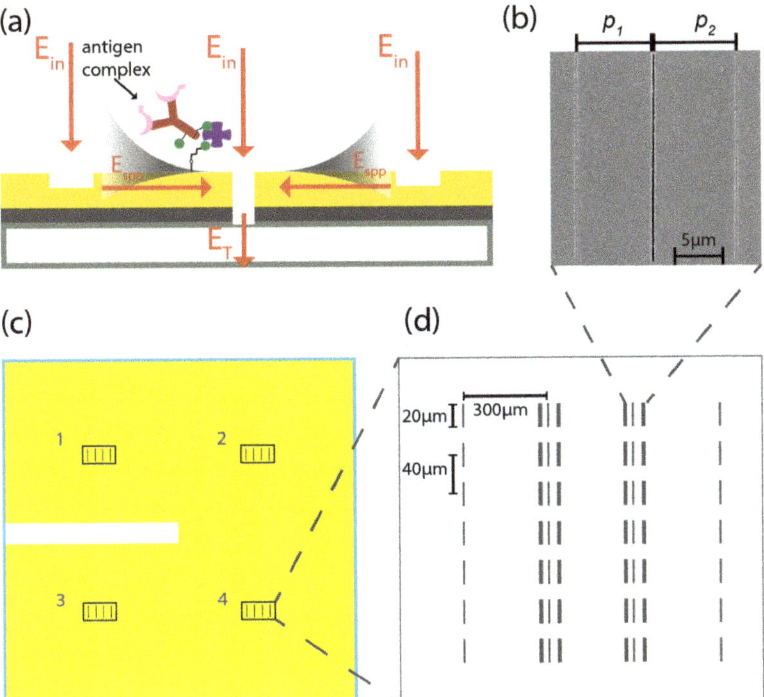

Figure 1. Design for TREM2 sensor chip based on plasmonic interferometry. (**a**) Cross-section schematic of the groove-slit-groove (GSG) architecture, which shows a slit flanked by two grooves, from which SPPs are excited by light diffraction and propagate towards the slit aperture, where they interfere and are then transmitted back into free space for far-field detection; diagram includes an example of an antigen complex, further described in Figure 2. The bottom slab represents quartz, the middle titanium, and the top layer gold. (**b**) Scanning electron micrograph (SEM) of a GSG interferometer with $p_1 = 7.65$ μm, $p_2 = 8.15$ μm. (**c**) Schematic of plasmonic interferometer sensor chip layout. The chip contains four nominally identical sensing spots enabling multiplex sensing applications. The yellow area indicates quartz covered by gold and the blank area is an uncoated quartz window used for optical alignment. (**d**) Schematic of a representative active sensing area. Each sensing area contains two columns of single slits and two columns of nominally identical asymmetric GSG interferometers with separation distance of 300 μm. The slit/grooves in each interferometer are ~20 μm long and, within each column, the distance between two adjacent interferometers is ~40 μm.

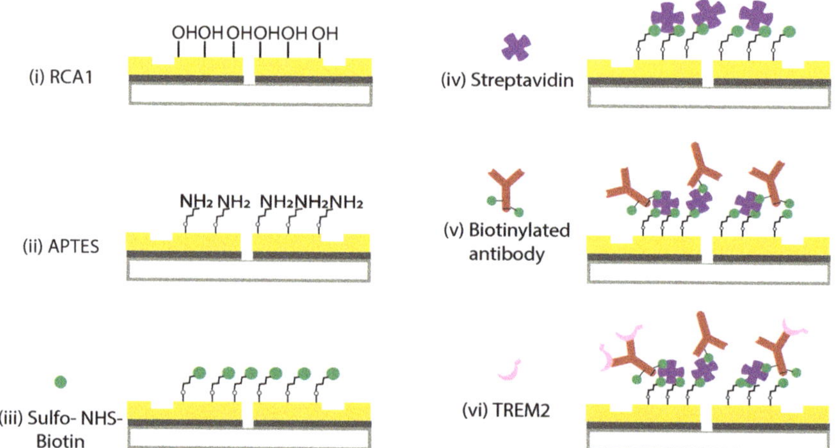

Figure 2. Surface immobilization protocol for capture of TREM2 in solution. Chip surface was treated with (**i**) an RCA1 cleaning procedure followed by (**ii**) (3-Aminopropyl)triethoxysilane (APTES) to form an amino-terminated surface. Sulfo-NHS-biotin (sulfo-N-Hydroxysulfosuccinimide biotin) covalently attaches to the amino groups of the surface (**iii**) and subsequently captures streptavidins (**iv**). Finally, the streptavidin functionalized chip is bound by the biotinylated TREM2 antibody (**v**) for sensing of the TREM2 molecule (**vi**). The green dot in (**v**) represents the sulfo-NHS ester of biotin that acts as the biotinylation reagent and allows to form a stable bond between the antibody and the streptavidin already bound to the sensor surface, as reported in (**iv**).

Compared to more conventional surface plasmon resonance (SPR) sensing platforms, sensors based on plasmonic interferometry (PI) retain high sensitivity and low detection limits whilst providing several advantages, including: (1) broadband (as opposed to single wavelength) operation that allows for spectroscopic capabilities and detection of refractive index changes at multiple wavelengths of interest, simultaneously [30,46]; (2) wide-angle excitation of SPPs (as opposed to the specific angle required to excite the surface plasmon resonance on the metal surface) that enables less stringent alignment requirements [30,43,46] and the use of incoherent light sources, which can also be integrated directly on the sensor surface [42]; (3) smaller sampling volumes and sensing areas, which lead to higher levels of device integration and multiplexing, since the same sensor chip can contain millions of individually addressable devices (over an area of just 1 cm²) that can be used to detect multiple analytes and perform screening on multiple patients at the same time [30,39,42].

Here, we employ plasmonic interferometry to develop and assess a surface functionalization protocol designed to detect TREM2 in solution. Specifically, we (i) monitor each step of the proposed functionalization protocol using the intensity change in the transmitted spectra of plasmonic interferometers and (ii) employ this surface functionalization scheme to detect biological levels of TREM2 in solution.

2. Biosensing Chip: Design, Fabrication and Implementation

2.1. Fabrication of Biosensing Chip Based on Plasmonic Interferometry

The proposed biosensing chip comprises four arrays of nominally identical GSG plasmonic interferometers that were designed and fabricated with asymmetrical arm lengths ($p_1 = 7.65$ μm, $p_2 = 8.15$ μm) to optimize the device sensitivity to refractive index change caused by TREM2 adsorbed to the sensor surface [30,39,42,44,46]. First, a 4 nm-thick titanium adhesion layer was deposited by electron-beam evaporation onto a previously cleaned 1 mm-thick fused quartz slide followed by a ~200 nm-thick gold layer. The thickness of the titanium layer was determined experimentally in order to cause strong surface adhesion of the gold layer, which would otherwise tend to delaminate and form blisters if directly deposited on glass. Four sensing spots were milled onto the metal film with a focused ion beam (FIB) using a gallium ion source. Each sensing spot contains two columns of seven nominally identical GSG interferometers and two columns of single slits for the purpose of statistical analysis and normalization, as shown in Figure 1c,d. The distance between two parallel columns and two adjacent interferometers in the same column are 300 μm and 40 μm, respectively. The area of each sensing spot is about 0.2 mm^2. Figure 1b shows a scanning electron micrograph of a representative GSG interferometer with left arm length 7.65 μm and right arm length 8.15 μm, within 2% fabrication error. Each groove is approximately 20 μm long, 200 nm wide, and ∼50 nm deep; each slit is 20 μm long, 180 nm wide, and ∼200 nm deep. These parameters were determined by analyzing SEM/FIB cross-sections (not reported). The actual values were chosen to optimize signal-to-noise light transmission ratio, SPP excitation efficiency, amplitude of SPP interference, and overall device sensitivity to refractive index change. Figure 1a shows a schematic illustration of a cross-section of the plasmonic interferometer, with two grooves flanking a slit in order to facilitate incoupling of incident light into SPPs (by optical scattering from each groove) and outcoupling of SPPs back into free space (through the slit).

2.2. Surface Functionalization of Optical Biochip

TREM2 antibodies were immobilized on the gold surface by using the protocol shown in Figure 2. The gold chip was first cleaned in an RCA1 solution, a mixture of 20 mL 29% ammonium hydroxide, 20 mL 30% hydrogen peroxide, and 100 mL deionized water (DI water), at 75 °C for 10 min. RCA1 cleaning removes organic residues from the gold surface and forms hydroxyl groups that facilitate silanol groups binding to the surface, as shown in Figure 2i. Next, the cleaned chip was soaked in 8 mL freshly prepared 2% (3-aminopropyl)triethoxysilane (APTES) solution for 1 h at room temperature to form an amino terminated surface, as shown in Figure 2ii. The 2% APTES solution was obtained by serial dilution of 99% APTES (Sigma-Aldrich, Burlington, MA, United States) in DI water. The APTES treatment time and concentration were chosen based on a standard surface plasmon resonance (SPR) functionalization protocol for a gold chip [48,49]. Then, the chip was rinsed thoroughly with DI water to remove loosely adsorbed APTES molecules that hadn't formed any covalent bonds with the hydroxyl terminated surface.

Next, biotinylation of the surface was achieved by soaking the chip in a 0.5 mg/mL sulfo-NHS-biotin solution (sulfo-N-Hydroxysuccinimide biotin ester sodium salt, Thermo Fisher Scientific, Waltham, MA, United States) for a duration of 2.5 h, followed by rinsing with phosphate buffer solution (PBS) and DI water. The sulfo-NHS-biotin solution was prepared in 0.01 M PBS (pH = 7.4, Sigma-Aldrich) immediately before using. On an amino-terminated gold surface, sulfo-NHS-biotin will covalently attach to the amino groups via ester linkage, as shown in Figure 2iii. Afterwards, the chip was treated with streptavidin by 1 h immersion in a 5 mg/mL fresh streptavidin (lyophilized powder, Sigma-Aldrich) PBS (0.01 M, pH = 7.4) containing 0.05% Tween 20, which was used to minimize the non-specific binding of streptavidin to the biotinlyated gold surface. The resulting streptavidin-coated chip was subsequently washed three times by washing buffer (0.01 M PBS with 0.05% Tween 20) to remove loosely adsorbed streptavidin molecules and then followed by PBS (0.01 M, pH = 7.4) to wash off the Tween 20 solution, leaving behind a streptavidin-functionalized metal surface, as shown in Figure 2iv.

In order to saturate the sulfo-NHS-biotin binding sites on the APTES-functionalized chip and to form a monolayer of streptavidins and antibodies on the surface, the chip was optically characterized after every hour of chemical treatment until no significant spectral peak shifts were observed in the optical transmission spectra through the GSG interferometers; the concentration ratio of sulfo-NHS-biotin and streptavidin was optimized to achieve a high biotinylated antibody covering rate. Subsequently, the chip was incubated with 0.1 mg/mL biotinylated TREM2 antibody PBS for 2.5 h, as shown in Figure 2v. Finally, human TREM2 biotinylated antibody was purchased from R&D (Cat. #BAF1828, R&D)

with 50 µg bovine serum albumin (BSA) per 1 µg as a carrier protein; the yielded solution contains 0.5% BSA in PBS. In sensing experiments, when TREM2 binds to the antibody the resulting chip surface is schematically shown in Figure 2vi.

2.3. Optical Characterization of Surface Functionalization Steps with Plasmonic Interferometry

After each functionalization step, the sensor chip was rinsed with DI water to remove unbound molecules/proteins and thoroughly dried under a stream of nitrogen gas to remove any residual water droplets/molecules that might affect the final wavelength shift measurement. This washing and drying protocol was enforced throughout the experiments to make sure that actual protein-protein binding events were measured, which allowed us to better validate the proposed surface functionalization and sensing methods. After surface rinsing and drying, the transmitted intensity of each plasmonic interferometer on the sensor chip was measured. To perform the optical characterization, a plasmonic interferometer sensor chip containing functionalized plasmonic interferometers was placed on the controllable moving stage of a Nikon Eclipse Ti Series inverted microscope. The collimated white light beam from a xenon arc lamp coupled with an optical lens system was focused onto the plasmonic interferometer sensor chip, transmitted through the slit, and collected by an objective lens (0.6 NA, 40×). The collected light was further dispersed by a monochromator and detected by a CCD camera. Transmitted spectra of the single slit accompanying each plasmonic interferometer were taken as well for spectral normalization.

Each solid line in Figure 3 corresponds to the mean value of seven normalized transmitted light intensities after each functionalization step,

$$I_{\text{n,mean}} = \frac{1}{7} \sum_{k=1}^{7} \frac{I_{\text{GSG},k}(\lambda)}{I_{\text{SS},k}(\lambda)}, \tag{2}$$

where $I_{\text{GSG},k}$ is the background-corrected light intensity transmitted through the kth GSG interferometer and $I_{\text{SS},k}$ is the background-corrected light intensity transmitted through the kth single slit, which serves as reference to estimate the light intensity baseline. The light gray shading represents the standard deviation that results from the proposed normalization and averaging procedure.

As shown in Figure 3, several maxima (minima) are observed as a result of constructive (destructive) interference between the counter-propagating SPPs and the original beam at the slit location. To fix the ideas and make it easier to follow the evolution of the various functionalization steps, we choose a specific peak as a reference, although the shift caused by the addition of reagents that are adsorbed to the surface can be tracked across the whole spectrum, which is an additional advantage of plasmonic interferometry compared to more conventional SPR that typically operates at single wavelength. After RCA1 cleaning of the sensor gold surface, the measured peak wavelength of the reference peak is around 588.1 nm (Figure 3i). After each functionalization step, the refractive index near the surface is modified by another layer of biomolecules, resulting in a spectral peak shift towards longer wavelengths. For the APTES and sulfo-NHS-biotin treatment steps (Figure 3ii,iii, respectively) the relative spectral peak shifts compared to their previous steps, are both ∼1.3 nm, while the relative spectral peak shift caused by the addition of streptavidin is ∼2.1 nm (Figure 3iv); finally, surface capture of the biotinylated TREM2 antibody produces an additional shift of ∼1.3 nm. A total spectral peak shift of ∼6 nm was observed after the entire functionalization procedure. The data shown in Figure 3 suggest that plasmonic interferometers can be effectively used to monitor the evolution of complex functionalization steps that involve adsorption of monolayer- and submonolayer-thick molecules to the sensor surface.

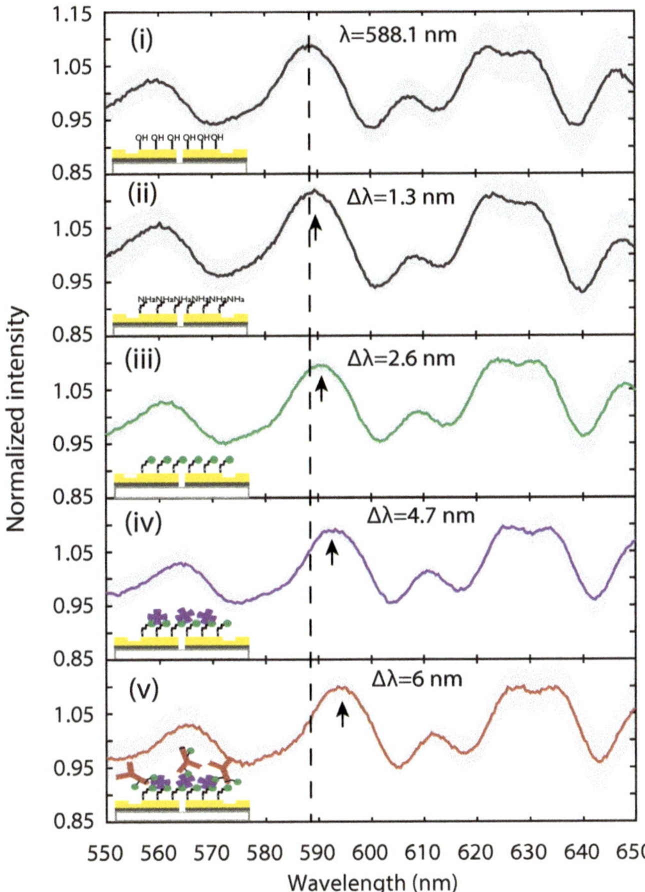

Figure 3. Tracking functionalization steps through plasmonic interference spectra. Measured results of transmitted intensity spectra after (**i**) RCA1, (**ii**) APTES, (**iii**) sulfo-NHS-biotin, (**iv**) streptavidin, and (**v**) biotinylated TREM2 antibody treatment. Solid lines represent the mean value of normalized intensity spectra averaged over seven nominally identical GSG interferometers after each functionalization step, as illustrated by the lower left insets. Light gray areas represent standard deviation. The vertical dashed line indicates the position of a representative transmission peak (588.1 nm) that results from constructive SPP interference after RCA1 cleaning. The black arrows mark the wavelength shift ($\Delta\lambda$) in this reference peak as the result of new constructive interference conditions after each functionalization step.

3. TREM2 Biosensing Experiment: Results and Analysis

3.1. Uniformity Study of Surface Functionalization Steps with Plasmonic Interferometry

The sensing chip was functionalized and stored in a refrigerator at 4 °C prior to testing. Immediately before TREM2 detection, the functionalized chip was blocked by 0.5~1% BSA PBS in order to saturate nonspecific binding sites and prevent false positive results [50]. Figure 4 shows the results of experiments carried out on each of the four sensing spots of the chip, tracking the spectral shift due to each functionalization step and after every hour of BSA blocking. Due to saturation of the blocking agent on the surface over time, the spectral shift due to BSA stabilized, enabling optimization of the BSA blocking time. We estimate that BSA almost saturated the surface of the chip after 4 h of blocking; therefore, the chip was blocked for 7 h at room temperature to ensure complete saturation. After the given blocking time, the chip was rinsed three times with a washing buffer (PBS with 0.05% Tween 20) followed by PBS and DI water, then dried under purified nitrogen gas flow.

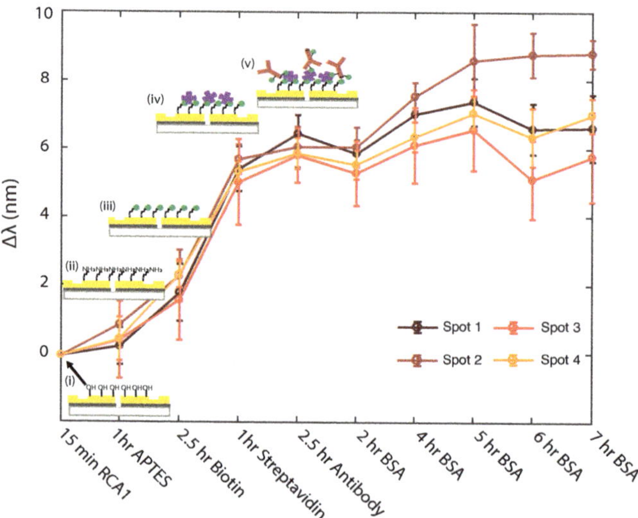

Figure 4. Uniformity study of surface functionalization steps across four sensing spots. Circles represent the mean value of wavelength shift ($\Delta\lambda$) measured from 7 GSG plasmonic interferometers after each functionalization step, labelled in the horizontal axis. Error bars represent the standard deviation. Lines and symbols with different colors indicate data measured from different sensing spots.

3.2. Sensing TREM2 Antigen-Antibody Binding Interaction with Plasmonic Interferometry

For sensing experiments, various concentrations of TREM2 in buffer solution were obtained by diluting a stock solution of TREM2 (0.27 mg/mL recombinant Human-TREM2 Fc Chimera PBS (Cat. #BAF1828, R&D)) in 0.5% BSA PBS (0.01 M, pH = 7.4). BSA serves as a protein stabilizer to maintain the integrity of TREM2 at low concentration.

TREM2 binding experiments were performed by directly dispensing 40 µL TREM2 BSA PBS on the chip and then drying it to verify that surface capture and adsorption had effectively taken place. More specifically, after a given binding time interval, the chip was rinsed three times with a washing buffer solution (PBS with 0.05% Tween 20) followed by PBS and then DI water, dried under nitrogen gas flow. Then, transmitted spectra through the slit of each GSG plasmonic interferometer were measured to assess the presence of TREM2 at the sensor surface. This process was repeated to obtain TREM2 kinetic binding curves as a function of time and for various initial TREM2 concentrations in buffer solution.

Figure 5 shows the observed peak shift $\Delta\lambda$ as a function of time in a binding experiment performed with a 2.7 ng/ml TREM2 0.5% BSA PBS. Blue circles and error bars represent the mean value and standard deviation, respectively, of $\Delta\lambda$ measured from seven GSG plasmonic interferometers. The inset of Figure 5 displays the average of normalized intensity spectra across all seven GSG plasmonic interferometers at different time steps, showing a spectral shift towards longer wavelengths as the reaction progresses. The time-domain sensing curve (or "sensorgram") reported in Figure 5 shows that by tracking the wavelength shift as a function of time we can indeed monitor the capture of TREM2 antigens from antibody binding sites and subsequent formation of a sub-monolayer of TREM2 on the sensor surface.

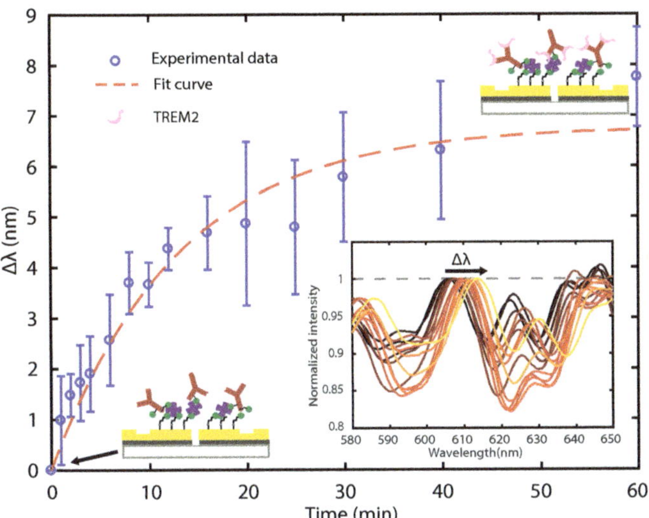

Figure 5. Sensing temporal evolution of TREM2 surface binding kinetics with plasmonic interferometry. Blue circles represent the mean peak shift ($\Delta\lambda$) averaged over seven nominally identical GSG plasmonic interferometers as the result of temporal evolution of antigen-antibody binding reaction for a 2.7 ng/ml TREM2 0.5% BSA PBS. Error bars represent the standard deviation. Bottom right inset illustrates the normalized transmitted spectra (averaged over seven identical GSG interferometers) measured at each time step. Color changing from dark red to yellow represents increasing reaction time from 0 to 60 min.

To understand the kinetic interaction of the binding between TREM2 and its antibody, we first consider an equilibrium model for the chemical reaction [39,51–56]. Ideally, the antibody-antigen interaction is a reversible reaction:

$$antibody + antigen \rightleftharpoons complex. \tag{3}$$

The time-dependent rate equation that governs this reaction can be expressed by:

$$\frac{d[complex]}{dt} = k_f[Ab][Ag] - k_b[complex] \tag{4}$$

where $[complex]$ is the molar concentration of antibody-antigen complex, $[Ab]$ is the molar concentration of unoccupied antibodies, $[Ag]$ is the molar concentration of antigen in the solution, k_f is the forward reaction constant and k_b is the backward reaction constant. When the reaction reaches equilibrium, we have:

$$k_f[Ab]_{eq}[Ag]_{eq} = k_b[complex]_{eq} \tag{5}$$

$$K_d \equiv \frac{k_b}{k_f} = \frac{[Ab]_{eq}[Ag]_{eq}}{[complex]_{eq}} \tag{6}$$

where K_d is the dissociation constant.

In addition, if we assume that the total number of antibodies immobilized at the sensor surface is fixed and the concentration of antigens is constant due to the large volume (or continuous dispensing) of the sample solution onto the surface, the following expression holds true at all times:

$$[Ab]_0 = [Ab] + [complex] \tag{7}$$

where $[Ab]_0$ is the initial concentration of the antibodies immobilized on the sensor surface. Applying Equations (6) and (7) to Equation (4) and taking a time integral, we can obtain:

$$[complex] = \frac{[Ag][Ab]_0}{[Ag] + K_d}[1 - e^{-(k_f[Ag]+k_b)t}]. \tag{8}$$

Equation (8) implies that: (a) the concentration of the antibody-antigen complex at the surface increases exponentially as a function of time; (b) the concentration of the antibody-antigen complex at equilibrium is always lower than the initial concentration of antibodies immobilized at the sensor surface; and (c) by increasing the antigen concentration in solution, the binding kinetics should occur with a faster rate and the equilibrium complex concentration should also increase. Figure 5 reports an exponential fit of the experimental data, where we assumed that the relative peak shift $\Delta\lambda$ was directly proportional to the number of TREM2 (the antigen) proteins captured by the immobilized antibodies and binding at the sensor surface over time. The data were fit by $\Delta\lambda = a(1 - e^{-t/\tau})$, where $a = 7.12$ nm and $\tau = 13.44$ min. Several studies have shown that protein adsorption is a very sophisticated process, strongly influenced by experimental parameters, such as surface wettability, pH, protein structure, and other factors [57,58]. Moreover, proteins could be adsorbed to the surface in a multilayer fashion, especially on a hydrophobic surface, which would produce a higher (and therefore more readily detectable) $\Delta\lambda$ than that caused by the formation of a single monolayer.

To further validate the model, we performed sensing experiments to detect TREM2 in solution with three different physiological concentrations (1.35 ng/ml, 2.7 ng/ml, and 8.1 ng/ml) by using three of the four sensing spots of the chip, separately, and investigated the concentration dependence of the binding kinetics. Note that the chip was first regenerated to bare gold by 15 min of RCA1 cleaning, then functionalized and blocked again with BSA PBS before TREM2 detection. After regeneration, we confirmed that the normalized transmission spectra went back to the initial spectra, well within the confidence intervals reported in Figure 3. Figure 6 shows the sensing results averaged over the three functionalized sensing spots after interaction with 0.5% BSA PBS spiked with different TREM2 concentrations. The $\Delta\lambda$ associated with the binding reaction corresponding to 1.35 ng/ml and 2.7 ng/ml TREM2 concentration showed a single-exponential time dependence $\Delta\lambda = a(1 - e^{-t/\tau})$, where $a = 2.8$ nm, $\tau = 41.08$ min for 1.35 ng/ml and $a = 6.89$ nm, $\tau = 15.04$ min for 2.7 ng/ml, which is consistent with the parameters identified for the same concentration in Figure 5. When the concentration of TREM2 increased from 1.35 ng/ml to 2.7 ng/ml, the binding time constant τ decreased, which confirms that the reaction occurred faster in the presence of a higher concentration of antigens, as predicted by the model. Additionally, the parameter a, which represents the total wavelength shift at equilibrium, increased with the concentration of antigens. These experimental findings are consistent with the simple model described by Equation (8).

Interestingly, the binding reaction data corresponding to the higher concentration (8.1 ng/ml TREM2) displayed a kinetic curve that has a second slow rising stage, which could be fit with a double-exponential function: $\Delta\lambda = a_1(1 - e^{-t/\tau_1}) + a_2(1 - e^{-t/\tau_2})$, where $a_1 = 5.83$ nm, $\tau_1 = 8.1$ min, $a_2 = 6.08$ nm, and $\tau_2 = 476.19$ min. This could indicate the presence of multiple adsorption mechanisms with distinct dissociation constants K_d that were active at higher analyte concentrations. These different values of K_d were manifested in the double-exponential curve by unique binding time constants τ_1 and τ_2 and associated wavelength shifts a_1 and a_2. Previous studies reported similar two-stage adsorption kinetics, which may be due to a number of possible mechanisms. For instance, the initial stage may be caused by formation of a single monolayer at the surface, while the second slow rising stage could correspond to multilayer condensation of proteins on the surface [57,59]. The longer time constant may also arise from more subtle adsorption mechanisms the various proteins may be subjected to: for example, for BSA, conformational changes may be the primary reason for multilayer conformation, while for the TREM2 antibody (immunoglobulin G), multilayer conformation behavior could be significantly influenced by long-range electrostatic interaction [60]. In addition, higher-order interaction effects during the adsorption process could cause the combined kinetics to differ significantly from a summation of two single component adsorption kinetics [61,62], especially at higher analyte concentrations.

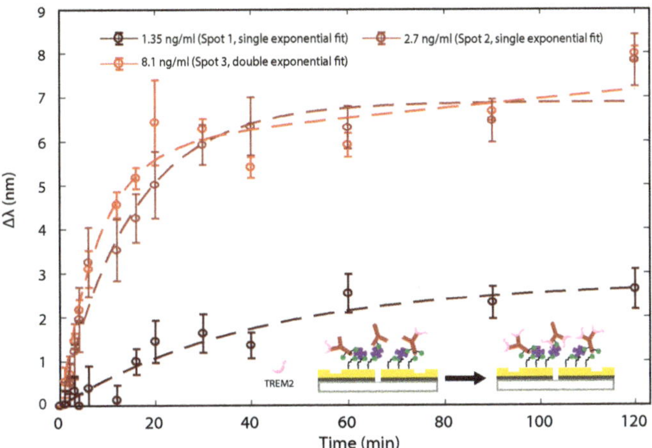

Figure 6. Binding times for different TREM2 concentrations. Temporal evolution of peak wavelength shifts measured from normalized transmission spectra for different TREM2 concentrations. Error bars represent standard deviation from 7 GSG interferometers. Dashed lines represent exponential fit using the kinetic model provided in the text.

Although a more complex model may be needed to fully describe all of the possible microscopic phenomena underlying the actual reaction kinetics, to first approximation the data presented in Figure 6 can be fit using the simple exponential model which allows us to generate reliable calibration curves that accurately describe the proposed sensing mechanism. For instance, by choosing a given incubation time of $t = 10$ min, the wavelength shifts observed at three different concentrations are statistically different, enabling us to infer the concentration based on the observed $\Delta\lambda$. However, we believe that a fit to the full time-resolved data set is an overall more statistically significant analysis of the acquired data and can lead to a better discrimination between different analyte concentrations, as shown in Figure 6. The full set of data presented so far validates the possibility of using plasmonic interferometry coupled with surface functionalization as a viable sensing scheme for TREM2 detection in solution.

4. Conclusions

We demonstrate a biophotonic sensing platform to monitor the chemical reactions associated with surface functionalization of a metal in time series through wavelength shifts in the transmission spectra of plasmonic interferometers. This work helps us devise and assess a functionalization protocol for detecting TREM2, a biomarker associated with the development of AD and other neurodegenerative diseases, in solution. We devise a chemical equilibrium model which we fit with the reported data. The fitting parameters from the data reinforce the theoretical model and are consistent between experiments, suggesting that the experimental process is repeatable. The chemical equilibrium model allows us to generate calibration curves for the reported data which are able to differentiate between different concentrations of TREM2. The results reported here open up the possibility to employ the proposed sensing platform based on plasmonic interferometry for the detection of physiological concentrations of TREM2 in CSF or other bodily fluids. In the future, plasmonic interferometry may be a promising method to test functionalization protocols for deployment on other scalable platforms, such as colloidal nanoparticles, which are widely used in biological testing, to help validate the importance of TREM2 to the development pathway of AD. Additionally, plasmonic interferometry may be adapted to other surface functionalization protocols for different antibody-antigen pairs to enable development of a wide array of testing schemes for a variety of clinically relevant biomarkers.

Author Contributions: Conceptualization, D.P., D.L. (Dingdong Li) and D.L. (Dongfang Li); data curation, D.L. (Dingdong Li), D.L. (Dongfang Li) and D.P.; formal analysis, D.L. (Dingdong Li), D.L. (Dongfang Li) and D.P.; investigation, D.L. (Dingdong Li); methodology, D.L. (Dingdong Li), D.L. (Dongfang Li) and D.P.; software, D.L. (Dongfang Li) and D.P.; validation, D.L. (Dingdong Li), D.L. (Dongfang Li) and D.P.; project administration, D.P.; resources, D.P.; visualization, D.L. (Dingdong Li), D.L. (Dongfang Li), R.O., and D.P.; writing—original draft preparation, D.L. (Dingdong Li); writing—review and editing, D.L. (Dongfang Li), D.L. (Dongfang Li), R.O. and D.P.; supervision, D.P.; project administration, D.P.; funding acquisition, D.P. All authors have read and agreed to the published version of the manuscript.

Funding: This research was funded by the National Science Foundation (NSF), CBET Div. Of Chem., Bioeng., Env., and Transp. Sys., ENG Directorate For Engineering, Grant. No. 1842605, "EAGER: Development of Surface Chemistry and Plasmonic Interferometers for Early-Onset Detection of Alzheimer Disease;" Program BIOSENS-Biosensing, Program Ref. Codes No. 7916, 9150.

Data Availability Statement: The data presented in this study are available on request from the corresponding author.

Acknowledgments: The authors would like to thank Jing Feng and Tinayi Shen for useful contributions and engaging discussions. D.P. would like to thank Alessandro, Paola, and Emma for their understanding and support throughout this research project.

Conflicts of Interest: The authors declare that there are no conflicts of interest related to this article.

Abbreviations

The following abbreviations are used in this manuscript:

AD	Alzheimer's disease
APTES	(3-Aminopropyl)triethoxysilane
BSA	Bovine serum albumin
CCD	Charged-coupled device
CSF	Cerebrospinal fluid
DI	Deionized
ELISA	Enzyme-linked immunosorbent assay
FIB	Focused ion beam
GSG	Groove-Slit-Groove
NA	Numerical aperture
NHS	N-Hydroxysulfosuccinimide
PBS	Phosphate buffer solution
RCA1	Radio Corporation of America Si wafer cleaning procedure, standard clean-1
SEM	Scanning Electron Micrograph/Microscopy
SPP	Surface Plasmon Polariton
SPR	Surface Plasmon Resonance
TREM2	Triggering Receptor Expressed on Myeloid Cells 2

References

1. Vos, T.; Allen, C.; Arora, M.; Barber, R.M.; Bhutta, Z.A.; Brown, A.; Carter, A.; Casey, D.C.; Charlson, F.J.; Chen, A.Z.; et al. Global, regional, and national incidence, prevalence, and years lived with disability for 310 diseases and injuries, 1990–2015: A systematic analysis for the Global Burden of Disease Study 2015. *Lancet* **2016**, *388*, 1545–1602. [CrossRef]
2. World Health Organization. *Towards a Dementia Plan: A WHO Guide*; World Health Organization: Geneva, Switzerland, 2018.
3. Alzheimer's Association. 2017 Alzheimer's disease facts and figures. *Alzheimer's Dement.* **2017**, *13*, 325–373. [CrossRef]
4. Calcoen, D.; Elias, L.; Yu, X. What does it take to produce a breakthrough drug? *Nat. Rev. Drug Discov.* **2015**, *14*, 161–162. [CrossRef]
5. Morris, J.C.; McKeel, D.W., Jr.; Fulling, K.; Torack, R.M.; Berg, L. Validation of clinical diagnostic criteria for Alzheimer's disease. *Ann. Neurol. Off. J. Am. Neurol. Assoc. Child Neurol. Soc.* **1988**, *24*, 17–22. [CrossRef]

6. Plenge, R.M. Disciplined approach to drug discovery and early development. *Sci. Transl. Med.* **2016**, *8*, 349ps15. [CrossRef] [PubMed]

7. Escott-Price, V.; Shoai, M.; Pither, R.; Williams, J.; Hardy, J. Polygenic score prediction captures nearly all common genetic risk for Alzheimer's disease. *Neurobiol. Aging* **2017**, *49*, 214.e7–214.e11. [CrossRef] [PubMed]

8. Van Cauwenberghe, C.; Van Broeckhoven, C.; Sleegers, K. The genetic landscape of Alzheimer disease: Clinical implications and perspectives. *Genet. Med.* **2016**, *18*, 421–430. [CrossRef] [PubMed]

9. Efthymiou, A.G.; Goate, A.M. Late onset Alzheimer's disease genetics implicates microglial pathways in disease risk. *Mol. Neurodegener.* **2017**, *12*, 43. [CrossRef]

10. Malik, M.; Parikh, I.; Vasquez, J.B.; Smith, C.; Tai, L.; Bu, G.; LaDu, M.J.; Fardo, D.W.; Rebeck, G.W.; Estus, S. Genetics ignite focus on microglial inflammation in Alzheimer's disease. *Mol. Neurodegener.* **2015**, *10*, 52. [CrossRef]

11. Yeh, F.L.; Hansen, D.V.; Sheng, M. TREM2, microglia, and neurodegenerative diseases. *Trends Mol. Med.* **2017**, *23*, 512–533. [CrossRef]

12. Suárez-Calvet, M.; Kleinberger, G.; Caballero, M.Á.A.; Brendel, M.; Rominger, A.; Alcolea, D.; Fortea, J.; Lleó, A.; Blesa, R.; Gispert, J.D.; et al. sTREM2 cerebrospinal fluid levels are a potential biomarker for microglia activity in early-stage Alzheimer's disease and associate with neuronal injury markers. *EMBO Mol. Med.* **2016**, e201506123. [CrossRef]

13. Piccio, L.; Deming, Y.; Del-Águila, J.L.; Ghezzi, L.; Holtzman, D.M.; Fagan, A.M.; Fenoglio, C.; Galimberti, D.; Borroni, B.; Cruchaga, C. Cerebrospinal fluid soluble TREM2 is higher in Alzheimer disease and associated with mutation status. *Acta Neuropathol.* **2016**, *131*, 925–933. [CrossRef]

14. Pimenova, A.A.; Marcora, E.; Goate, A.M. A tale of two genes: Microglial Apoe and Trem2. *Immunity* **2017**, *47*, 398–400. [CrossRef] [PubMed]

15. Kober, D.L.; Alexander-Brett, J.M.; Karch, C.M.; Cruchaga, C.; Colonna, M.; Holtzman, M.J.; Brett, T.J. Neurodegenerative disease mutations in TREM2 reveal a functional surface and distinct loss-of-function mechanisms. *Elife* **2016**, *5*. [CrossRef]

16. Suárez-Calvet, M.; Caballero, M.Á.A.; Kleinberger, G.; Bateman, R.J.; Fagan, A.M.; Morris, J.C.; Levin, J.; Danek, A.; Ewers, M.; Haass, C.; et al. Early changes in CSF sTREM2 in dominantly inherited Alzheimer's disease occur after amyloid deposition and neuronal injury. *Sci. Transl. Med.* **2016**, *8*, 369ra178. [CrossRef]

17. Gispert, J.D.; Suárez-Calvet, M.; Monté, G.C.; Tucholka, A.; Falcon, C.; Rojas, S.; Rami, L.; Sánchez-Valle, R.; Lladó, A.; Kleinberger, G.; et al. Cerebrospinal fluid sTREM2 levels are associated with gray matter volume increases and reduced diffusivity in early Alzheimer's disease. *Alzheimer Dementia J. Alzheimer Assoc.* **2016**, *12*, 1259–1272. [CrossRef]

18. Jonsson, T.; Stefansson, H.; Steinberg, S.; Jonsdottir, I.; Jonsson, P.V.; Snaedal, J.; Bjornsson, S.; Huttenlocher, J.; Levey, A.I.; Lah, J.J.; et al. Variant of TREM2 associated with the risk of Alzheimer's disease. *N. Engl. J. Med.* **2013**, *368*, 107–116. [CrossRef] [PubMed]

19. Guerreiro, R.; Wojtas, A.; Bras, J.; Carrasquillo, M.; Rogaeva, E.; Majounie, E.; Cruchaga, C.; Sassi, C.; Kauwe, J.S.K.; Younkin, S.; et al. TREM2 variants in Alzheimer's disease. *N. Engl. J. Med.* **2013**, *368*, 117–127. [CrossRef] [PubMed]

20. Gratuze, M.; Leyns, C.E.; Holtzman, D.M. New insights into the role of TREM2 in Alzheimer's disease. *Mol. Neurodegener.* **2018**, *13*, 1–16. [CrossRef]

21. Jay, T.R.; von Saucken, V.E.; Landreth, G.E. TREM2 in neurodegenerative diseases. *Mol. Neurodegener.* **2017**, *12*, 1–33. [CrossRef]

22. Yilmaz, A.; Geddes, T.; Han, B.; Bahado-Singh, R.O.; Wilson, G.D.; Imam, K.; Maddens, M.; Graham, S.F. Diagnostic biomarkers of Alzheimer's disease as identified in saliva using 1H nmR-based metabolomics. *J. Alzheimer Dis.* **2017**, *58*, 355–359. [CrossRef] [PubMed]

23. Heslegrave, A.; Heywood, W.; Paterson, R.; Magdalinou, N.; Svensson, J.; Johansson, P.; Öhrfelt, A.; Blennow, K.; Hardy, J.; Schott, J.; et al. Increased cerebrospinal fluid soluble TREM2 concentration in Alzheimer's disease. *Mol. Neurodegener.* **2016**, *11*, 3. [CrossRef] [PubMed]

24. Wilson, E.N.; Swarovski, M.S.; Linortner, P.; Shahid, M.; Zuckerman, A.J.; Wang, Q.; Channappa, D.; Minhas, P.S.; Mhatre, S.D.; Plowey, E.D.; et al. Soluble TREM2 is elevated in Parkinson's disease subgroups with increased CSF tau. *Brain* **2020**, *143*, 932–943. [CrossRef]

25. Piccio, L.; Buonsanti, C.; Cella, M.; Tassi, I.; Schmidt, R.E.; Fenoglio, C.; Rinker, J.; Naismith, R.T.; Panina-Bordignon, P.; Passini, N.; et al. Identification of soluble TREM-2 in the cerebrospinal fluid and its association with multiple sclerosis and CNS inflammation. *Brain* **2008**, *131*, 3081–3091. [CrossRef]

26. Haun, J.B.; Yoon, T.J.; Lee, H.; Weissleder, R. Magnetic nanoparticle biosensors. *Wiley Interdiscip. Rev. Nanomed. Nanobiotechnol.* **2010**, *2*, 291–304. [CrossRef]

27. Vamvakaki, V.; Chaniotakis, N.A. Carbon nanostructures as transducers in biosensors. *Sens. Actuators B Chem.* **2007**, *126*, 193–197. [CrossRef]

28. Zhang, C.Y.; Yeh, H.C.; Kuroki, M.T.; Wang, T.H. Single-quantum-dot-based DNA nanosensor. *Nat. Mater.* **2005**, *4*, 826–831. [CrossRef]

29. Li, Y.; Schluesener, H.J.; Xu, S. Gold nanoparticle-based biosensors. *Gold Bull.* **2010**, *43*, 29–41. [CrossRef]

30. Feng, J.; Siu, V.S.; Roelke, A.; Mehta, V.; Rhieu, S.Y.; Palmore, G.T.R.; Pacifici, D. Nanoscale plasmonic interferometers for multispectral, high-throughput biochemical sensing. *Nano Lett.* **2012**, *12*, 602–609. [CrossRef]

31. Graydon, O. Sensing: Plasmonic interferometry. *Nat. Photonics* **2012**, *6*, 139. [CrossRef]

32. Martín-Becerra, D.; Armelles, G.; González, M.; García-Martín, A. Plasmonic and magnetoplasmonic interferometry for sensing. *New J. Phys.* **2013**, *15*, 085021. [CrossRef]

33. Yeung, K.Y.; Yoon, H.; Andress, W.; West, K.; Pfeiffer, L.; Ham, D. Two-path solid-state interferometry using ultra-subwavelength two-dimensional plasmonic waves. *Appl. Phys. Lett.* **2013**, *102*, 021104. [CrossRef]

34. Gao, Y.; Xin, Z.; Gan, Q.; Cheng, X.; Bartoli, F.J. Plasmonic interferometers for label-free multiplexed sensing. *Opt. Express* **2013**, *21*, 5859–5871. [CrossRef] [PubMed]

35. Gao, Y.; Xin, Z.; Zeng, B.; Gan, Q.; Cheng, X.; Bartoli, F.J. Plasmonic interferometric sensor arrays for high-performance label-free biomolecular detection. *Lab Chip* **2013**, *13*, 4755–4764. [CrossRef]

36. Gan, Q.; Gao, Y.; Bartoli, F.J. Vertical plasmonic Mach-Zehnder interferometer for sensitive optical sensing. *Opt. Express* **2009**, *17*, 20747–20755. [CrossRef]

37. Gao, Y.; Gan, Q.; Xin, Z.; Cheng, X.; Bartoli, F.J. Plasmonic Mach–Zehnder interferometer for ultrasensitive on-chip biosensing. *ACS Nano* **2011**, *5*, 9836–9844. [CrossRef]

38. Li, X.; Tan, Q.; Bai, B.; Jin, G. Non-spectroscopic refractometric nanosensor based on a tilted slit-groove plasmonic interferometer. *Opt. Express* **2011**, *19*, 20691–20703. [CrossRef]

39. Siu, V.S.; Feng, J.; Flanigan, P.W.; Palmore, G.T.R.; Pacifici, D. A "plasmonic cuvette": Dye chemistry coupled to plasmonic interferometry for glucose sensing. *Nanophotonics* **2014**, *3*, 125–140. [CrossRef]

40. Cetin, A.E.; Coskun, A.F.; Galarreta, B.C.; Huang, M.; Herman, D.; Ozcan, A.; Altug, H. Handheld high-throughput plasmonic biosensor using computational on-chip imaging. *Light. Sci. Appl.* **2014**, *3*, e122. [CrossRef]

41. Sun, X.; Dai, D.; Thylén, L.; Wosinski, L. High-sensitivity liquid refractive-index sensor based on a Mach-Zehnder interferometer with a double-slot hybrid plasmonic waveguide. *Opt. Express* **2015**, *23*, 25688–25699. [CrossRef]

42. Li, D.; Feng, J.; Pacifici, D. Nanoscale optical interferometry with incoherent light. *Sci. Rep.* **2016**, *6*, 20836. [CrossRef]

43. Li, D.; Feng, J.; Pacifici, D. Higher-order surface plasmon contributions to passive and active plasmonic interferometry. *Opt. Express* **2016**, *24*, 27309–27318. [CrossRef] [PubMed]

44. Feng, J.; Li, D.; Pacifici, D. Circular slit-groove plasmonic interferometers: A generalized approach to high-throughput biochemical sensing. *Opt. Mater. Express* **2015**, *5*, 2742–2753. [CrossRef]

45. Raether, H. Surface plasmons on smooth surfaces. In *Surface Plasmons on Smooth and Rough Surfaces and on Gratings*; Springer: Berlin, Germany, 1988; pp. 4–39.

46. Feng, J.; Pacifici, D. A spectroscopic refractometer based on plasmonic interferometry. *J. Appl. Phys.* **2016**, *119*, 083104. [CrossRef]

47. Morrill, D.; Li, D.; Pacifici, D. Measuring subwavelength spatial coherence with plasmonic interferometry. *Nat. Photonics* **2016**, *10*, 681–687. [CrossRef]

48. Vashist, S.K.; Dixit, C.K.; MacCraith, B.D.; O'Kennedy, R. Effect of antibody immobilization strategies on the analytical performance of a surface plasmon resonance-based immunoassay. *Analyst* **2011**, *136*, 4431. [CrossRef]

49. Williams, E.H.; Davydov, A.V.; Motayed, A.; Sundaresan, S.G.; Bocchini, P.; Richter, L.J.; Stan, G.; Steffens, K.; Zangmeister, R.; Schreifels, J.A.; et al. Immobilization of streptavidin on 4H–SiC for biosensor development. *Appl. Surf. Sci.* **2012**, *258*, 6056–6063. [CrossRef]

50. Xiao, Y.; Isaacs, S.N. Enzyme-linked immunosorbent assay (ELISA) and blocking with bovine serum albumin (BSA)—not all BSAs are alike. *J. Immunol. Methods* **2012**, *384*, 148–151. [CrossRef]

51. Malmqvist, M. Surface plasmon resonance for detection and measurement of antibody-antigen affinity and kinetics. *Curr. Opin. Immunol.* **1993**, *5*, 282–286. [CrossRef]

52. O'Shannessy, D.J.; Brigham-Burke, M.; Soneson, K.K.; Hensley, P.; Brooks, I. Determination of rate and equilibrium binding constants for macromolecular interactions using surface plasmon resonance: Use of nonlinear least squares analysis methods. *Anal. Biochem.* **1993**, *212*, 457–468. [CrossRef]

53. O'Shannessy, D.J.; Brigham-Burke, M.; Soneson, K.K.; Hensley, P.; Brooks, I. Determination of rate and equilibrium binding constants for macromolecular interactions by surface plasmon resonance. In *Methods in Enzymology*; Elsevier: Amsterdam, The Netherlands, 1994; Volume 240, pp. 323–349.

54. O'Shannessy, D.J. Determination of kinetic rate and equilibrium binding constants for macromolecular interactions: A critique of the surface plasmon resonance literature. *Curr. Opin. Biotechnol.* **1994**, *5*, 65–71. [CrossRef]

55. Myszka, D.G. Kinetic analysis of macromolecular interactions using surface plasmon resonance biosensors. *Curr. Opin. Biotechnol.* **1997**, *8*, 50–57. [CrossRef]

56. Cornish-Bowden, A. *Fundamentals of Enzyme Kinetics*; John Wiley & Sons: Hoboken, NJ, USA, 2013.

57. Teichroeb, J.; Forrest, J.; Jones, L.; Chan, J.; Dalton, K. Quartz crystal microbalance study of protein adsorption kinetics on poly(2-hydroxyethyl methacrylate). *J. Colloid Interface Sci.* **2008**, *325*, 157–164. [CrossRef]

58. Holmberg, M.; Hou, X. Competitive Protein Adsorption—Multilayer Adsorption and Surface Induced Protein Aggregation. *Langmuir* **2009**, *25*, 2081–2089. [CrossRef] [PubMed]

59. Ball, V.; Ramsden, J.J. Analysis of hen egg white lysozyme adsorption on Si (Ti) O_2 aqueous solution interfaces at low ionic strength: A biphasic reaction related to solution self-association. *Colloids Surfaces B Biointerfaces* **2000**, *17*, 81–94. [CrossRef]

60. Haynes, C.A.; Norde, W. Globular proteins at solid/liquid interfaces. *Colloids Surfaces B Biointerfaces* **1994**, *2*, 517–566. [CrossRef]

61. Rabe, M.; Verdes, D.; Zimmermann, J.; Seeger, S. Surface Organization and Cooperativity during Nonspecific Protein Adsorption Events. *J. Phys. Chem. B* **2008**, *112*, 13971–13980. [CrossRef] [PubMed]
62. Minton, A.P. Effects of excluded surface area and adsorbate clustering on surface adsorption of proteins: I. Equilibrium models. *Biophys. Chem.* **2000**, *86*, 239–247. [CrossRef]

biosensors

Article

SPR-Based Kinetic Analysis of the Early Stages of Infection in Cells Infected with Human Coronavirus and Treated with Hydroxychloroquine

Petia Genova-Kalou [1], Georgi Dyankov [2,*], Radoslav Marinov [1], Vihar Mankov [2], Evdokiya Belina [2], Hristo Kisov [2], Velichka Strijkova-Kenderova [2] and Todor Kantardjiev [1]

[1] National Center of Infectious and Parasitic Diseases, 44A "Gen. Stoletov" Blvd., 1233 Sofia, Bulgaria; petia.d.genova@abv.bg (P.G.-K.); r.m.r@mail.bg (R.M.); nrlcell_rik_herp@ncipd.org (T.K.)
[2] Institute of Optical Materials and Technologies "Acad. J. Malinowski" (IOMT), Bulgarian Academy of Sciences (BAS), 109 "Acad. G. Bonchev" Str., 1113 Sofia, Bulgaria; viharmankov@gmail.com (V.M.); evdokiyabelina@yahoo.de (E.B.); christokissov@yahoo.com (H.K.); vily@iomt.bas.bg (V.S.-K.)
* Correspondence: gdyankov@iomt.bas.bg; Tel.: +359-897-771-945

Abstract: Cell-based assays are a valuable tool for examination of virus–host cell interactions and drug discovery processes, allowing for a more physiological setting compared to biochemical assays. Despite the fact that cell-based SPR assays are label-free and thus provide all the associated benefits, they have never been used to study viral growth kinetics and to predict drug antiviral response in cells. In this study, we prove the concept that the cell-based SPR assay can be applied in the kinetic analysis of the early stages of viral infection of cells and the antiviral drug activity in the infected cells. For this purpose, cells immobilized on the SPR slides were infected with human coronavirus HCov-229E and treated with hydroxychloroquine. The SPR response was measured at different time intervals within the early stages of infection. Methyl Thiazolyl Tetrazolium (MTT) assay was used to provide the reference data. We found that the results of the SPR and MTT assays were consistent, and SPR is a reliable tool in investigating virus–host cell interaction and the mechanism of action of viral inhibitors. SPR assay was more sensitive and accurate in the first hours of infection within the first replication cycle, whereas the MTT assay was not so effective. After the second replication cycle, noise was generated by the destruction of the cell layer and by the remnants of dead cells, and masks useful SPR signals.

Keywords: SPR; cell-based assay; viral growth kinetics; human coronavirus; hydroxychloroquine

Citation: Genova-Kalou, P.; Dyankov, G.; Marinov, R.; Mankov, V.; Belina, E.; Kisov, H.; Strijkova-Kenderova, V.; Kantardjiev, T. SPR-Based Kinetic Analysis of the Early Stages of Infection in Cells Infected with Human Coronavirus and Treated with Hydroxychloroquine. *Biosensors* **2021**, *11*, 251. https://doi.org/10.3390/bios11080251

Received: 20 June 2021
Accepted: 21 July 2021
Published: 26 July 2021

Publisher's Note: MDPI stays neutral with regard to jurisdictional claims in published maps and institutional affiliations.

1. Introduction

The cell is the minimum functional unit of living organisms. Knowledge of the basic cellular components and the way cells work is fundamental to life sciences, including molecular biology, cell biology, cell physiology, etc. With the traditionally used cell-based assays, it is a common practice to use radioactivity, chemiluminescence, or fluorescence to produce a measurable signal. Label-free cell-based assays have sparked interest due to their ability to measure cell responses without additional reporter compounds. Among the different label-free techniques, optical methods have been widely adopted for cell-based assays. The most effective one—surface plasmon resonance (SPR)—has been applied in the study of a variety of cellular processes.

In its conventional approach, SPR detects the binding of molecules in the detection volume on a sensor chip in real-time without any labeling. The signals are generated by a change in the biomolecule layers and are linearly related to their thickness. This is true in the first approximation since the layers are uniform and much smaller than the penetration depth of the plasmon wave. The situation is different in cell-based SPR assays where cells are immobilized on the sensing surface and serve as sensing elements. Nevertheless, SPR sensing has been extended into a powerful method for sensing large biological objects such as cells [1].

1.1. Penetration Depth and Detection Depth

An essential concept in SPR sensing is the nature of the evanescent field of the plasmon wave. This is especially relevant in functional cell-based assays because the penetration depth of 150–500 nm from the metal surface is only a fraction of the height (vertical dimension) of commonly used cells, which is in the range of several micrometers. Thus, it was suggested that the SPR signal is provoked by biological events in the area near the plasma membrane, whereas events inside cells, especially in their upper area, cannot be detected [2–4].

Two approaches have been used to achieve a greater penetration depth of the plasmon wave: excitation of long-range surface plasmon in a specific SPR biochip [5] and plasmon excitation in the UV range [6,7]. The latter approach seems to be more effective—the penetration depth could reach several microns. Even though these modifications are clearly advantageous, it may turn out that the more distant cell regions can be detected without applying them.

Although in the majority of cases the penetration depth is within the range of several hundred nanometers, SPR sensing is not necessarily applied to this limited range. SPR has been successfully used in detecting cell responses to external triggers such as drugs [8–11] and external stimuli [12].

It has been demonstrated [1,13,14] that a refractive index (RI) near the plasma membrane might reflect the accumulation and rearrangement of proteins activated by intracellular signal transduction provoked by exogenous stimuli. The SPR signals generated by the cellular response originate from complex biological events that have a local impact on RI. Additional experiments are required to find out what biological matter elicits the SPR signal.

1.2. Application in Drug Research

SPR technology has been widely applied in studying drugs. These studies have been generally limited to bimolecular binding assays outside living cells where purified biomolecules have been immobilized and a binding reaction with the target drug molecules has been detected [15–17].

Instead of immobilizing purified biomolecules in binding assays, a whole-cell adhesion to a sensing surface would provide on-site signals from drug–living cell interactions. Therefore, the pharmacokinetic parameters obtained by the cell-based assays would be more accurate and reliable than those obtained by the biochemical binding assays. A number of research groups have reported cell-based SPR assays used in evaluating the efficiency of a variety of drugs. A comprehensive review of cell-based SPR assay can be found elsewhere [18].

1.3. Cell-Based SPR Assay in Virus Research

The SPR technique has been widely used in studying viruses. Its well-known advantages are as follows: (i) the fact that it is label-free, thus eliminating functionalization of multiple antibodies, which occurs with ELISA; (ii) dynamic measurement of binding–unbinding kinetics; and (iii) high sensitivity, providing reliable virus detection. Reasonably, more research has been focused on viral diagnosis. The recognition elements used have included antibodies, antigens, DNA, and aptamers. Viral kinetic analysis of cells infected with SARS [19], SARS-CoV [20], and SARS-CoV-2 [21] has been performed. Surprisingly, the cell-based SPR assay has never been used in virus research so far.

1.4. Aim of the Present Study

Coronaviruses are disease-causing agents that infect many species of mammals and birds. Some, such as HKU1, OC43, 229E, and NL63, circulate seasonally and cause respiratory diseases in children and adults, which are not life-threatening. At the end of 2019, the identification of SARS-CoV-2 as the causal agent of atypical pulmonary diseases was the latest example of these emerging coronaviruses. It is essential to investigate the way in which the virus replication cycle occurs. MTT and immunofluorescence have been widely applied in investigating virus kinetics. However, the necessity of studying virus kinetics in the first hours of infection requires the application of new methods. Although the cell-based SPR assay has been successfully used to study intracellular processes, it has never been applied in examining the kinetics of the ultrastructure of virus-infected cells. The aim of this study was to prove that the cell-based SPR assay can be applied in the kinetic analysis of the early stages of viral infection of cells and the antiviral drug activity in the infected cells.

2. Materials and Methods

There are seven human coronaviruses (HCoV), highly diverse and causing respiratory diseases with mild to severe outcomes [22]. Currently, no specific antiviral drugs to treat HCoV infection are available, although hydroxychloroquine (HCQ) has been suggested as appropriate [23]. The way in which HCQ exerts its effect on time-dependent HCoV growth is well known and helps us analyze the SPR signal and compare it with other methods.

Human cell line culture: HCoV, strain 229E (HCoV-229E) was isolated using Vero E6 (African green monkey kidney) cell line, obtained from the National Center of Infectious and Parasitic Diseases (NCIPD), Bulgaria. The cells were cultured in Dulbecco's Modified Eagle's Medium (DMEM, Sigma-Aldrich, Sent Luis, MO, USA), supplemented with 10% fetal bovine serum (FBS, Gibco^{TM} by Life Technology, Darmstadt, Germany) and antibiotics at 37 °C, 5% CO_2 atmosphere.

Virus propagation: The HCoV-229E (from the NCIPD viral collection) was propagated in Vero E6 cells that reached 70% confluence in DMEM media, supplemented with 2% FBS at 37 °C, 5% CO_2 atmosphere.

Virus titration: Confluent Vero cells (3×10^4 cells/well) were cultured in 96-well plates (100 µL/well). Serial 10-fold dilutions of the HCoV-229E stock (10^{-1} to 10^{-8}) were prepared in DMEM supplemented with 2% FBS, at 37 °C and 5% CO_2 atmosphere for 4 days. Cell viability and yields of virus progeny were measured post-infection (*p.i.*) every 24 h for a 96 h period of total incubation time. The infected cells were monitored microscopically daily for cytopathic effects (CPE) in the infected cells caused by HCoV-229. The titer of the purified HCoV-229E was $10^{4.5}$ (high-titer) 50% tissue culture infection doses (TCID(50)/mL.

MTT assay: The MTT assay (3-(4,5-dimethyl-2-thiazolyl)-2,5-diphenyl-2H-tetrazolium bromide (Methyl Thiazolyl Tetrazolium; MTT) is used to measure cellular metabolic activity as an indicator of cell viability, proliferation, and cytotoxicity. MTT is reduced by mitochondrial dehydrogenases to the water-insoluble pink compound formazan, depending on the viability of cells. Vero cells were seeded in 96-well microtiter plates (3×104 cells/mL) and infected with HCoV-229E, multiplicity of infection (MOI) 0.1, and treated with different HCQ non-toxic concentrations at different hours. Measuring the optical density (OD) by the MTT assay has been used as a sensitive method to quantify the density of the HCQ-treated infected cells [24]. The OD values have been measured at a wavelength of 540 nm using ELISA reader (Sunrise Basic Tecan, Männedorf, Switzerland), whereby the concentration of viable cells is found [25]. The same approach was used in our study as well.

Cell-based SPR assay: The SPR slides were derived from a recordable compact disc (CD-R). A gold layer with thickness 80–100 nm was deposited onto the polycarbonate substrate by vacuum thermal evaporation.

Before seeding the cells, the slides were immersed in isopropyl alcohol and cleaned ultrasonically for 10 min. Then, they were rinsed thoroughly with high purity water, dried, and illuminated by UV light for 24 h.

The Vero E6 cells were cultured in DMEM (Dulbecco's Modified Eagle Medium), supplemented with 10% heat-inactivated fetal bovine serum (FBS) and antibiotics at a density 3×10^4 cells/mL and incubated for 24 h at 37 °C and 5% CO_2 conditions to allow cell adhesion to the SPR slide surface. When the cells achieved appropriate density (about 70% confluence on the SPR slide surface), a monolayer was washed twice with phosphate buffer solution (PBS), pH = 7.3, the supernatant was carefully removed, and the cell culture medium was supplemented with 2% FBS and HCoV-229E with multiplicity of infection (MOI) 0.1. After virus adsorption, the infected cells were treated with a non-cytotoxic concentration (1 mg/mL) of the antiviral drug HCQ. The infected and treated cultures were incubated at 33 °C in a humidified 5% CO_2 atmosphere. Cell morphology was observed every 6 h for microscopically detectable morphological alterations, such as loss of confluency, cell rounding and shrinking, and cytoplasm granulation and vacuolization. The viability of the infected and treated cells from each well of the 96-well culture plate was determined every 6 h by an MTT-assay, and the SPR spectral shift of the cell-based SPR assay was also measured.

SPR measurement: The gilded diffraction grating is part of a continuous CD-R spiral groove with a period of 1.55 μm. A Θ-2θ goniometer with a 0.01 deg accuracy was used for the SPR excitation and registration. Spectral interrogation was used for the SPR registration. A collimated beam of p-polarized white light under angle of incidence in the range 35–42 degrees excited resonances between 710 nm and 610 nm in a bare grating. A spectrometer registered the spectrum in the zero-order reflection. The optical setup is depicted in Figure 1.

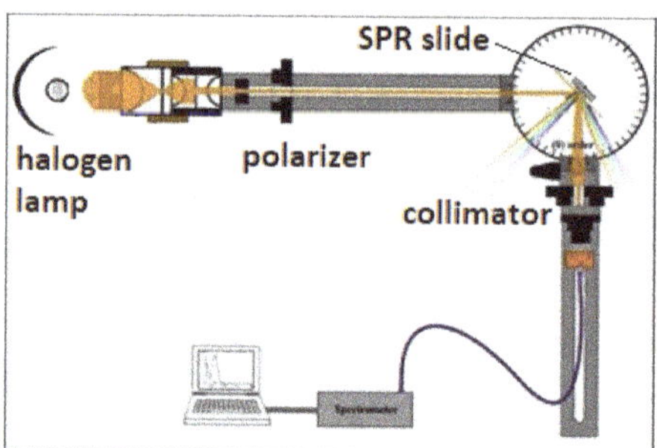

Figure 1. Optical setup for the SPR measurements.

Figure 2 shows the experimentally observed resonances in the bare SPR slides and the cell-based SPR assays: a slide with cells obtained at 12 h after seeding and a slide with infected cells at 12 h *p.i.* The changes in cellular morphology, which in turn led to a variation of the effective refractive index at the interface between the cell membrane and the metal layer, caused a well observable spectral shift, marked in Figure 2. Reference resonances were established for bare grating—the black curve in Figure 2. The spectral shift of the slides with non-infected cells was evaluated against reference resonances—marked as "A" in Figure 2. The spectral shift established in this way is referred to as "cell control". The SPR shift for infected cells was evaluated against the cell control—marked as "B" in Figure 2. The spectral shift established in this way is referred to as "virus control". The SPR spectral shift of the treated cells was evaluated against the virus control, after which it was compensated for by dividing the difference between cell control and virus control. The signal established was referred to as "SPR compensated signal". The SPR responses were measured at different time intervals between 4 and 48 h.

AFM examination: An Atomic Force Microscope (AFM) (Asylum Research MFP-3D (Oxford Instruments) was supplied with silicon nitride probes: frequency of 30 kHz, spring constant of 0.27 N/m, and radius <15 nm. The experiments were carried out at ambient conditions using the AFM contact mode. For the purposes of scanning, the cell-based SPR assays were fixed with 2.5% glutaraldehyde.

In the next section, we describe what type of biological events occur at different moments and determine the registered SPR shift.

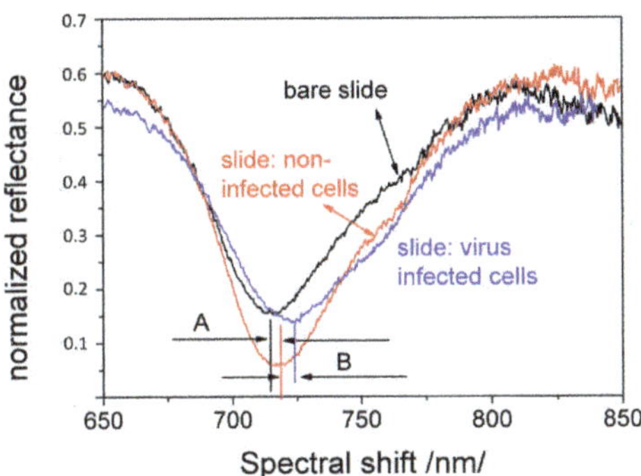

Figure 2. Experimentally observed resonances of bare grating and cell-based SPR assays.

3. Results

3.1. Cells Growth Kinetics

Non-human primate kidney cell line Vero E6 was seeded at concentration 3×10^4 cells/mL on SPR slides and on glass plates. The cells on the glass plates were counted by MTT. Every 6 h, the SPR spectral shift of the cell-based SPR assay was measured. The results are presented as a dotted line in Figure 3. A reference measurement with an MTT assay was provided. The data obtained—the mean values of three indep endent experiments—are presented in Figure 3: an SPR assay measurement (dotted line) and an MTT measurement (solid line). Obviously, the SPR signal follows the temporal change of cell viability established by the MTT assay. We also observed that the Vero E6 cells grew rapidly, producing a confluent monolayer. Even after a prolonged period, the Vero E6 cells showed the typical morphological characteristics of spindle-shaped fibroblasts—flat, without prominence in their shape and surface, with intact cytoplasm and oval nucleus, 15–20 μm long, and about 5 μm high—all of which was microscopically proven by the AFM examination (Figure 4). The AFM study showed that the cells had adhered tightly to the grating surface, which ensured an effective penetration of the plasmon wave into the cells. The population doubling per day (Pd/D) and cell density were found to increase with the prolongation of cell cultivation expressed as a steeper SPR curve in the range 25–38 h (Figure 3). The growth curves constructed from both the SPR and MTT assays showed the typical pattern of a growth curve. The SPR curve clearly indicated a lag phase (until 20 h), an exponential phase (25–38 h), a plateau (around 40 h), and reaching a death phase (40–48 h). This suggested that the cell density used was appropriate for running the experiments with the present virus.

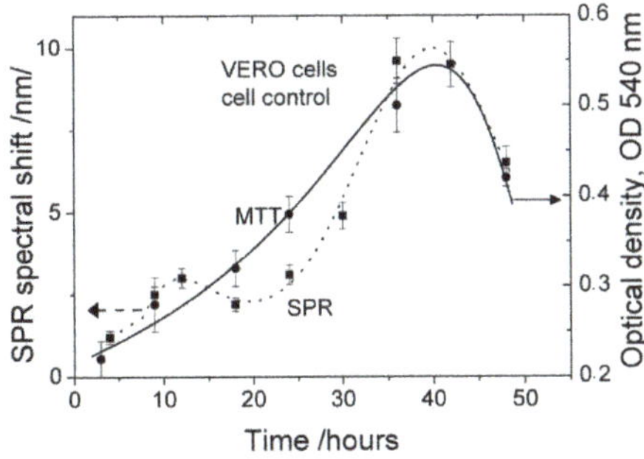

Figure 3. Growth kinetics of Vero cells: dotted line—SPR results (cell control); solid line—MTT results.

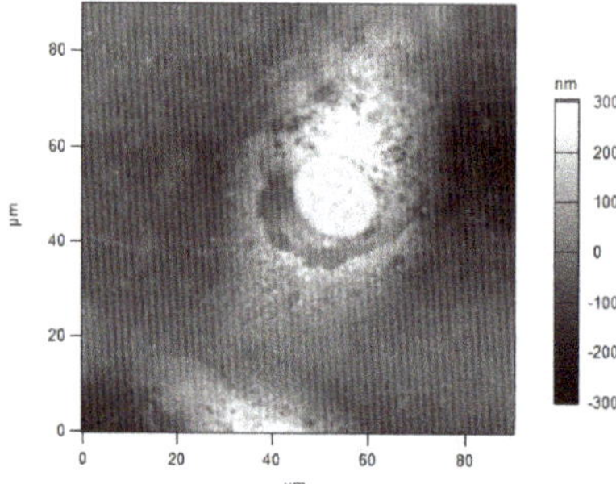

Figure 4. Vero cell at 24 h *p.i.*; AFM scan of the diffraction grating.

3.2. Viral Growth Kinetics

To estimate the viral growth curve characteristics, the infectivity titer of HCOV-229E was determined every 6 h *p.i.* by MTT (solid curve in Figure 5). The MTT assay measurement revealed the main phases of the viral growth kinetics. The initial increase in viability by 10 h is due to an increase in the number of uninfected cells as a result of the cell growth process.

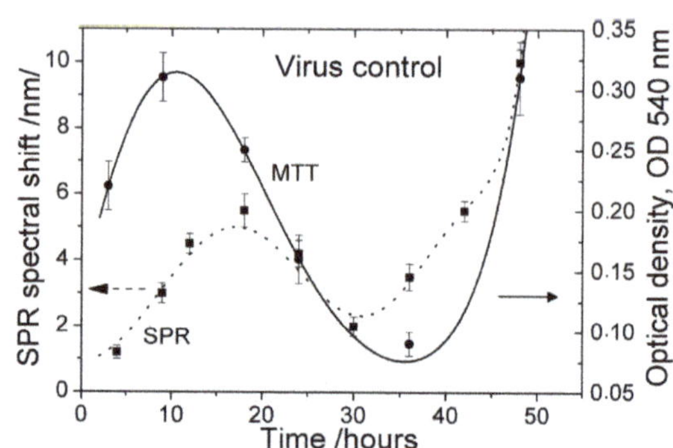

Figure 5. Growth kinetics of HCoV-229E-infected cells: dotted line—SPR results (virus control); solid line—MTT results.

The increase in the SPR spectral shift (dotted curve in Figure 3) lasted until 18 h, probably due not only to the increased cell density on the grating surface but also to the attachment of the viruses to the cell membrane. Then, the SPR assay measurement accurately indicates the infection efficiency.

We would like to point out that the measurement time interval of MTT assay was almost twice as long as the SPR time interval. This is due to the lower time resolution of MTT. The highest time resolution of SPR assay can explain observed local maximums of the cell control around 12 h (Figure 3) and of virus control around 18 h (Figure 4).

MTT showed a substantial decrease in cell viability in the interval 20–24 h (Figure 5), which corresponded to dramatic ultrastructural changes—marked granulation of cell cytoplasm, particularly around the nucleus with the fragmentation of the latter, and proliferation of pseudopodia at the cell periphery. This was confirmed by the AFM examination (Figure 6).

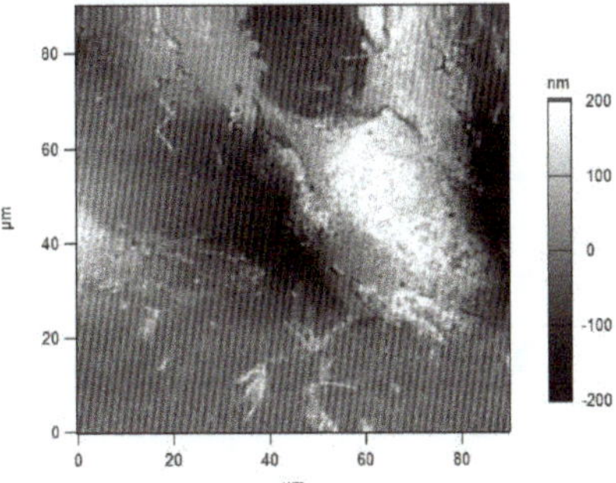

Figure 6. Infected Vero cell at 42 h *p.i.*; AFM scan of the diffraction grating.

The SPR signal also decreased in this interval and reached its minimum around 30 h. This is close to the minimum of cell viability (MTT curve—Figure 5), well pronounced in the interval 30–40 h, confirming the cell monolayer destruction, cell deterioration, and detachment from the surface as a result of the increased virus production.

At the end of the first replication cycle (24–30 h), the virus was expressed in the intercellular space. This was clearly observed by the AFM study as shown in Figure 7, which represents a magnified part of an area (shown in Figure 4) located near the cell membrane. Viruses (marked with arrows) have just been expressed from the host cell at 24 h p.i and are still close to the cell membrane. After this moment, the second replication cycle starts: virus attachment to the cell membranes of new cells. As a result of this, compaction of the cell membrane occurred, and the refractive index increased. This coincided with the increase in the SPR signal after 30 h.

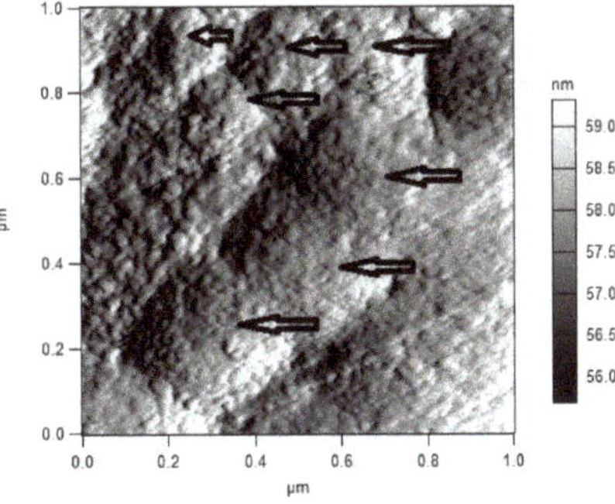

Figure 7. Viruses expressed in the intercellular space at 24 h *p.i.*

The change in the MTT signal was more inert—it increased after 36 h due to the competing processes of cell growth and viral replication.

3.3. Kinetics of Antiviral Activity of HCQ

Hydroxychloroquine (HCQ), used to treat malaria and some autoimmune disorders, might be of certain use in the clinical management of infections caused by HCoV, especially SARS coronavirus (SARS-CoV-1) and SARS-CoV-2, by potently inhibiting the infection, a fact that has been found in cell culture studies [26]. Here, we report an in vitro kinetic analysis of the antiviral activity of HCQ against the HCoV-229E strain. We would like to point out here that such a study has not been carried out so far.

The cytotoxicity of HCQ in Vero E6 cells was measured in advance for the purposes of determining the concentrations that would not cause injury or death to the treated cells (data not shown). This experiment was conducted three times and the results obtained are shown in Figure 8. To gain an initial insight into the stages of the viral replication cycle, at which HCQ is likely to exert its antiviral activity, time-of-drug-addition assays, such as SPR and MTT, were elaborated.

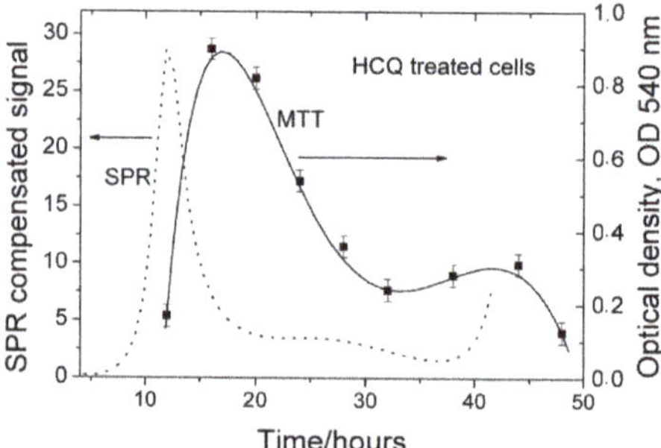

Figure 8. Kinetics of HCQ activity: dotted line—SPR results; solid line—MTT results.

The experiment involved Vero E6 cells (3×10^4 cells/mL) infected with HCoV-229E (MOI = 0.1) and treated with an HCQ standard (Sigma-Aldrich) at a concentration of 1 mg/mL (maximum non-toxic concentration). The replication cycle of HCoV-229E has demonstrated rapid viral propagation inside host cells, reaching maximum levels at 24 h *p.i.* [27]. This was confirmed by the AFM scans performed in our study—Figure 7, as well.

The inhibition of post-translational glycosylation with subsequent reduction in SARS-CoV-2 binding to and fusion with the angiotensin-converting enzyme 2 (ACE2) receptor of the host cell is an important antiviral effect of HCQ used in the treatment of SARS-CoV-2 infections [28]. Cleavage of SARS-CoV-2 spike (S) proteins by HCQ in the autophagosomes has also been reported [29]. Thus, the highest antiviral activity has to be expected at the stage of virus attachment to cells.

Indeed, our MTT assay confirmed the same mechanism of action in HCoV-229E: HCQ inhibited its replication in Vero E6 cells until 18 h *p.i.* A maximum antiviral activity was observed around 18 h—Figure 8. This coincided with the local maximum of virus control established by SPR (Figure 5), which confirmed that HCQ effectively inhibited virus replication at the stage of its attachment to the membrane and penetration into the cells.

The SPR study showed that maximum antiviral activity was reached around 12 h. This is also the stage of the virus replication cycle corresponding to the attachment to the membrane and penetration into the cells. The temporal shift against the MMT results could be explained by the method of compensation of SPR signals generated by the processes of cell growth and virus growth, as expanded in Section 2. The compensation procedure accumulates the error of surface plasmon resonance measurements (about 1.5 nm) and influences the peak position. However, the SPR signal is well pronounced at the expense of decreased accuracy.

4. Conclusions

A cell-based SPR assay was used to study cell growth, virus growth kinetics, and hydroxychloroquine antiviral kinetics. The MTT method was used as a reference since it is widely adopted for assessing cell metabolic activity. The kinetics revealed by the cell-based SPR assay was consistent with the findings of the MTT assay. Although the principles of the SPR and MTT methods are very different, the results obtained were very similar. All that showed that cell-based SPR is a reliable tool in investigating virus–host cell kinetics and antiviral drug activity. As expected, we found that the SPR assay provides better time resolution than MTT.

To the best of our knowledge, the present study is the first one focusing on the inhibiting effect of HCQ on the HCoV-229E virus. Both the SPR and MTT assay revealed that the antiviral efficiency is highest at the first stages of infection.

However, a few points have to be considered for the correct SPR measurement. First, the cells have to be seeded on the SPR slides with almost uniform density, which would ensure a reliable SPR signal across the slide. Additionally, the cell density has to prevent light scattering so that the reflection from the grating can be detected reliably. In addition, the method for compensating the signals generated by the processes of cell and virus growth has to be carefully considered.

There is a significant limitation to the cell-based SPR assay in investigating virus–host kinetics—it cannot be applied for a period lasting more than two virus replication cycles. After this period, the SPR signal is masked by destruction of the cell monolayer, detachment from the grating surface, and presence of remnants of the destroyed cells on the surface. As a result, the SPR signal is not unambiguously defined by the virus–host interaction.

In the present research, we showed that the cell-based SPR assay is applicable for in vitro studies. However, an extension of the SPR assay for in vivo application is a matter of engineering solutions.

Author Contributions: Conceptualization, P.G.-K.; methodology, P.G.-K. and G.D.; validation, P.G.-K. and R.M.; formal analysis, R.M. and E.B.; investigation, P.G.-K. and G.D.; resources, P.G.-K. and V.M.; data curation, E.B. and V.S.-K.; writing—original draft preparation, review and editing, P.G.-K. and G.D.; visualization, H.K.; supervision, T.K.; project administration, P.G.-K.; funding acquisition, G.D. All authors have read and agreed to the published version of the manuscript.

Funding: This research was funded by Bulgarian National Science Fund, grand number PN33/1, and by the Bulgarian Ministry of Education and Science, grant number D01-392.

Institutional Review Board Statement: Not applicable.

Informed Consent Statement: Not applicable.

Data Availability Statement: Data is contained within the article.

Acknowledgments: The authors thank Veneta Koceva, Institute of Optical Materials and Technology, for SPR slides elaboration.

Conflicts of Interest: The authors declare no conflict of interest.

References

1. Hide, M.; Tsutsui, T.; Sato, H.; Nishimura, T.; Morimoto, K.; Yamamoto, S.; Yoshizato, K. Real-Time analysis of ligand-induced cell surface and intracellular reactions of living mast cells using a surface plasmon resonance-based biosensor. *Anal. Biochem.* **2002**, *302*, 28–37. [CrossRef] [PubMed]
2. Méjard, R.; Griesser, H.; Thierry, B. Optical biosensing for label-free cellular studies. *TrAC Trends Anal. Chem.* **2014**, *53*, 178–186. [CrossRef]
3. Abadian, P.N.; Kelley, C.; Goluch, E. Cellular analysis and detection using surface plasmon resonance techniques. *Anal. Chem.* **2014**, *86*, 2799–2812. [CrossRef] [PubMed]
4. Yanase, Y.; Hiragun, T.; Ishii, K.; Kawaguchi, T.; Yanase, T.; Kawai, M.; Sakamoto, K.; Hide, M. Surface Plasmon Resonance for Cell-Based Clinical Diagnosis. *Sensors* **2014**, *14*, 4948–4959. [CrossRef]
5. Méjard, R.; Thierry, B. Systematic study of the surface plasmon resonance signals generated by cells for sensors with different characteristic lengths. *PLoS ONE* **2014**, *9*, e107978. [CrossRef] [PubMed]
6. Golosovsky, M.; Lirtsman, V.; Yashunsky, V.; Davidov, D.; Aroeti, B. Midinfrared surface-plasmon resonance: A novel biophysical tool for studying living cells. *J. Appl. Phys.* **2009**, *105*, 102036. [CrossRef]
7. Lirtsman, V.; Golosovsky, M.; Davidov, D. Surface plasmon excitation using a Fourier-transform infrared spectrometer: Live cell and bacteria sensing. *Rev. Sci. Instruments* **2017**, *88*, 103105. [CrossRef]
8. Wang, W.; Yin, L.; Gonzalez-Malerva, L.; Wang, S.; Yu, X.; Eaton, S.; Zhang, S.; Chen, H.-Y.; LaBaer, J.; Tao, N. In situ drug-receptor binding kinetics in single cells: A quantitative label-free study of anti-tumor drug resistance. *Sci. Rep.* **2014**, *4*, 6609. [CrossRef]
9. Yin, L.; Yang, Y.; Wang, S.; Wang, W.; Zhang, S.; Tao, N. Measuring binding kinetics of antibody-conjugated gold nanoparticles with intact cells. *Small* **2015**, *11*, 3782–3788. [CrossRef]

10. Zhang, F.; Wang, S.; Yin, L.; Yang, Y.; Guan, Y.; Wang, W.; Xu, H.; Tao, N. Quantification of epidermal growth factor receptor expression level and binding kinetics on cell surfaces by surface plasmon resonance imaging. *Anal. Chem.* **2015**, *87*, 9960–9965. [CrossRef]

11. Berthuy, O.I.; Blum, L.J.; Marquette, C.A. Cancer-cells on chip for label-free detection of secreted molecules. *Biosensors* **2016**, *6*, 2. [CrossRef] [PubMed]

12. Yanase, Y.; Hiragun, T.; Kaneko, S.; Gould, H.; Greaves, M.W.; Hide, M. Detection of refractive index changes in individual living cells by means of surface plasmon resonance imaging. *Biosens. Bioelectron.* **2010**, *26*, 674–681. [CrossRef]

13. Yanase, Y.; Suzuki, H.; Tsutsui, T.; Hiragun, T.; Kameyoshi, Y.; Hide, M. The SPR signal in living cells reflects changes other than the area of adhesion and the formation of cell constructions. *Biosens. Bioelectron.* **2007**, *22*, 1081–1086. [CrossRef]

14. Chabot, V.; Miron, Y.; Charette, P.; Grandbois, M. Identification of the molecular mechanisms in cellular processes that elicit a surface plasmon resonance (SPR) response using simultaneous surface plasmon-enhanced fluorescence (SPEF) microscopy. *Biosens. Bioelectron.* **2013**, *50*, 125–131. [CrossRef] [PubMed]

15. Kumar, P.K. Systematic screening of viral entry inhibitors using surface plasmon resonance. *Rev. Med. Virol.* **2017**, *27*, e1952. [CrossRef]

16. Sohrabi, F.; Saeidifard, S.; Ghasemi, M.; Asadishad, T.; Hamidi, S.M.; Hosseini, S.M. Role of plasmonics in detection of deadliest viruses: A review. *Eur. Phys. J. Plus* **2021**, *136*, 1–71. [CrossRef]

17. Hong, M.; Lee, S.; Clayton, J.; Yake, W.; Li, J. Genipin suppression of growth and metastasis in hepatocellular carcinoma through blocking activation of STAT-3. *J. Exp. Clin. Cancer Res.* **2020**, *39*, 1–17. [CrossRef] [PubMed]

18. Su, Y.-W.; Wang, W. Surface plasmon resonance sensing: From purified biomolecules to intact cells. *Anal. Bioanal. Chem.* **2018**, *410*, 3943–3951. [CrossRef]

19. Park, T.J.; Hyun, M.S.; Lee, H.J.; Lee, S.Y.; Ko, S. A self-assembled fusion protein-based surface plasmon resonance biosensor for rapid diagnosis of severe acute respiratory syndrome. *Talanta* **2009**, *79*, 295–301. [CrossRef]

20. Chen, H.; Gill, A.; Dove, B.K.; Emmett, S.R.; Kemp, C.F.; Ritchie, M.A.; Dee, M.; Hiscox, J.A. Mass spectroscopic characterization of the coronavirus infectious bronchitis virus nucleoprotein and elucidation of the role of phosphorylation in rna binding by using surface plasmon resonance. *J. Virol.* **2005**, *79*, 1164–1179. [CrossRef] [PubMed]

21. Shang, J.; Ye, G.; Shi, K.; Wan, Y.; Luo, C.; Aihara, H.; Geng, Q.; Auerbach, A.; Li, F. Structural basis of receptor recognition by SARS-CoV-2. *Nat. Cell Biol.* **2020**, *581*, 221–224. [CrossRef]

22. Zhao, X.; Ding, Y.; Du, J.; Fan, Y. 2020 update on human coronaviruses: One health, one world. *Med. Nov. Technol. Devices* **2020**, *8*, 100043. [CrossRef] [PubMed]

23. Gautret, P.; Lagier, J.C.; Parola, P.; Hoang, V.T.; Meddeb, L.; Mailhe, M.; Doudier, B.; Courjon, J.; Giordanengo, V.; Vieira, V.E.; et al. Hydroxychloroquine and azithromycin as a treatment of COVID-19: Results of an open-label non-randomized clinical trial. *Int. J. Antimicrob. Agents* **2020**, *56*, 105949. [CrossRef]

24. Mosmann, T. Rapid colorimetric assay for cellular growth and survival: Application to proliferation and cytotoxicity assays. *J. Immunol. Methods* **1983**, *65*, 55–63. [CrossRef]

25. Sieuwerts, A.M.; Klijn, J.G.M.; Peters, H.A.; Foekens, J.A. The MTT Tetrazolium Salt Assay Scrutinized: How to Use this Assay Reliably to Measure Metabolie Activity of Cell Cultures in vitro for the Assessment of Growth Characteristics, IC50-Values and Cell Survival. *Clin. Chem. Lab. Med.* **1995**, *33*, 813–824. [CrossRef] [PubMed]

26. Hashem, A.M.; Alghamdi, B.S.; Algaissi, A.A.; Alshehri, F.S.; Bukhari, A.; Alfaleh, M.; Memish, Z.A. Therapeutic use of chloroquine and hydroxychloroquine in COVID-19 and other viral infections: A narrative review. *Travel Med. Infect. Dis.* **2020**, *35*, 101735. [CrossRef] [PubMed]

27. Friedman, N.; Jacob-Hirsch, J.; Drori, Y.; Eran, E.; Kol, N.; Nayshool, O.; Mendelson, E.; Rechavi, G.; Mandelboim, M. Transcriptomic profiling and genomic mutational analysis of Human coronavirus (HCoV)-229E -infected human cells. *PLoS ONE* **2021**, *16*, e0247128. [CrossRef] [PubMed]

28. Lu, H. Drug treatment options for the 2019-new coronavirus (2019-nCoV). *Biosci. Trends* **2020**, *14*, 69–71. [CrossRef] [PubMed]

29. Mauthe, M.; Orhon, I.; Rocchi, C.; Zhou, X.; Luhr, M.; Hijlkema, K.-J.; Coppes, R.P.; Engedal, N.; Mari, M.; Reggiori, F. Chloroquine inhibits autophagic flux by decreasing autophagosome-lysosome fusion. *Autophagy* **2018**, *14*, 1435–1455. [CrossRef]

biosensors

Communication

Surface Plasmon Resonance for Protease Detection by Integration of Homogeneous Reaction

Ning Xia [1], Gang Liu [1,2] and Xinyao Yi [3,*

[1]   College of Chemistry and Chemical Engineering, Anyang Normal University, Anyang 455000, China; ningxia@aynu.edu.cn (N.X.); liugang08215@163.com (G.L.)
[2]   College of Chemistry and Chemical Engineering, Henan University of Technology, Zhengzhou 450011, China
[3]   College of Chemistry and Chemical Engineering, Central South University, Changsha 410083, China
*   Correspondence: yixinyao@csu.edu.cn

Abstract: The heterogeneous assays of proteases usually require the immobilization of peptide substrates on the solid surface for enzymatic hydrolysis reactions. However, immobilization of peptides on the solid surface may cause a steric hindrance to prevent the interaction between the substrate and the active center of protease, thus limiting the enzymatic cleavage of the peptide. In this work, we reported a heterogeneous surface plasmon resonance (SPR) method for protease detection by integration of homogeneous reaction. The sensitivity was enhanced by the signal amplification of streptavidin (SA)-conjugated immunoglobulin G (SA-IgG). Caspase-3 (Cas-3) was determined as the model. A peptide labeled with two biotin tags at the N- and C-terminals (bio-GDEVDGK-bio) was used as the substrate. In the absence of Cas-3, the substrate peptide was captured by neutravidin (NA)-covered SPR chip to facilitate the attachment of SA-IgG by the avidin-biotin interaction. However, once the peptide substrate was digested by Cas-3 in the aqueous phase, the products of bio-GDEVD and GK-bio would compete with the substrate to bond NA on the chip surface, thus limiting the attachment of SA-IgG. The method integrated the advantages of both heterogeneous and homogeneous assays and has been used to determine Cas-3 inhibitor and evaluate cell apoptosis with satisfactory results.

Keywords: surface plasmon resonance; protease; caspase; avidin-biotin interaction

Citation: Xia, N.; Liu, G.; Yi, X. Surface Plasmon Resonance for Protease Detection by Integration of Homogeneous Reaction. *Biosensors* **2021**, *11*, 362. https://doi.org/10.3390/bios11100362

Received: 7 September 2021
Accepted: 27 September 2021
Published: 29 September 2021

1. Introduction

Proteases play an important role in a wide variety of biological processes, including protein digestion, wound healing, apoptosis, fertilization, growth differentiation, and immune system activation [1]. In the human body, at least 1.7% of human genes are encoded by proteases. The activities of proteases are closely related to many diseases, such as cancer, cardiovascular disease, Alzheimer's disease, human immunodeficiency virus (HIV), thrombosis, and diabetes [2]. Thus, extensive efforts have been made to screen protease inhibitors as potential drugs. This provides a powerful motivation for the development of sensitive, selective, and robust methods to detect protease and discover potential inhibitors.

Until now, many homogeneous and heterogeneous biosensors have been reported for the detection of proteases and screening of their inhibitors [2,3]. In homogeneous analysis, the substrate and protease sample are present in the aqueous phase. For instance, in the fluorescence resonance energy transfer (FRET) assay, the commonly used method for protease activity detection, the peptides labeled with two different fluorophores at two ends are digested by protease in the aqueous phase [4]. The activity of protease can be measured by monitoring the change of fluorescence signal after the cleavage of the peptide. On the contrary, the peptide substrate is anchored on a solid surface in the heterogeneous assay, and the enzymatic reaction happens at the solid-liquid interface [5,6]. Both the homogeneous and heterogeneous methods have their own advantages and disadvantages. Usually, homogeneous biosensors have the advantages of easy operation, rapid response, excellent sensitivity, and high throughput, but they show poor anti-interference ability and require large sample volumes and complex sample handling procedures. Conversely, heterogeneous assays exhibit the advantages of less sample consumption, ultra-high sensitivity and selectivity, and low instrument investment. Overall, the heterogeneous biosensors provide tremendous advantages over conventional homogeneous assays since numerous peptide substrates are immobilized at a discrete location on the solid interface [2]. However, immobilization of peptides on the solid surface will cause a steric hindrance to prevent the interaction between the substrate and the active center of protease [7], thus limiting the enzymatic cleavage of the peptide. Although the steric hindrance can be reduced by the use of nanomaterials-modified interface and the well-design of peptide substrate [8–10], the surface chemistry and coverage of peptide on the solid surface demands laborious optimization. Therefore, it is of importance to integrate the advantages of both heterogeneous and homogeneous assays for the design of general protease biosensors.

Surface plasmon resonance (SPR) is a simple, label-free technology to monitor the protein-protein interactions by measuring the refractive index change at the sensor surface [11–14]. The technology can be used to monitor the cleavage of protein or peptide fixed on the chip surface, providing a label-free detection method for protease analysis due to the advantages of fast response, real-time detection, high signal-to-noise, and good compatibility with the microfluidic system. For example, Steinrücke and co-workers suggested that cleavage of the helical protein with 78 amino acids by protease caused a detectable SPR signal [15]. However, the cleavage of low molecular weight peptides leads to a small, undetectable change in the refractive index [16–18]. Thus, it usually requires a signal amplification strategy to detect protease by labeling the peptide substrate with nanomaterials or specific groups [19–24]. Biotin is usually used to label peptide substrate for the design of heterogeneous biosensors. It can interact with avidin or its analogs of neutravidin (NA) and streptavidin (SA) with a binding coefficient as high as ~10^{15} M^{-1}. Such an interaction allows for the immobilization and recognition of peptide substrate at the solid-liquid interface [25–28]. By integrating the advantages of homogeneous assays, herein, we proposed a novel SPR method for protease detection by the signal amplification of SA-conjugated immunoglobulin G (SA-IgG). Caspases, a family of cysteine proteases, play an important role in apoptosis. To demonstrate the feasibility of the method, caspase-3 (Cas-3) that can specifically recognize and cleave the C-terminal of the peptide with the DEVD sequence was determined as the model. A peptide labeled with two biotin tags at the N- and C-terminals (bio-GDEVDGK-bio) was used as the substrate (Scheme 1). In the absence of Cas-3, the peptide substrate can be captured by the NA-covered chip through the avidin-biotin interaction (Channel 1). The biotin group at the other end of the peptide allows for the capture of SA-IgG, thus resulting in a strong SPR signal. When the peptide substrate was digested by Cas-3 in the aqueous phase, the biotinylated products (bio-GDEVD and GK-bio) would compete with the substrate to bond NA on the chip surface (Channel 2). This prevents the attachment of bio-GDEVDGK-bio and the follow-up capture of SA-IgG by the avidin-biotin interaction. However, when the activity of Cas-3 was suppressed by inhibitor, more bio-GDEVDGK-bio substrates would be anchored on the chip surface, which facilitating the capture of SA-IgG. The method was used to evaluate the inhibition efficiency of the inhibitor and monitor the activity of Cas-3 in cell lysates.

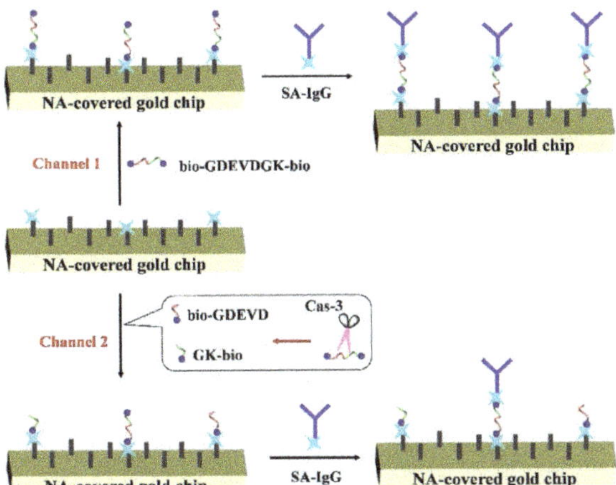

Scheme 1. Schematic representation of SPR method for the detection of Cas-3 using NA-covered gold chip. The signal was amplified by SA-IgG conjugates.

2. Materials and Methods

2.1. Chemicals and Materials

NA protein was purchased from Thermo Fisher Scientific (Shanghai, China). Cas-3 was obtained from New England BioLabs (Ipswich, MA, USA). Thrombin, beta-secretase, prostate-specific antigen (PSA), and bovine serum protein (BSA) were acquired from Sigma-Aldrich (Shanghai, China). SA-IgG and glutathione (GSH) were purchased from Sangon Biotech (Shanghai, China). Peptides were provided by China Peptide Co., Ltd. (Shanghai, China). Other reagents were ordered from Aladdin Reagent Co., Ltd. (Shanghai, China). All aqueous solutions were prepared daily with ultrapure water collected from a Milli-Q purification system.

2.2. Preparation of SPR Chips

The gold chips were annealed in a hydrogen flame to eliminate the surface contaminant. Then, the cleaned gold chips were incubated with 1 µM NA protein in carbonate buffer (pH 10) for 2 h. NA was capped on the gold surface through the hydrophobic and Au-S interactions [29]. The unbound NA proteins were removed by rinsing the chip with the carbonate buffer. The unreacted gold surface were blocked by incubation of the chip with 10 µM BSA and 100 µM GSH. Finally, the NA-covered chips were thoroughly washed with the carbonate buffer.

2.3. Procedure for Cas-3 Detection

The peptide bio-GDEVDGK-bio was digested by Cas-3 in the HEPES buffer with the optimized reaction conditions. The reaction mixture was delivered to the flow cell by a syringe pump. When the baseline is stable, SA-IgG in phosphate buffer (10 mM, pH 7.4) was injected into the SPR channel by the pump. The signal was collected by measuring the change of SPR dip shift on a BI-SPR 3000 system (Biosensing Instrument Inc., Tempe, AZ, USA).

2.4. Inhibitor Detection and Cell Lysate Analysis

For the detection of Cas-3 inhibitor, DEVD-FMK at different concentrations was mixed with a given concentration of Cas-3 for 10 min. Then, the resultant solution was incubated with the peptide substrate. For real sample assays, HeLa cells were cultured and the cell lysates were extracted with our reported procedures [30,31]. Then, the peptide substrate was incubated with the diluted cell lysates to react for 2 h. Finally, the reaction mixture was delivered to the flow cell, followed by injection of SA-IgG to the channel after the baseline stable was attained.

3. Results and Discussion

3.1. Feasibility of the Strategy

Based on the avidin-biotin interactions, NA or SA-modified magnetic beads, chromatography columns and solid surfaces have been widely used for the immobilization and separation of biotinylated biomolecules. In this work, an NA-covered gold chip was used to capture the biotinylated peptide. Figure 1 depicts the SPR responses when injecting SA-IgG, SA, and IgG to the sensor channel. A negligible change in the dip shift was observed after injecting SA-IgG conjugate to the NA-covered chip (curve a), demonstrating that SA-IgG showed no interaction with the sensor chip. Interestingly, the SPR dip shift reached 207 mD when injecting the conjugate to the chip treated by bio-GDEVDGK-bio (curve b). No significant change was observed when injecting IgG onto the bio-GDEVDGK-bio treated channel (curve c) and a smaller SPR dip shift (59 mD) was attained when injecting SA onto the channel (curve d). Thus, the change in curve b should be attributed to the avidin-biotin interaction and the signal was greatly amplified by IgG due to its large molecular weight (~150000 Da). We also found that the signal was intensified with the increase in bio-GDEVDGK-bioconcentration from 0.01 to 20 nM and began to level off beyond 5 nM. To attain higher sensitivity and a wider linear range, 5 nM bio-GDEVDGK-bio was used as the substrate for the assays of Cas-3.

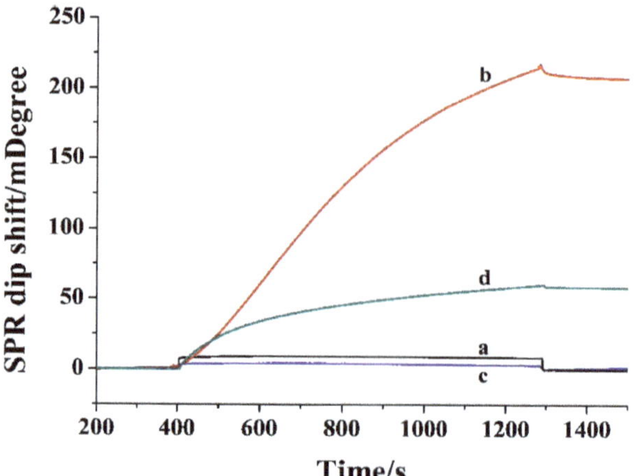

Figure 1. SPR sensorgrams when injecting 0.1 mg/mL SA-IgG to the fluidic channel wherein the NA-covered chip had been treated by blank buffer (curve a) and 20 nM bio-GDEVDGK-bio (curve b). Curves c and d were acquired when injecting 0.1 mg/mL IgG and 0.05 mg/mL SA onto the NA-covered chip treated by bio-GDEVDGK-bio.

3.2. Detection of Cas-3 and Its Inhibitor

The analytical performances were first investigated by determining different concentrations of Cas-3. In Figure 2A, the SPR signal decreased gradually with the increase in Cas-3 concentration in the range of 0~2000 pg/mL. The plateau beyond 1000 pg/mL is indicative of the completion of the enzymatic hydrolysis (Figure 2B). The signal did not decrease to the background value, indicating that not all the substrate peptides were cleaved by Cas-3 even at a higher concentration with a very long reaction time. The detection limit was estimated to be 0.5 pg/mL by measuring the sensor response to a dilution series and determining the smallest target concentration at which the signal was clearly distinguishable from the response to the blank solution. The value is lower than that attained by the homogeneous methods, such as fluorescence (128 pg/mL) [32], colorimetric assay (5 ng/mL) [33], differential pulse voltammetry (27.4 ng/mL) [34], and mass spectrometry (3.02 ng/mL) [35]. The value is comparable to or even lower than that achieved by heterogeneous methods based on the signal amplification of enzymes and nanomaterials (Table 1). The high sensitivity of the method should be attributed to the "immobilization-free" hydrolysis reaction and the large molecular weight of SA-IgG.

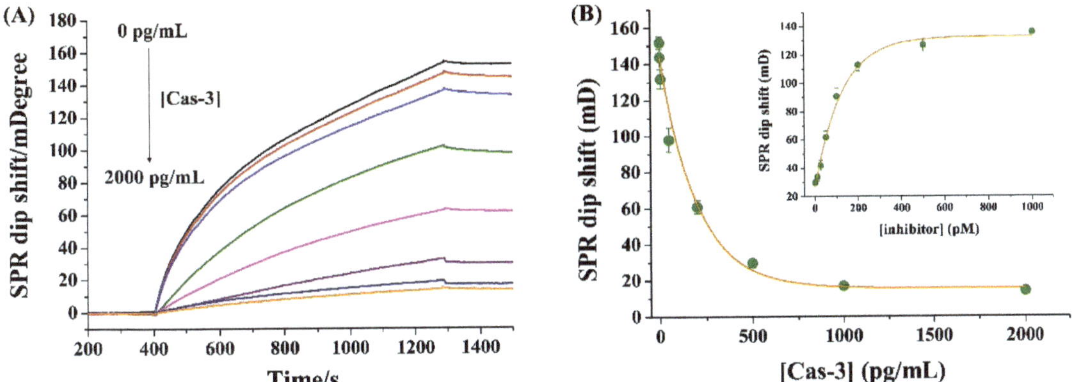

Figure 2. (**A**) SPR sensorgrams when injecting 0.02 mg/mL SA-IgG onto the NA-covered chips treated by the mixture of 5 nM bio-GDEVDGK-bio and a given concentration of Cas-3 (from top to bottom: 0, 0.5, 5, 50, 200, 500, 1000, and 2000 pg/mL). (**B**) Dependence of SPR signal on the concentration of Cas-3. The inset shows the effect of inhibitor concentration on the SPR signal wherein the concentration of Cas-3 was 500 pg/mL.

Table 1. An overview of heterogeneous methods for Cas-3 detection based on the signal amplification of enzymes and nanomaterials.

Method	Signal Label	Linear Range	Detection Limit	Ref.
EIS	Biotin-FNP network	1–125 pg/mL	1 pg/mL	[30]
EIS	SA-peptide network	0–50 pg/mL	0.2 pg/mL	[36]
DPV	Methylene blue/GO	0.1–100 pg/mL	0.06 pg/mL	[37]
DPV	Calixarene-rGO	10–100 pg/mL	0.0167 pg/mL	[38]
DPV	AuNPs-MCM/MB-HRP	10 fM–10 nM	5 fM	[39]
LSV	CB(8)-capped AgNPs	1–10 ng/mL	24.62 pg/mL	[40]
SWV	MB/HRP	0.1–1 nM	56 pM	[41]
ECL	Ru@SiO$_2$	0.2–200 pg/mL	0.07 pg/mL	[42]
ECL	HRP-SA-MB	0.5–100 fM	0.5 fM	[43]

Abbreviations: DPV, differential pulse voltammetry; LSV, linear sweep voltammetry; SWV, square wave voltammetry; EIS, electrochemical impedance spectroscopy; ECL, electrochemiluminescence; GO, graphene oxide; rGO, reduced graphene oxide; AuNPs-MCM, gold nanoparticle-coated silica-based mesoporous materials; MB, magnetic bead; HRP, horseradish peroxidase; AgNPs, silver nanoparticles; FNP, self-assembled biotin-phenylalanine nanoparticle.

As a proof-of-concept experimental for evaluation of Cas-3 activity, the inhibitor DEVD-FMK at different concentrations was incubated with 500 pg/mL Cas-3. The inset in Figure 2B shows the dependence of SPR dip shift on the concentration of inhibitor. The increase in inhibitor concentration induced the enhancement in the SPR signal, indicating that DEVD-FMK is an effective Cas-3 inhibitor. The half-maximal inhibitory concentration (IC$_{50}$) was estimated to be 98 nM, which is in agreement with that measured by other methods [30,36]. Thus, the method has bright prospects for the screening of protease inhibitors.

3.3. Selectivity

To evaluate the specificity of the method, the method was first challenged by determining other proteases (e.g., thrombin, beta-secretase, and PSA) to replace Cas-3. As shown in Figure 3, the tested proteases did not induce a significant decrease in the SPR dip shift, suggesting that the method shows high selectivity toward Cas-3 (*cf.* curves 1~4). However, trace biotin or other molecules in real samples may interact with biotin, thereby limiting the practical application of the technique. For this consideration, the interferences from avidin and biotinylated biomolecule such as bio-GLRRASLG were examined. As envisaged, both avidin and bio-GLRRASLG caused a significant decrease in the SPR signal (curves 5~6). The result is understandable as avidin can bind to the peptide substrate (bio-GDEVDGK-bio) and the biotinylated peptide can compete with the substrate to bind NA on the chip surface, thus preventing the attachment of bio-GDEVDGK-bio on the chip. To resolve this problem, a certain amount of biotin was added to the sample in advance to eliminate the interference of avidin. The free biotin or biotinylated peptide was then removed by the commercial SA-modified magnetic beads. As a result, the interferences from avidin and biotinylated peptide have been well eliminated (curves 7~8).

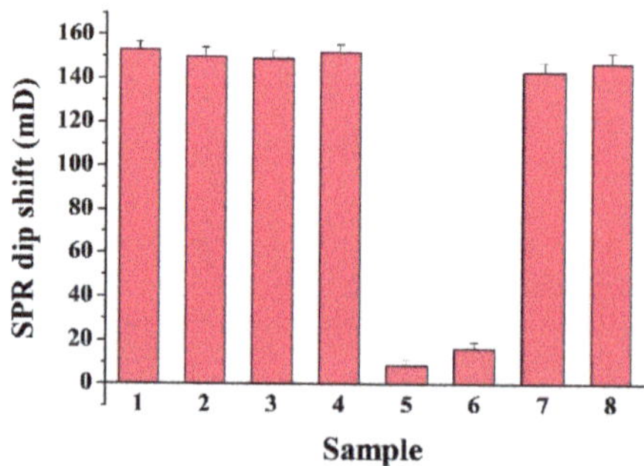

Figure 3. Selectivity of the method: bar 1, thrombin; bar 2, beta-secretase; bar 3, PSA, bar 4, Cas-3; bar 5, avidin; bar 6, bio-GLRRASLG; bar 7, the mixture of avidin and biotin pretreated by SA-modified magnetic beads; bar 8, bio-GLRRASLG pretreated by SA-modified magnetic beads. The concentrations of thrombin, beta-secretase, PSA, Cas-3, avidin, biotin, and bio-GLRRASLG were 5 ng/mL, 5 ng/mL, 5 ng/mL, 500 pg/mL, 200 ng/mL, 12.5 nM, and 5 nM, respectively.

3.4. Evaluation of Cell Apoptosis

Apoptosis is a highly regulated physiological process, which is of great significance in the life cycle of organisms. However, the imbalance of apoptosis may directly lead to the occurrence of many diseases. Therefore, the death caused by apoptosis has attracted extensive attention from experts in pathology, pharmacology, and toxicology. Among various types of caspases, Cas-3 is the central molecule to mediate the apoptotic pathway inside and outside cells. Therefore, Cas-3 has been regarded as the biomarker and therapeutic target for the diagnosis and treatment of apoptosis-related diseases. To verify the feasibility of this method for monitoring cell apoptosis, HeLa cells were used as the models. As shown in Figure 4A, when the peptide substrates were incubated with the cell lysates extracted from normal HeLa cells, the SPR signals were high and no significant changes were observed with the increase in cell number. However, when the cells were treated by STS (a common apoptosis inducer), the SPR signals decreased gradually with the increase in cell number. This indicated that the apoptosis was triggered by STS and the activity of Cas-3 was activated during apoptosis. STS-induced apoptosis was also confirmed by characterizing the cell morphology with a microscope (Figure 4B). The result is consistent with that obtained by other methods, indicating that the method can be used for the evaluation of apoptosis by monitoring the Cas-3 activity.

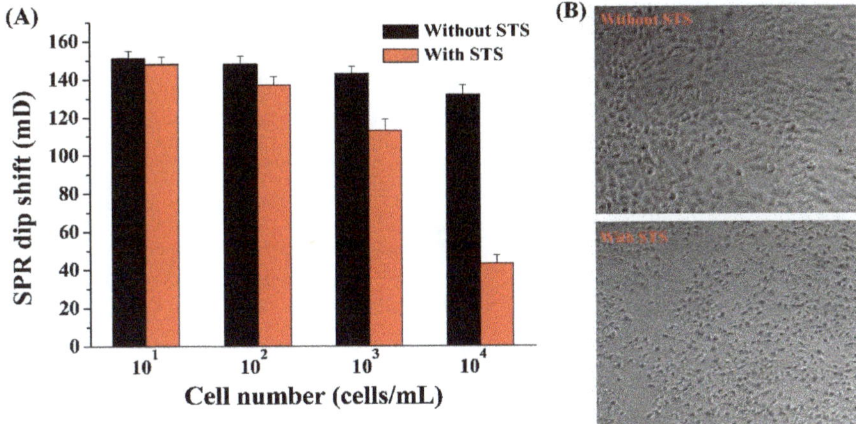

Figure 4. (**A**) Dependence of SPR signal on the concentration of normal and STS-treated HeLa cells. (**B**) Confocal images of normal and STS-treated HeLa cells.

4. Conclusions

In summary, we reported a heterogeneous SPR method for protease detection by integration of homogeneous enzymatic hydrolysis reaction. The signal was amplified by SA-IgG because of its large molecular weight. The method was used to determine Cas-3 activity and evaluate cell apoptosis with satisfactory results. The method exhibited high sensitivity and obviated the use of enzymes or nanomaterial for signal amplification. The "immobilization-free" strategy for the enzymatic reaction should be valuable for the design of novel heterogeneous biosensors to eliminate the effect of steric hindrance.

Author Contributions: Conceptualization, G.L. and N.X.; methodology, N.X.; investigation, G.L.; writing—original draft preparation, N.X.; writing—review and editing, X.Y.; project administration, N.X.; funding acquisition, X.Y. All authors have read and agreed to the published version of the manuscript.

Funding: This research was funded by the Program for Innovative Research Team of Science and Technology in the University of Henan Province (21IRTSTHN005), and the National Natural Science Foundation of China (21705166).

Institutional Review Board Statement: Not applicable.

Informed Consent Statement: Not applicable.

Data Availability Statement: Not applicable.

Conflicts of Interest: The authors declare no conflict of interest.

References

1. Suaifan, G.A.; Shehadeh, M.; Al-Ijel, H.; Ng, A.; Zourob, M. Recent progress in prostate-specific antigen and HIV proteases detection. *Expert Rev. Mol. Diagn.* **2013**, *13*, 707. [CrossRef] [PubMed]
2. Ong, I.L.H.; Yang, K.-L. Recent developments in protease activity assays and sensors. *Analyst* **2017**, *142*, 1867. [CrossRef] [PubMed]
3. Kaur, J.; Singh, P.K. Trypsin detection strategies: A review. *Crit. Rev. Anal. Chem.* **2020**. [CrossRef] [PubMed]
4. Dos Santos, M.C.; Algar, W.R.; Medintz, I.L.; Hildebrandt, N. Quantum dots for Forster Resonance Energy Transfer FRET. *TrAC-Trend Anal. Chem.* **2020**, *125*, 115819. [CrossRef]

5. La, M.; Zhao, X.; Peng, Q.; Chen, C.; Zhao, G. Electrochemical biosensors for probing of protease activity and screening of protease inhibitors. *Int. J. Electrochem. Sci.* **2015**, *10*, 3329.

6. Pavan, S.; Berti, F. Short peptides as biosensor transducers. *Anal. Bioanal. Chem.* **2012**, *402*, 3055. [CrossRef]

7. Xuan, F.; Fan, T.W.; Hsing, I.-M. Electrochemical interrogation of kinetically-controlled dendritic DNA/PNA assembly for immobilization-free and enzyme-free nucleic acids sensing. *ACS Nano* **2015**, *9*, 5027. [CrossRef] [PubMed]

8. Song, Y.; Fan, H.; Anderson, M.J.; Wright, J.G.; Hua, D.H.; Koehne, J.; Meyyappan, M.; Li, J. Electrochemical activity assay for protease analysis using carbon nanofiber nanoelectrode arrays. *Anal. Chem.* **2019**, *91*, 3971–3979. [CrossRef]

9. Mahshid, S.S.; Ricci, F.; Kelley, S.O.; Vallée-Bélisle, A. Electrochemical DNA-based immunoassay that employs steric hindrance to detect small molecules directly in whole blood. *ACS Sens.* **2017**, *2*, 718–723. [CrossRef]

10. Anderson, M.J.; Song, Y.; Fan, H.; Wright, J.G.; Ren, Z.; Hua, D.H.; Koehne, J.E.; Meyyappan, M.; Li, J. Simultaneous, multiplex quantification of protease activities using a gold microelectrode array. *Biosens. Bioelectron.* **2020**, *165*, 112330. [CrossRef]

11. Homola, J. Present and future of surface plasmon resonance biosensors. *Anal. Bioanal. Chem.* **2003**, *377*, 528. [CrossRef] [PubMed]

12. Jean-Francois, M. Surface plasmon resonance clinical biosensors for medical diagnostics. *ACS Sens.* **2017**, *2*, 16–30.

13. Chang, C.-C. Recent advancements in aptamer-based surface plasmon resonance biosensing strategies. *Biosensors* **2021**, *11*, 233. [CrossRef] [PubMed]

14. Kenshin, T. Surface plasmon resonance (SPR)- and localized SPR (LSPR)-based virus sensing systems: Optical vibration of nano- and micro-metallic materials for the development of next-generation virus detection technology. *Biosensors* **2021**, *11*, 250.

15. Steinrücke, P.; Aldinger, U.; Hill, O.; Hillisch, A.; Basch, R.; Diekmann, S. Design of helical proteins for real-time endoprotease assays. *Anal. Biochem.* **2000**, *286*, 26. [CrossRef]

16. Chen, H.; Mei, Q.; Hou, Y.; Zhu, X.; Koh, K.; Li, X.; Li, G. Fabrication of a protease sensor for caspase-3 activity detection based on surface plasmon resonance. *Analyst* **2013**, *138*, 5757. [CrossRef]

17. Tomassetti, M.; Conta, G.; Campanella, L.; Favero, G.; Sanzò, G.; Mazzei, F.; Antiochia, R. A flow SPR immunosensor based on a sandwich direct method. *Biosensors* **2016**, *6*, 22. [CrossRef]

18. Nakamura, S.; Yatabe, R.; Onodera, T.; Toko, K. Sensitive detection of capsaicinoids using a surface plasmon resonance sensor with anti-homovanillic acid polyclonal antibodies. *Biosensors* **2013**, *3*, 374–384. [CrossRef]

19. Esseghaier, C.; Ng, A.; Zourob, M. A novel and rapid assay for HIV-1 protease detection using magnetic bead mediation. *Biosens. Bioelectron.* **2013**, *41*, 335. [CrossRef]

20. Gao, Y.M.; Zou, F.; Wu, B.P.; Wang, X.X.; Zhang, J.J.; Koh, K.; Chen, H.X. CB [7]-mediated signal amplification approach for sensitive surface plasmon resonance spectroscopy. *Biosens. Bioelectron.* **2016**, *81*, 207. [CrossRef]

21. Huang, Y.; Sun, T.; Liu, L.; Xia, N.; Zhao, Y.; Yi, X. Surface plasmon resonance biosensor for the detection of miRNAs by combining the advantages of homogeneous reaction and heterogeneous detection. *Talanta* **2021**, *234*, 122622. [CrossRef] [PubMed]

22. Sun, T.; Zhang, Y.; Zhao, F.; Xia, N.; Liu, L. Self-assembled biotin-phenylalanine nanoparticles for the signal amplification of surface plasmon resonance biosensors. *Microchim. Acta* **2020**, *187*, 473. [CrossRef] [PubMed]

23. Zeng, S.; Baillargeat, D.; Ho, H.-P.; Yong, K.-T. Nanomaterials enhanced surface plasmon resonance for biological and chemical sensing applications. *Chem. Soc. Rev.* **2014**, *43*, 3426. [CrossRef] [PubMed]

24. Tabasi, O.; Falamaki, C. Recent advancements in the methodologies applied for the sensitivity enhancement of surface plasmon resonance sensors. *Anal. Methods* **2018**, *10*, 3906. [CrossRef]

25. Wu, L.; Wang, Y.; Xu, X.; Liu, Y.; Lin, B.; Zhang, M.; Zhang, J.; Wan, S.; Yang, C.; Tan, W. Aptamer-based detection of circulating targets for precision medicine. *Chem. Rev.* **2021**. [CrossRef] [PubMed]

26. Park, J.; Kim, G.B.; Lippitz, A.; Kim, Y.M.; Jung, D.; Unger, W.E.S.; Kim, Y.-P.; Lee, T.G. Plasma-polymerized antifouling biochips for label-free measurement of protease activity in cell culture media. *Sens. Actuators B Chem.* **2019**, *281*, 527. [CrossRef]

27. Liu, L.; Deng, D.; Wu, D.; Hou, W.; Wang, L.; Li, N.; Sun, Z. Duplex-specific nuclease-based electrochemical biosensor for the detection of microRNAs by conversion of homogeneous assay into surface-tethered electrochemical analysis. *Anal. Chim. Acta* **2021**, *1149*, 338199. [CrossRef]

28. Chang, Y.; Ma, X.; Sun, T.; Liu, L.; Hao, Y. Electrochemical detection of kinase by converting homogeneous analysis into heterogeneous assay through avidin-biotin interaction. *Talanta* **2021**, *234*, 122649. [CrossRef]

29. Liu, X.; Wang, Y.; Chen, P.; Wang, Y.; Zhang, J.; Aili, D.; Liedberg, B. Biofunctionalized gold nanoparticles for colorimetric sensing of botulinum neurotoxin a light chain. *Anal. Chem.* **2014**, *86*, 2345. [CrossRef]

30. Xia, N.; Huang, Y.; Cui, Z.; Liu, S.; Deng, D.; Liu, L.; Wang, J. Impedimetric biosensor for assay of caspase-3 activity and evaluation of cell apoptosis using self-assembled biotin-phenylalanine network as signal enhancer. *Sens. Actuators B Chem.* **2020**, *320*, 128436. [CrossRef]

31. Liu, L.; Wu, D.; Zhen, S.; Lu, K.; Yi, X.; Sun, Z. Electrochemical detection of telomerase in cancer cells based on the in-situ formation of streptavidin-biotin-DNA-biotin networks for signal amplification. *Sens. Actuators B Chem.* **2021**, *334*, 129659. [CrossRef]

32. Yang, Y.; Liang, Y.; Zhang, C. Label-free and homogenous detection of caspase-3-like proteases by disrupting homodimerization-directed bipartite tetracysteine display. *Anal. Chem.* **2017**, *89*, 4055. [CrossRef] [PubMed]

33. Pan, Y.; Guo, M.; Nie, Z.; Huang, Y.; Peng, Y.; Liu, A.; Qing, M.; Yao, S. Colorimetric detection of apoptosis based on caspase-3 activity assay using unmodified gold nanoparticles. *Chem. Commun.* **2012**, 997. [CrossRef] [PubMed]

34. Takano, S.; Shiomoto, S.; Inoue, K.Y.; Ino, K.; Ino, H.; Matsue, T. Electrochemical approach for the development of a simple method for setecting cell apoptosis based on caspase-3 activity. *Anal. Chem.* **2014**, *86*, 4723. [CrossRef] [PubMed]
35. Ouyang, F.; Yu, T.; Gu, C.; Wang, G.; Shi, R.; Lv, R.; Wu, E.; Ma, C.; Guo, R.; Li, J.; et al. Sensitive detection of caspase-3 enzymatic activities and inhibitor screening by mass spectrometry with dual maleimide labelling quantitation. *Analyst* **2019**, *144*, 6751. [CrossRef]
36. Xia, N.; Sun, Z.; Ding, F.; Wang, Y.; Sun, W.; Liu, L. Protease biosensor by conversion of a homogeneous assay into a surface-tethered electrochemical analysis based on streptavidin-biotin interactions. *ACS Sens.* **2021**, *6*, 1166. [CrossRef]
37. Chen, H.; Zhang, J.; Gao, Y.; Liu, S.; Koh, K.; Zhu, X.; Yin, Y. Sensitive cell apoptosis assay based on caspase-3 activity detection with graphene oxide-assisted electrochemical signal amplification. *Biosens. Bioelectron.* **2015**, *68*, 777. [CrossRef]
38. Song, S.; Shang, X.; Zhao, J.; Hu, X.; Koh, K.; Wang, K.; Chen, H. Sensitive and selective determination of caspase-3 based oncalixarene functionalized reduction of graphene oxide assisted signal amplification. *Sens. Actuators B Chem.* **2018**, *267*, 357. [CrossRef]
39. Khalilzadeh, B.; Charoudeh, H.N.; Shadjou, N.; Mohammad-Rezaei, R.; Omidi, Y.; Velaei, K.; Aliyari, Z.; Rashidi, M.R. Ultrasensitive caspase-3 activity detection using an electrochemical biosensor engineered by gold nanoparticle functionalized MCM-41: Its application during stem cell differentiation. *Sens. Actuators B Chem.* **2016**, *231*, 561. [CrossRef]
40. Song, S.; Hu, X.; Li, H.; Zhao, J.; Koh, K.; Chen, H. Guests involved CB [8] capped silver nanoparticles as a means of electrochemical signal enhancement for sensitive detection of Caspase-3. *Sens. Actuators B Chem.* **2018**, *274*, 54. [CrossRef]
41. Khalilzadeh, B.; Shadjou, N.; Eskandani, M.; Charoudeh, H.N.; Omidi, Y.; Rashidi, M.-R. A reliable self-assembled peptide based electrochemical biosensor fordetection of caspase 3 activity and apoptosis. *RSC Adv.* **2015**, *5*, 58316. [CrossRef]
42. Dong, Y.P.; Chen, G.; Zhou, Y.; Zhu, J.J. Electrochemiluminescent sensing for caspase-3 activity based on Ru(bpy)$_3{}^{2+}$-doped silica nanoprobe. *Anal. Chem.* **2016**, *88*, 1922. [CrossRef] [PubMed]
43. Khalilzadeh, B.; Shadjou, N.; Afsharan, H.; Eskandani, M.; Charoudeh, H.N.; Rashidi, M.-R. Reduced graphene oxide decorated with gold nanoparticle as signal amplification element on ultra-sensitive electrochemiluminescence determination of caspase-3 activity and apoptosis using peptide based biosensor. *BioImpacts* **2016**, *6*, 135. [CrossRef] [PubMed]

Article

Effect of Graphene vs. Reduced Graphene Oxide in Gold Nanoparticles for Optical Biosensors—A Comparative Study

Ana P. G. Carvalho [1,*], Elisabete C. B. A. Alegria [1,2], Alessandro Fantoni [3,4], Ana M. Ferraria [5,6], Ana M. Botelho do Rego [5,6] and Ana P. C. Ribeiro [2]

[1] Departamento de Engenharia Química, ISEL, Instituto Politécnico de Lisboa, 1949-014 Lisbon, Portugal; elisabete.alegria@isel.pt

[2] Centro de Química Estrutural, Instituto Superior Técnico, Universidade de Lisboa, 1049-001 Lisbon, Portugal; apribeiro@tecnico.ulisboa.pt

[3] Departamento de Engenharia Eletrónica e Telecomunicações e de Computadores, ISEL, Instituto Politécnico de Lisboa, 1949-014 Lisbon, Portugal; afantoni@deetc.isel.ipl.pt

[4] Centro de Tecnologias e Sistemas, UNINOVA, Faculdade de Ciências e Tecnologia, 2829-517 Caparica, Portugal

[5] iBB—Institute for Bioengineering and Biosciences and Departamento de Engenharia Química, Instituto Superior Técnico, Universidade de Lisboa, Av. Rovisco Pais, 1049-001 Lisbon, Portugal; ana.ferraria@tecnico.ulisboa.pt (A.M.F.); amrego@tecnico.ulisboa.pt (A.M.B.d.R.)

[6] Associate Laboratory i4HB—Institute for Health and Bioeconomy at Instituto Superior Técnico, Universidade de Lisboa, Av. Rovisco Pais, 1049-001 Lisbon, Portugal

* Correspondence: ana.carvalho.ana@gmail.com

Citation: Carvalho, A.P.G.; Alegria, E.C.B.A.; Fantoni, A.; Ferraria, A.M.; do Rego, A.M.B.; Ribeiro, A.P.C. Effect of Graphene vs. Reduced Graphene Oxide in Gold Nanoparticles for Optical Biosensors—A Comparative Study. *Biosensors* **2022**, *12*, 163. https://doi.org/10.3390/bios12030163

Received: 22 January 2022
Accepted: 2 March 2022
Published: 4 March 2022

Publisher's Note: MDPI stays neutral with regard to jurisdictional claims in published maps and institutional affiliations.

Abstract: Aiming to develop a nanoparticle-based optical biosensor using gold nanoparticles (AuNPs) synthesized using green methods and supported by carbon-based nanomaterials, we studied the role of carbon derivatives in promoting AuNPs localized surface plasmon resonance (LSPR), as well as their morphology, dispersion, and stability. Carbon derivatives are expected to work as immobilization platforms for AuNPs, improving their analytical performance. Gold nanoparticles (AuNPs) were prepared using an eco-friendly approach in a single step by reduction of $HAuCl_4 \cdot 3H_2O$ using phytochemicals (from tea) which act as both reducing and capping agents. UV–Vis spectroscopy, transmission electron microscopy (TEM), zeta potential (ζ-potential), and X-ray photoelectron spectroscopy (XPS) were used to characterize the AuNPs and nanocomposites. The addition of reduced graphene oxide (rGO) resulted in greater dispersion of AuNPs on the rGO surface compared with carbon-based nanomaterials used as a support. Differences in morphology due to the nature of the carbon support were observed and are discussed here. AuNPs/rGO seem to be the most promising candidates for the development of LSPR biosensors among the three composites we studied (AuNPs/G, AuNPs/GO, and AuNPs/rGO). Simulations based on the Mie scattering theory have been used to outline the effect of the phytochemicals on LSPR, showing that when the presence of the residuals is limited to the formation of a thin capping layer, the quality of the plasmonic resonance is not affected. A further discussion of the application framework is presented.

Keywords: biosensors; AuNPs; metal–graphene hybrid; simulations; Mie theory

1. Introduction

Plasmonic biosensors are widely explored as promising sensing tools due to their low cost, simplicity, and short response time. These devices can be useful in a variety of situations, such as medical emergencies in more isolated populations or/and with poor access to health services [1]. For these applications, plasmonic structures can be integrated in point-of-care systems with optical, electrical, or thermal signals delivered in response to certain stimuli, mediated by biomolecules immobilized on biosensor surfaces—biological recognition elements (BREs)—thereby allowing the selective detection of analytes of interest [2].

Gold nanoparticles (AuNPs) have been widely used in the development of optical biosensors [3]. Their utilization in optical signal analysis is based on their characteristic surface plasmon resonance (SPR) effects. No additional material is required for the generation of surface plasmons after interaction with light [4], making their use advantageous in label-free sensing devices.

When AuNPs are used as optical transducers, the region of confinement in which the evanescent wave is propagated is smaller than the wavelength of the incident light. In this case, the phenomenon is called localized surface plasmon resonance (LSPR), where the collective oscillation of the electrons originates a dipolar or multipolar moment in the nanoparticle. LSPR is excited at a specific light wavelength, which is determined by the morphology, size, and composition of the nanoparticle [5]. Any modification on the surface of AuNPs affects the behavior of the LSPR, allowing the exploration of this phenomenon in the development of biosensors through the functionalization of its surface with biomolecules.

When the analyte of interest is recognized by the BRE, the surface refractive index of the AuNPs changes, promoting a shift in the LSPR wavelength [4].

The interest in AuNPs for the development of biomedical diagnostic devices is related to their sensitivity and the possibility of controlling and optimizing the limit of detection (LOD) [5].

Moreover, these interesting properties may act synergistically with graphene when this carbon nanomaterial is added to form a nanocomposite [6]. For example, Banerjee [7] highlighted the efficiency of nanocomposites formed with metallic nanoparticles and graphene in biomedical applications as compared to conventional materials. This greater efficiency is due to their small size and high surface-to-volume ratio, which improves their responses to external stimuli. Carbon allotropes, such as graphene (G), graphene oxide (GO), and reduced graphene oxide (rGO), have similar optical, electronic, and electrochemical properties, and have therefore been investigated as potential materials for biosensor design, drug delivery, bioimaging, and tissue engineering applications. The unique properties of graphene derivatives are also related to their 2D dimensionality, high electrical conductivity and thermal properties, malleability, and functionalization potential. The properties of graphene, which is composed of a single 2D sheet of carbon atoms forming vertices of hexagons with covalent bonds and sp^2 hybridization [7], can be manipulated by chemical modification either through oxidation (from G to GO) or reduction (from GO to rGO) reactions, increasing its functionalization capabilities and broadening its range of applications.

The oxidation of graphene, yielding graphene oxide (GO), allows the addition of oxygen functional groups to its surface, reducing electrical conductivity and malleability, and increasing solubility in polar solvents, as well as reducing the aggregation of carbon-based nanomaterials in aqueous solutions.

Graphene oxide shows a greater functionalization capacity due to the presence of a larger number of oxygen groups, but for the same reason there is also restriction of the mobility of electrons on its surface. Reduction of the number of oxygen groups, yielding reduced graphene oxide, allows the restructuring of electrical and thermal conductivity towards pristine graphene, maintaining functionalization capacity and the distance between graphene sheets [7]. The maintenance of this distance is mandatory to prevent agglomeration between the sheets and avoid the graphite state, since this material has different properties to graphite-derivative nanomaterials [8]. The AuNPs act synergistically with these nanoscale carbon-based nanomaterials, not only in optical properties but also in the dispersion of both nanoparticles and graphene sheets. Studies report that nanoparticles can function as nano-spacers of these materials [9], enhancing AuNPs' stability after the addition of rGO.

In this work, the influence of a variety of carbon-based materials on gold nanoparticles' (AuNPs) localized surface plasmon resonance (LSPR) has been studied and, considering environmental concerns, an ecofriendly approach in the synthesis of gold nanoparticles was chosen, using tea extract as a reducing/capping agent. As a matter of fact, the phytochemicals present in tea, such as polyphenols and tannins, have the interesting capacity of reducing metallic salts, such as chloroauric acid ($HAuCl_4$) [10–12].

Although the use of composite materials based on metal nanoparticles and graphene allotropes to explore plasmonic properties in sensing applications has already been reported in the literature [13,14], a green method of synthesis for nanoparticle fabrication represents a novel approach in the preparation of hybrid materials. Since a residual layer of polyphenols remains on the surface of the nanoparticles, this approach results in the introduction of an additional factor in the set of parameters necessary to obtain the desired LSPR tuning which deserves to be investigated. In this paper, the frequency and intensity of the LSPRs for three different types of graphene are analyzed, compared, and discussed, combined with a green protocol for the synthesis of gold nanoparticles as well as different sequences in the fabrication of the composites.

Application Framework

Sensors based on the local surface plasmon resonance of metal nanoparticles are characterized by simple structures and good sensitivities. Despite the good sensing properties of LSPR structures, full commercialization has been prevented by the production costs associated with the bio-functionalization and the high-precision systems necessary to extract the optoelectronic output. There is a great interest in new strategies for bringing the excellent detection properties of LSPR sensors into play in low-cost devices made with low-cost materials [15]. The combination of carbon-based nanomaterials (CNMs) with MNPs has been demonstrated to enhance LSPR response [16] and facilitate functionalization with specific and selective antibodies [17]. In addition, the introduction of CNMs in the plasmonic layer allows a tuning of the LSPR central frequency. With the double dependence of the LSPR on MNP size and the presence of CNMs, it is possible to create a set of plasmonic layers whose LSPR wavelengths are distributed in a spectral range of a few tenths of a nanometer. This consideration paves the way to an LSPR sensor with an arrayed structure, where each element maximizes its specific LSPR at its own wavelength. Illumination with a broad light source will produce a different response in each one of the elements and the biomarkers' immobilization in the surrounding medium will cause a transition to a different state. In such a configuration, the output can be extracted by the application of an image analysis approach based on a color-recognition algorithm [18]. The experimental characterization presented in this work represents a first step toward the development of an arrayed LSPR sensor whose elements are composed by metal nanoparticles with different dimensions and supported by different CNMs, joined to a reading scheme provided by a CCD imager, supported by an image processing algorithm. The use of low-cost materials together with a simplified interrogation scheme aims to overcome the elevated costs related to high-precision mechanical systems and wavelength selective light sources. The result will be a proof of concept for a low-cost LSPR sensor with potential for large-scale biosensing applications in environmental monitoring or the medical stratification of diseases.

The main advantages of these nanomaterials are their biocompatibility, high chemical stability, and high surface areas. The reduced dimensions and weight, high resistance, simplicity of use, and low cost makes graphene and its derivatives excellent candidates for the development of biosensors.

An optical sensor has, by definition, the ability to convert an external stimulus into an optical output signal.

The π–π binding capacity of graphene with biomolecules allows the use of this carbon-derived nanomaterial as a substrate, and the SPR technique can be used to detect the interaction with the analyte of interest, in which the angle of an incident polarized light is adjusted as a result of the change in the refractive index caused by the interaction between them. rGO has advantages over graphene, since its oxygen functional groups improve the interaction of this nanomaterial with biomolecules [19].

Optical sensors, such as those based on fluorescence, surface-enhanced Raman scattering (SERS), optical fiber biological sensors, among other kinds of optical sensors, are being developed due to the amazing properties of graphene and its derivatives. Various sensing applications, such as single-cell detection, cancer diagnosis, and protein and DNA sensing, have been reported in recent years [19].

Due to its unique optical and electrical properties, graphene and its derivatives are widely used in photonic and optoelectronic devices as they have displayed several ideal properties, including broadband light absorption, the ability to quench fluorescence, excellent biocompatibility, and strong polarization-dependent effects, making them very popular platforms for optical sensors. Graphene and its derivatives-based optical sensors have numerous advantages, such as high sensitivity, low-cost, fast response time, and small dimensions. The use of metallic nanoparticles, namely, gold nanoparticles, may improve the response of biosensors, amplifying the signals obtained and increasing the sensitivity of these devices [19].

Several applications of graphene and its derivatives-based optical sensors are summarized in the following Table 1:

Table 1. Several applications of graphene and its derivatives-based optical sensors.

Optical Sensors	Material	Application	Ref.
Fluorescence sensing	GO	Effect of pH on fluorescence	[20]
Fluorescence sensing	GO	Fluorescence quencher	[21]
Fluorescence sensing	GO	Two-photon multi-color bio-imaging of multiple drug-resistant bacteria (MDRB)	[22]
Fluorescence sensing	GO	Fluorescence imaging	[23]
Fluorescence sensing	GO	High-sensitivity detection of miRNA in cells	[24]
Graphene-Based SERS Sensing	G	Adsorbed molecules	[25]
Graphene-Based SERS Sensing	G	Detection of biomarkers and biomolecules	[26]
Graphene-Based SERS Sensing	G	Bio-imaging, cancer diagnostics	[27]
Graphene-Based SERS Sensing	GO	Effects of pH values on SERS intensities of some aromatic molecules	[28]
Graphene-Based SERS Sensing	RGO	SERS effects of RGO with different degrees of reduction	[29]
Graphene-Based Optical Fiber Sensing	G	Biochemical sensing	[30]
Graphene-Based Optical Fiber Sensing	G	Gas sensor	[31]
Graphene-Based Optical Fiber Sensing	G	Biomolecule detector	[32]
Graphene-Based Optical Fiber Sensing	GO and RGO	Sensors for volatile organic compounds	[33]
Other Kind of Graphene-Based Optical Sensors	G	Detection of cancer cells	[34]
Other Kind of Graphene-Based Optical Sensors	RGO	Detection of cancer cells	[35]
Other Kind of Graphene-Based Optical Sensors	G and RGO	Photothermal detection (PTD)	[36]

It is undeniable that there are still many challenges in this area. The need to synthesize high-quality graphene, to achieve a low-cost, environmentally friendly method for synthesizing graphene and its derivatives are issues still to be addressed. With this study, the use of a simple green method to produce light-responsive material is our aim and contribution.

Addressing the above challenges, we hope to show the potential of green methods as well as the importance of graphene and its derivatives in the development of optical sensing technologies, which will ultimately increase the quality of future life.

2. Results and Discussion

2.1. Sequence 1—Addition of Carbon-Based Derivatives to AuNPs (SQ1)

After the addition of chloroauric acid ($HAuCl_4 \cdot 3H_2O$) to a tea extract (5% w/w), a change of color from yellow to red was observed, which is consistent with the formation of AuNPs. The AuNPs' characteristic LSPR was observed only after one week (t_{1w}), the absence of LSPR at the initial time (t_0) possibly being associated with the period of reduction required to originate AuNPs or due to the fact that in the case of hybrid nanostructures, carbon materials may be responsible for radiation absorption, inhibiting the detection of AuNPs, as described by Biris et al. [37]. Figure 1 shows the UV–Vis spectrum of the aqueous solution containing AuNPs with the corresponding LSPR occurring at 556 nm. The presence of a single band suggests a spherical morphology for the synthesized AuNPs [37]. The solution remained stable for several weeks without any signal of aggregation.

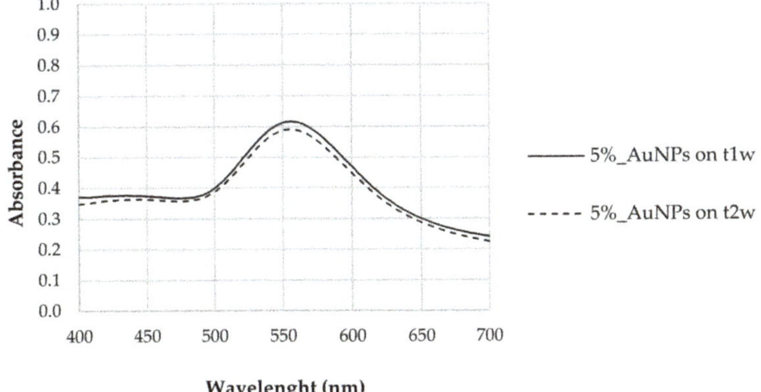

Wavelenght (nm)

Figure 1. LSPR of AuNP samples synthesized with 5% tea extract (*Thea sinensis*) at t_{1w} (after 1 week) and t_{2w} (after 2 weeks).

The stability of the synthesized nanoparticles was evaluated. We found that between the first (t_{1w}) and second week (t_{2w}) after synthesis no significant differences in LSPR response was observed, confirming that polyphenols guarantee a certain stability to the synthesized AuNPs. As proven previously by the authors of [10], polyphenols prevailed as reducing agents of the metallic salt and capping agents of the produced AuNPs [10], maintaining their capacity for at least two weeks after the synthesis. This is due to the ionic force of polyphenols that promote good dispersion by efficiently counteracting the mutual attraction caused by Van der Waals forces and which are responsible for the aggregation of nanoparticles [38].

Figure 2 reports the LSPR of AuNPs synthesized with 5% tea extract after the addition of G, GO, and rGO. For the naked AuNPs (5%_AuNPs), the LSPR at t_{1w} occurred at 556 nm, with a shift to higher frequencies for the mixtures containing graphene (G) (LSPR = 541 nm) and reduced graphene oxide (rGO) (LSPR = 542 nm). The composite containing graphene oxide (AuNPs/GO) did not show any LSPR shift after the addition of carbon-based nanomaterials. For the AuNPs/G hybrid nanostructure, band amplitude decreased by ~10% when compared to the sample without any of the carbon-based nano-material, while for AuNPs to AuNPs/rGO, it increased by ~0.29, which corresponds to 46%

of the naked/unsupported AuNPs' absorbance. This result may be ascribed to a dispersion of the AuNPs [38] in the presence of the polyphenols [37].

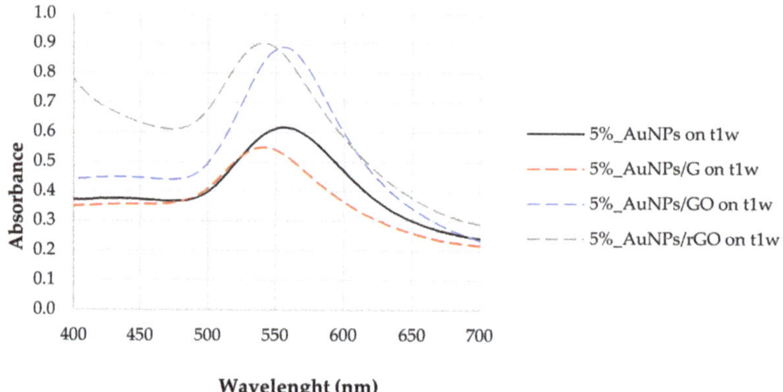

Figure 2. LSPR of AuNPs synthesized with 5% tea (w/w) extract and after the addition of G, GO, and rGO, one week after their synthesis (t_{1w}).

LSPR properties depend not only on the AuNPs' shape, size, and dispersion but also on the surrounding dielectric constant. TEM characterization showed the formation of spherical AuNPs (Figure 3). Although the addition of rGO caused a LSPR shift to lower wavelengths, the AuNPs supported on rGO exhibited a larger diameter (61 nm $\pm$ 3 nm) when compared to AuNPs only encapsulated by polyphenols (30 nm $\pm$ 2 nm). The shift to lower wavelengths after the carbon-based material addition may be related to the increased dispersion of AuNPs, with no change in the diameter of the nanoparticles, and to charge transfer interactions between graphene and AuNPs [39]. By acting as nano-spacers [6], when graphene is used as a support, the nanoparticles benefit from increased dispersion, while at the same time they avoid graphene sheet agglomeration. Additionally, the LSPR deviation can also be caused by the surface energy of neighboring nanoparticles [40].

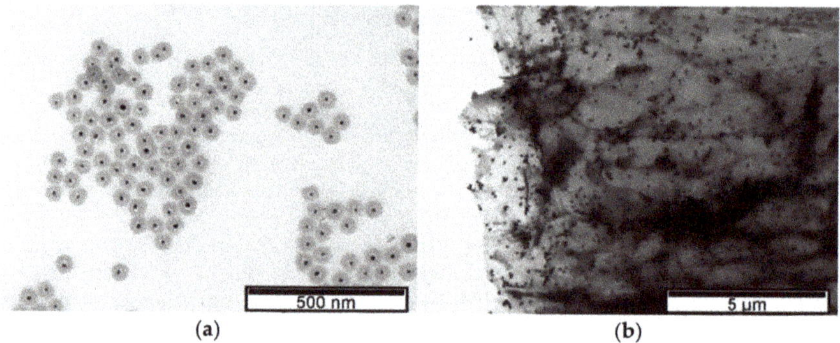

Figure 3. TEM images. (**a**) AuNPs in 5% tea extract. (**b**) rGO-supported AuNPs in 5% tea extract.

The probable formation of π–π bonds between AuNPs and the functional oxygen groups of rGO [6] may be responsible for the stable dispersion confirmed by the measured zeta potential value -20.17 mV when compared to -15.59 mV obtained for the sample containing AuNPs dispersed in 5% tea extract (Table 2). Since LSPR did not occur at t_0, we assumed that the nucleation of metallic salts occurred near the rGO functional oxygen groups [6] which may have promoted a higher stability due to the dispersion of the AuNPs as well as the increase in the diameter of the nanoparticles [41].

Table 2. Zeta potential values of AuNPs synthesized before and after rGO addition.

Sample	Zeta Potential (mV)	Standard Deviation (mV)
5%_AuNPs	−15.59	0.595
5%_AuNPs/rGO	−20.17	0.868

The reason for using a pre-prepared solution of AuNPs stabilized with polyphenols then mixed with rGO is to help anchor the polyphenol-protected AuNPs to the rGO surface. This process could be assisted by: (i) $\pi-\pi$ interactions between AuNPs and functional oxygen groups of rGO; (ii) $\pi-\pi$ interactions between benzene rings of the polyphenols and the surface of rGO; and (iii) electrostatic interactions between the OH on the surface of rGO and the polyphenols capping the AuNPs. After mixing, the shape of the AuNPs appeared to be the same when observed by TEM (Figure 3), but the presence of the rGO seemed to have anchored them, as expected. Schematically (Figure 4), the introduction of rGO leads to the settling of the AuNPs [42].

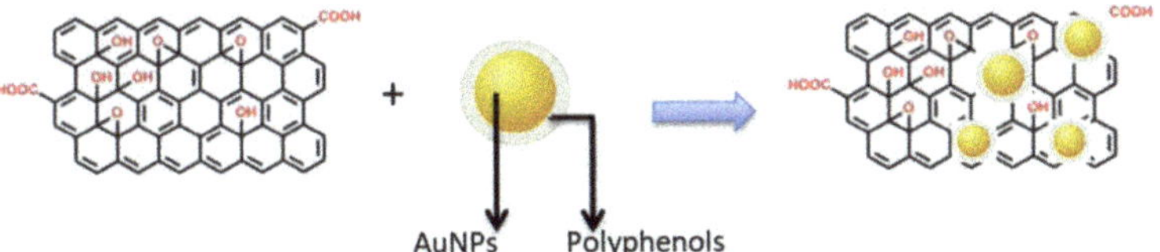

Figure 4. Schematic illustration of AuNPs anchored on the surface of reduced graphene oxide (rGO).

In the case where GO is added to the AuNPs, no significant deviation in the LSPR occurs relative to the value observed prior to the addition of this carbon-based material (λ = 556 nm). TEM images (Figure 3) show a greater dispersion of AuNPs on the surface of the carbon-based nanomaterial which is probably responsible for the increase in the LSPR band amplitude observed by UV–Vis [38,43,44]. As with the addition of rGO, nucleation may have occurred next to the functional oxygen groups [41]. To confirm this hypothesis, Figure 5 shows dispersed nanoparticle clusters on the GO surface. The wide number of oxygen functional groups on the GO may have promoted AuNP agglomeration and significant dispersion of diameters. These results are in accordance with those reported by Parnianchi et al. [45] related to the difficulty of controlling both the morphology and the homogeneous distribution of AuNPs when GO is present. The frequency versus size distributions for the three TEM images are presented in ESI (Figure S1).

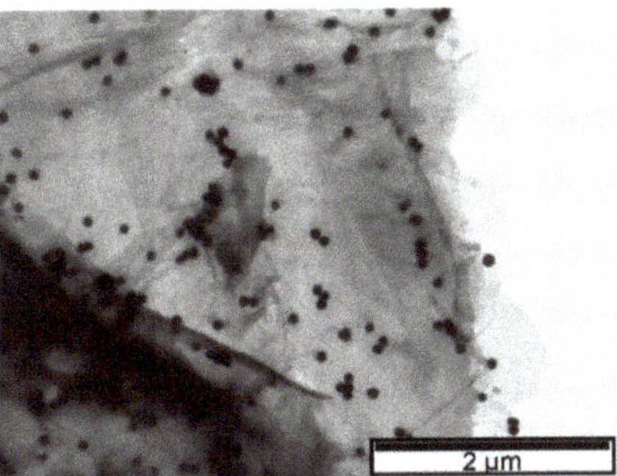

Figure 5. TEM image. GO supported on AuNPs synthesized with 5% tea extract.

After the addition of both GO and rGO, sharper resonance bands as well as an increase in the resonance intensity or amplitude were observed (Figure 2). In fact, both enhancements of plasmonic resonance intensity and shift in the position are indicators of enhanced sensitivity of the LSPR sensor.

The partial reduction of oxygen groups in GO driven by tea polyphenols [46] may contribute to reducing the availability of these phytochemicals both as reduction and capping agents. The dispersion of diameters verified after GO addition and the fact that this nanomaterial is insulating (due to the functional groups of oxygen, the electronic mobility is reduced) compromises the use of this nanocomposite in the development of biosensors.

We also measured the electrostatic potential at the electrical double layer surrounding a nanoparticle in solution, commonly referred to as the zeta potential. Nanoparticles with a zeta potential between -10 and $+10$ mV are considered approximately neutral, while nanoparticles with zeta potentials of greater than $+30$ mV or less than -30 mV are considered strongly cationic and strongly anionic, respectively [47]. Since most cellular membranes are negatively charged, zeta potential can affect the nanoparticle tendency to permeate membranes, with cationic particles generally displaying more toxicity associated with cell wall disruption. It can be observed that the initial 5%_AuNPs are weakly anionic. With the addition of rGO, they became strongly anionic due, most likely, to the increase in acidity by the introduction of rGO.

According to the classical physical theory for the electromagnetics of metals, the plasma frequency (ω_p) of the free electron gas depends linearly on the density of the electrons (N_e). Such a consideration leads, in the Drude model, to a linear dependence of the metal dielectric function on the value of N_e. Within the dipole approximation, i.e., when the nanoparticle size is much smaller than the light wavelength, and under the Fröhlich condition, it can be shown that the LSPR frequency (ω_{LSPR}) can be directly related to ω_p by the following equation:

$$\omega_{LSPR} = \frac{\omega_p}{\sqrt{1 + 2\varepsilon_m}}$$

where ε_m is the dielectric constant of the surrounding medium [48]. Therefore, a change in the electron density is expected, leading to a blue or red shift of the localized plasmon resonance. This effect has been reported in the literature, supported by electrochemical experiments [49] and exploited for sensing applications [38]. Optical gas sensors based on gold nanoparticles and carbon nanomaterials have also been demonstrated to rely on the reactions for both reducing and oxidizing gases and the correspondent injection or subtraction of electrons to and from graphene oxide, implying a shift in the observed LSPR [50]. The charge transfer interaction between AuNPs and graphene has been also demonstrated to be suitable for active modulation of surface plasmon resonance. These considerations are in agreement with our experimental findings, namely, the blue shift observed in the LSPR of the composites (Figure 2) and the negative charge transfer between the graphene allotropes and the nanoparticles observed in the zeta potential measurements (Table 2).

The degree of oxidation of graphene before and after the addition of AuNPs, the type of oxygen functional groups, the oxidation state of gold at AuNP surfaces within the AuNPs/G, AuNPs/GO, or AuNPs/rGO composites, as well as relevant relative atomic amounts were assessed by XPS. The characterization of AuNPs prior to any graphene addition has already been reported [10]. Here, the three different types of graphene and the corresponding gold composites are studied.

XPS confirms that samples are composed mainly of carbon and oxygen, and, where expected, gold. GO-based samples also contain some sulphur, and some samples have a residual or low relative amount of silicon. Table 3 shows the corrected binding energies (BE) of the peaks fitted in the different XPS regions: C 1s, O 1s, Au 4f, S 2p, and Si 2p.

Table 3. Corrected BE $\pm$ 0.1 eV and corresponding assignments.

	AuNPs/ rGO	AuNPs/ GO	AuNPs/ G	rGO	GO	G	Assignments [51,52]
C 1s	284.4	284.4	284.4	284.4	284.4 [1]	284.4	C-C and C-H sp^2
	285.5	285.1	285.7	285.4	285.0	285.2	C-C and C-H sp^3
	286.3	287.0	286.7	286.4	287.3	286.2	C-O or epoxide
	287.6	288.3	287.9	287.7		287.4	C=O
	288.8		289.1	288.8	289.2	288.8	XO-C=O (X=H or C)
	290.2	289.7	290.3	289.9		290.1	
	291.3		291.6	291.3	292.0	291.3	
	293.3		292.6	294.1		293.8	π-π*
	295.6		294.9				
O 1s	531.5	531.6	530.7	531.3			O in electropositive vicinity
	532.7	532.9	532.3	532.9	532.9	532.3	O bonded to C
Au 4f$_{7/2}$	84.1	84.6	84.1				Au^0; in "AuNPs/GO": Au$^+$?
Au 4f$_{5/2}$	87.8	88.3	87.8				(see text)
S 2p$_{3/2}$		168.6			169.0		SO$_4^{2-}$
S 2p$_{1/2}$		169.8			170.0		
Si 2p$_{3/2}$	101.7	101.7		101.6	102.0		silicone
Si 2p$_{1/2}$	102.3	102.4		102.2	102.6		

[1] see text.

C 1s regions of AuNPs/rGO and AuNPs/G are dominated by the peak assigned to carbon atoms in C-C and C-H sp^2 bonds, centred at 284.4 $\pm$ 0.1 eV, and by a long tail at the high BE side, detected roughly between 287 eV and 297 eV (Figure 6a–c), corresponding to π–π* excitations, typical of extended delocalized systems, such as graphene. In addition, at BE > 297 eV plasmon loss features are detected. Included in the C 1s envelope, peaks attributed to carbon atoms bonded to oxygen in different functional groups are identified in Table 3. The latter are particularly intense in GO and AuNPs/GO, where the loss of electron delocalization due to the oxidation of graphene is evident: the sp^2 carbon peak has a lower relative intensity compared with AuNPs/rGO and AuNPs/G, and the energy loss features are hardly detected (Figure 6d). In GO, a peak at 284.4 eV was fitted, but it is almost cancelled with the fitting, leaving only the peak centred at 285 eV (Figure 6d).

Au 4f regions are doublet peaks (Figure 7) with a spin–orbit energy separation of 3.7 eV. Au 4f$_{7/2}$ is centred at 84.1 $\pm$ 0.1 eV in AuNPs/rGO and AuNPs/G, which is attributed to Au^0. In AuNPs/GO, Au 4f$_{7/2}$ is centred at 84.6 $\pm$ 0.1 eV, which has been identified as Au$^+$ in HAuCl [53]. However, in this case, no chlorine was detected to corroborate this assignment. Still, positive BE shifts for Au 4f photoelectrons have been reported for very small gold nanoparticles (diameter $\leq$20 nm) [54]. In addition, a contact potential effect between the metal nanoparticles and the organic substrate may be present, leading to an underestimation of the charge shift. It is noteworthy that the full widths at half maximum, for all Au 4f fitted peaks, are very similar to each other (1.1 $\pm$ 0.1 eV), which also sustains the hypothesis of having reduced gold nanoparticles in all the samples. Finally, sulphur is only present in GO-based samples. S 2p is a doublet with a spin–orbit split of 1.1 $\pm$ 0.1 eV, and the main component, S 2p$_{3/2}$, centred at 168.8 $\pm$ 0.2 eV, is assigned to sulphate groups. The silicon detected in some of the samples may come from the polysiloxane-based tape used to mount the powder for XPS analysis. Table 4 shows the XPS quantitative analysis.

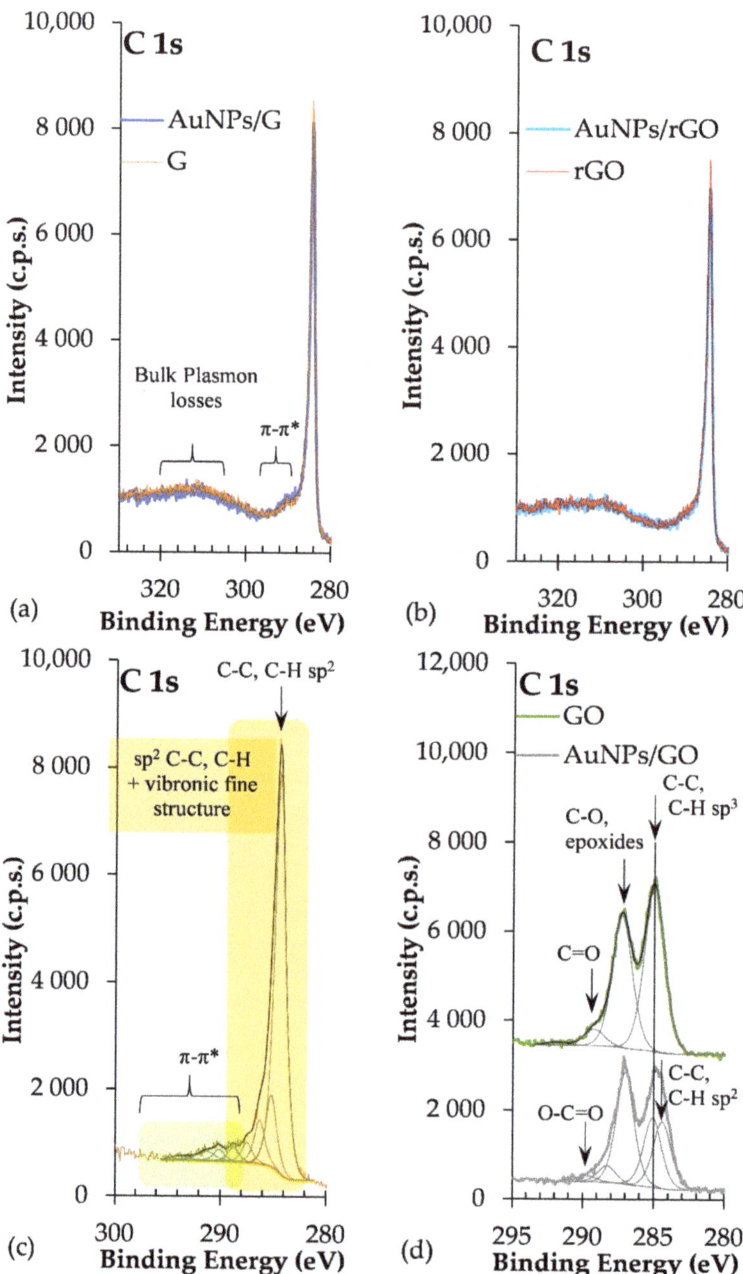

Figure 6. C 1s regions of (**a**) AuNPs/G and G; (**b**) AuNPs/rGO and rGO; (**c**) G with fitting (similar to rGO and AuNPs/rGO and AuNPs/G); and (**d**) GO and AuNPs/GO.

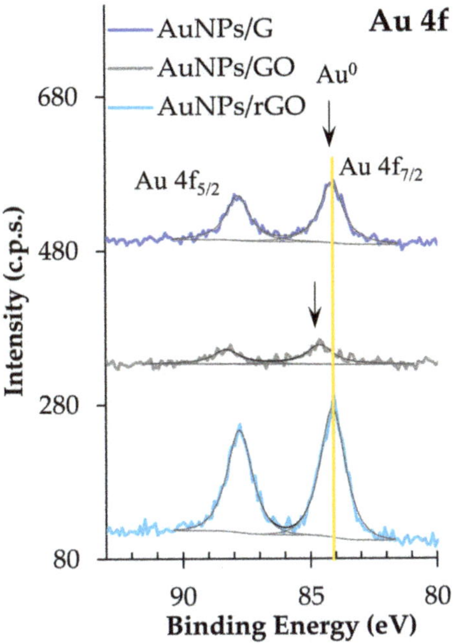

Figure 7. Au 4f XPS regions.

Table 4. XPS atomic concentrations (%) and relevant atomic ratios.

	AuNPs/ rGO	AuNPs/ GO	AuNPs/ G	rGO	GO	G
Atomic Concentrations (%)						
C	86.2	66.5	92.8	87.2	71.2	93.9
O	13.4	32.3	7.1	12.6	27.3	6.1
Au	0.12	0.03	0.06			
S		0.8			1.0	
Si	0.2	0.4		0.2	0.6	
Atomic ratios						
Au/C	0.0014	0.0004	0.0006			
O/C	0.15	0.44	0.08	0.14	0.33	0.06

It is interesting to note that the larger relative amount of Au in AuNPs/rGO compared with AuNPs/G, computed from XPS data, is compatible with the UV–Vis absorbance spectra shown in Figure 2: AuNP LSPR absorbance is much larger for AuNPs/rGO than for AuNPs/G. Moreover, the relative amount of Au, detected by XPS, is much lower in AuNPs/GO than in AuNPs/rGO. Actually, as attested by Table 3 and Figure 6, graphene oxide has a very different chemical composition from G or rGO, with many more oxygen functional groups than G or rGO. These oxidized carbonaceous groups establish stronger intermolecular interactions with polyphenols surrounding the AuNPs, allowing for the formation of sandwich-like GO/AuNPs/GO, significantly attenuating the Au 4f photo-electron signal detected by XPS. It is also clear from the quantification results that samples with gold nanoparticles are slightly more oxidized than the parent samples with no gold. Actually, since AuNPs are capped with phenolic functional groups, a larger relative amount of oxygen is expected in the samples with AuNPs. In addition, a further reduction of gold may occur when in contact with G, GO, or rGO with the simultaneous oxidation of graphene. In Table 4, the ratios computed were obtained after subtracting the contribution of sulphate and silicone, these being, exclusively, the ratios in the graphene-based samples discarding the contaminations. Other effects of the introduction of Au nanoparticles can be found in the C 1s spectral differences presented in ESI (Figure S2).

2.2. Sequence 2—Addition of Carbon-Based Derivatives Prior to AuNP Formation (SQ2)

The addition of carbon-based nanomaterials prior to AuNP formation seems to compromise phytochemicals' reducing and capping capacities. This may be related to the adsorption of phytochemicals by carbon-based nanomaterials [6] which compromise their availability to reduce the metallic salt. Some of these samples revealed a slight LSPR at t_0 (Figures S4–S6, Supplementary Materials). We believe that the addition of these derivatives initially increases the contact between the metallic salt and polyphenols, promoting the synthesis of AuNPs. However, at t_{1w} and t_{2w} the LSPR did not occur. We believe that the adsorption of polyphenols by carbon-based nanomaterials through electrostatic bonds and Van der Waals interactions [6] compromises the efficiency of these species as capping agents. In this case, we did not verify stability between t_{1w} and t_{2w} in any of the samples. No further testing was performed.

2.3. AuNP Stability Study

The stability of the synthetized nanoparticles was also evaluated by the sample's characterization, 1 week (t_{1w}) and 2 weeks (t_{2w}) after synthesis. Figure 8 reports the absorbance spectra for the different composites taken at time intervals of one week. We found that the resonance of the surface plasma had a red shift one week after synthesis (t_{1w} vs. t_0). These data show that the synthesis of the nanoparticles did not cease after 20 min of vigorous stirring of chloroauric acid with the tea extract. This shift could mean that the diameters of the AuNPs increased between t_0 and t_{1w}. As mentioned above, the increased band amplitude after a week suggests that there was an increase in the concentration of nanoparticles, indicating that the formation of AuNPs was still happening [44].

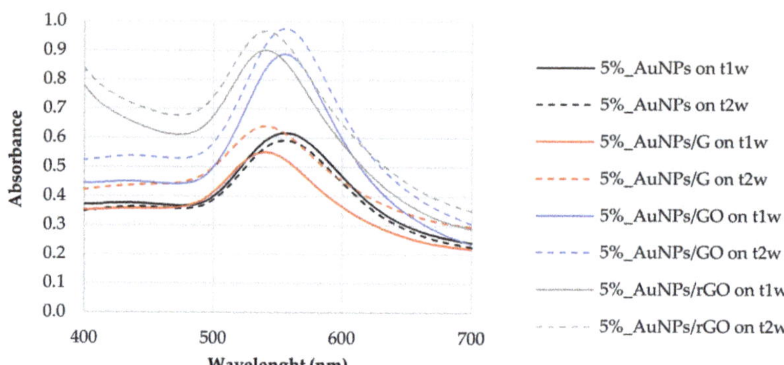

Figure 8. AuNP samples stability (t_{1w}–t_{2w}). Carbon-based nanomaterials added after AuNP formation.

UV–Vis analysis of the samples after 2 weeks (t_{2w}) showed that LSPR did not shift in comparison to that observed after 1 week (t_{1w}) (Figure 8). The polyphenols acted as capping agents [10] of the AuNPs contrary to the Van der Waals forces that promote the agglomeration of AuNPs. This result leads us to conclude that polyphenols can be good candidates for green synthesis of AuNPs, not only because they were efficient as reducing agents but also because they gave stability to synthesized nanoparticles. The synthesis with tea extract allows the use of phytochemicals for the production and stabilization of AuNPs, simplifying the process.

Between t_{1w} and t_{2w}, we can observe a slight increase in the absorbance value without a change in the behavior of the plasmonic response. Tea contains a multitude of different chemicals. Some of these, e.g., tannins, are fairly dark to begin with, but, if they are allowed to react with the oxygen [55] in the air, they oxidize, producing other compounds that are even darker in color. We can observe this phenomenon by UV spectroscopy because, in general, bigger molecules absorb more light, and as the oxidized tannins tend to aggregate over time [56], creating bigger molecules, this leads to a change in tea spectra. GO and rGO oxygen functional groups contribute to a more efficient oxidation of samples. The difference between t_{1w} and t_{2w} is more obvious in the 5%_AuNPs/GO in comparison with the 5%_AuNPs/rGO sample, since GO has more oxygen groups on its surface.

2.4. Simulation (Mie Theory)

The variation in the transmission spectra caused by the plasmonic resonance of the nanoparticles can be calculated by recourse to the Mie Theory [57], and the intensity of the LSPR effect can be correlated with the material properties of the surrounding medium [58]. Regarding gold spherical nanoparticles, as a bottom line of the Mie analysis, it can be stated, as a rule, that increasing AuNP size results in an LSPR shift towards the red part of the spectrum. Such a behavior can also be observed in experimental measurements [59]. Moreover, as the ratio between AuNP radius and light wavelength increases, multipolar behavior is to be expected and a widening of the peak waist observed. Increasing the values of the refractive index of the surrounding medium will also lead to a red shift of the LSPR peak, accompanied by a significant enhancement of its maximum value. For the analysis of the specific case of the AuNPs, produced by combining $HAuCl_4$ with the phytochemicals present in tea extract as a reducing agent, we have considered the gold nanosphere capped by a thin uniform layer of tea polyphenols, immersed in pure water. This approach agrees with the morphology observed in Figure 3a. The most abundant polyphenol encountered in tea is epigallocatechin gallate (EGCG) [60], whose reported refractive index is 1.857 [61], which was the value used in the simulations.

Figure 9 reports the light extinction profile calculated for a AuNP with increasing radius and capped by a thin layer of EGCG (1–30 nm). From the analysis of this Figure, the LSPR wavelength (560 nm for a 10 nm radius of the AuNP) is only slightly affected by the thickness of the EGCG layer, but if it is too thick, LSPR intensity is reduced. Anyway, when compared with the LSPR produced by AuNPs in pure water, without EGCG capping, the LSPR peak always suffers a red shift. This red shift is observable for any thickness of the capping layer.

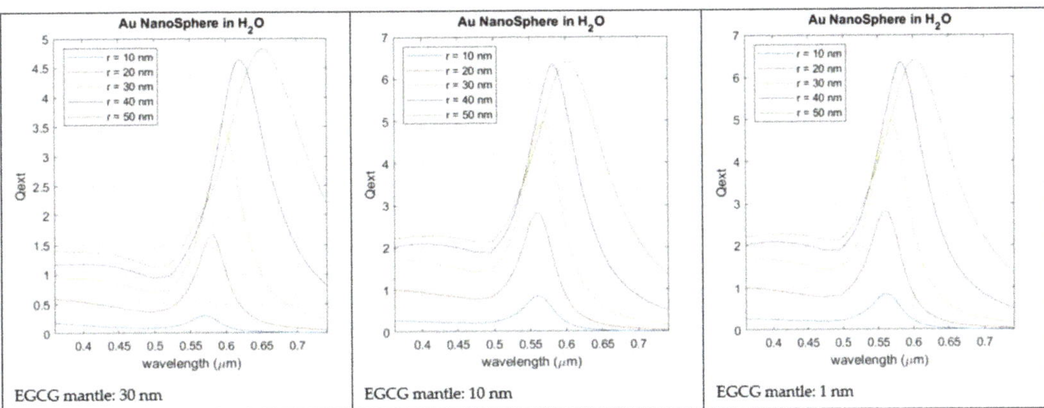

Figure 9. Simulated LSPR intensity for AuNPs with increasing dimensions (radius between 10 and 50 nm). Gold nanospheres are immersed in pure water and have a capping layer of EGCG with a thickness between 1 and 30 nm.

Figure 10 shows the results for the simulation where the light extinction is calculated for a AuNP with a radius of 30 nm and a capping layer thickness between 1 and 100 nm.

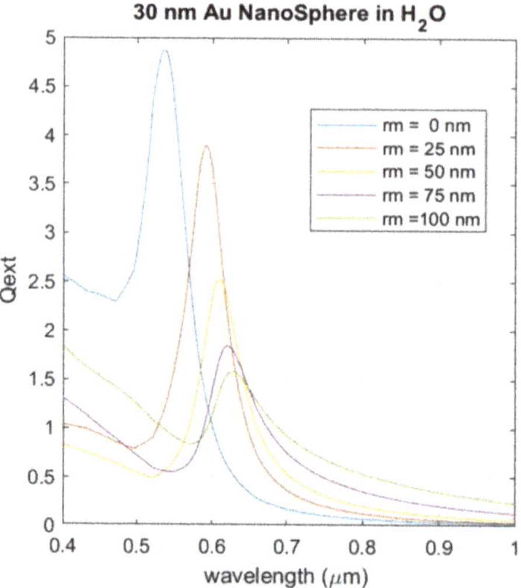

Figure 10. Simulated LSPR intensity for AuNPs with fixed dimensions (radius 30 nm). Gold nanospheres are immersed in pure water and have a capping layer of EGCG with thickness between 0 and 100 nm.

The effect of increasing the thickness of the cover layer results in the LSPR shifting towards the red region. Simultaneously, the intensity of the LSPR shows a marked reduction as the capping thickness increases.

We can conclude from this simulation analysis that the residual polyphenols which remain after AuNP fabrication have a negligible effect on the quality of the plasmonic response. An excessive accumulation of the residual polyphenols on the AuNP surface would reduce the LSPR intensity, but the simulation shows that for the range of the capping thickness observed in the TEM images, such a level of EGCD accumulation is not reached. The red shift foreseen by the simulation has no impact on the operation mechanism of a sensor device built with these materials. At the same time, the presence of the capping layer can be expected to physically separate the nanoparticles, preventing their aggregation. Thus, AuNPs synthesized by this green method combine the advantage of a simplified fabrication method that avoids aggregation with a reliable plasmonic resonance that can be successfully exploited in a sensing device.

The wavelength and intensity of LSPR are strongly dependent on the refractive index of the surrounding medium. An alteration of this parameter provokes a shift of the LSPR peak which can be used as the output value of a sensing system. To evaluate the sensing efficiency of the AuNPs, the spectral shift ($\Delta\lambda$) of the resonance wavelength, as a function of the variation in the refractive index (Δn) of the surrounding medium, can be translated into a sensitivity parameter (S):

$$S = \Delta\lambda/\Delta n$$

Figure 11 reports the position of the LSPR central wavelength as a function of the refractive index of the surrounding medium and the corresponding sensitivity. The presence of the EGCG capping layer also acts as a separation layer between the AuNPs and the surrounding medium, reducing the sensitivity of the AuNPs. The simulation results show that when the thickness of the EGCG layer remains below a few tenths of a nanometer, even if reduced, the sensitivity value is maintained within the limits described in the literature [62,63]. The variation of the refractive index also produces a modification in the LSPR peak intensity. The combined analysis of peak wavelength and intensity can be used to improve the sensor signal-to-noise ratio (SNR), allowing a higher tolerance to the sensitivity reduction introduced by the EGCG capping layer.

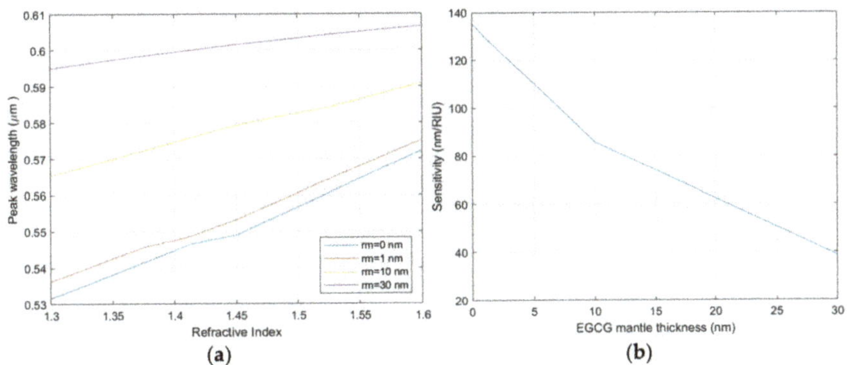

Figure 11. (**a**) Variation of the central wavelength for the LSPR resonance as a function of the medium refractive index for different thickness of the EGCG capping layer. (**b**) Sensitivity of the NPs' LSPR as a function of the EGCG capping layer thickness.

3. Materials and Methods

Chloroauric acid ($HAuCl_4 \cdot 3H_2O$) was purchased from Sigma-Aldrich (Sigma-Aldrich, Munich, Germany); the black tea leaves (*Thea sinensis*) for the tea extract preparation used as a reducing agent for chloroauric acid was purchased from Pingo Doce (Pingo Doce, Portugal, tea brand—batch 1832); graphene [64] and derivatives were synthetized using the modified Hummers method [64].

Black tea extracts with 5% concentration were prepared as reported previously [10], the main difference being that the resides used in the addition of graphene and its derivatives were added in a different order to see if this would produce different UV spectra (different transitions). This led to the following sequences (Figure 12):

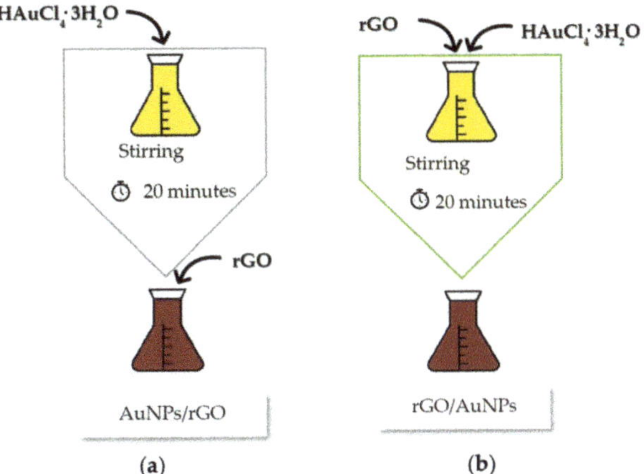

Figure 12. Preparation of samples by SQ1-AuNPs/rGO (**a**) or SQ2-rGO/AuNPs (**b**).

Sequence 1 (SQ1): 1 mL of chloroauric acid (0.1 M) was added to each tea extract *Thea sinensis* concentration (5%) with a 6 mL total volume. This mixture was stirred for 20 min at 400 rpm at room temperature. After that, 1 mg of graphene or derivative was added to every 2 mL of the samples previously obtained (Figure 12a).

Sequence 2 (SQ2): The same weight of graphene or derivatives was added before the addition of chloroauric acid (Figure 12b). The remaining synthesis process of nanoparticles was maintained as described in SQ1.

Both nanoparticles and nanocomposites were characterized by UV–Vis spectroscopy (UV-2501PC Schimadzu, Waltham, MA, USA) and transmission electron microscopy (Hitachi 8100, Tokyo, Japan) with a ThermoNoran EDS light and zeta potential (Litesizer 500—Anton-Paar, Gratz, Austria). The characterization by UV–Vis was undertaken at 3 distinct moments, t_0, t_{1w} and t_{2w}, to determine the stability of the AuNPs and the nanocomposites over time, t_0 being the starting moment, immediately after stirring the samples, t_{1w} one week after, and t_{2w} two weeks after. All samples were preserved in cold conditions and protected from light. TEM images were obtained between t_0 and $t_{1w,}$ as was the zeta potential of the samples.

Modified and unmodified graphene were analyzed by X-ray photoelectron spectroscopy with a XSAM800 spectrometer from KRATOS. Non-monochromatic radiation from a Mg Kα source was used (hν = 1253.6 eV). Powdered samples were fixed on the XPS holder with a double face tape and analyzed at UHV, at TOA = 45°. The BE was corrected considering the charge shift observed for the sp^2 C-C and C-H peak set at 284.4 eV [52]. Other operational conditions and data treatment details were as published elsewhere [65].

4. Conclusions

The black tea extract (*Thea sinensis*) showed reducing capacity of chloroauric acid, allowing the synthesis of spherical gold nanoparticles. The production of AuNPs through this green synthetic approach proved to be sustainable, not only due to its low cost but also because of the reducing capacity of the tea and its coating agent function, conferring stability to the synthesized AuNPs. We found that 2 weeks after stirring of the reagents, polyphenols prevailed as coating agents, sustaining the stability of the AuNPs. Simulation results based on the Mie theory for the LSPR effects support the conclusion that, even with a thin capping layer of residuals, no significant reduction of plasmonic resonance should be expected.

The addition of carbon-based nanomaterials before stirring of the tea extract with chloroauric acid proved not to be efficient with respect to the stability of synthesized AuNPs, and, in some cases, LSPR did not take place at any of the moments of characterization. The reverse order of material addition, i.e., the addition of carbon-based nanomaterials after the stirring of the tea extract with metallic salt, proved to be more efficient, both in terms of the synthesis and stability of the AuNPs.

Of the three materials studied, rGO proved to be the most efficient carbon-based nanomaterial used as a support for AuNPs to be applied in biosensors. The stability revealed by the zeta potential, the greater dispersion of AuNPs, and the conductivity of this nanomaterial reported by several authors support this statement. Khalil et al. [6] corroborate our conclusion by stating that rGO has advantages as a support of AuNPs, since this conjugation enhances stability, inhibiting agglomeration through a closer contact between rGO and AuNPs, as also suggested by XPS results. The possibility of producing a metal–graphene hybrid nanostructure, composed of AuNPs and graphene allotropes, opens an interesting avenue for the exploration of biosensing applications, as these composites can be tuned to a specific wavelength of resonance, while at the same time they are known to simplify functionalization with bioelements (antibodies or antigens) for selective detection of specific biomarkers or disease carriers. Once joined to a low-cost optoelectronic setup for output extraction, a LSPR sensing element fabricated with these graphene–metal hybrid nanostructures (AuNPs-G-GO-rGO) could be of great use in a situation where a large-scale, low-cost, and timely disease screening action is needed, as, for example, in a future pandemic crisis or in Third World countries, where access to laboratory facilities is problematic.

The topic of this paper has not been widely explored in the literature and it will be necessary to carry out more exhaustive research in order to reach conclusions about certain questions raised in our study. Nevertheless, we believe that this work can be a good starting point for this investigation. Regarding future work, it would be interesting to monitor the synthesis of AuNPs between t_0 and t_{1w} so that we can determine when this reaction ceases. In addition, trying other brands of black tea could supply additional valuable information, since different conditions, both soil and climatic, may be associated with different antioxidant properties.

Supplementary Materials: The following supporting information can be downloaded at: https://www.mdpi.com/article/10.3390/bios12030163/s1: Figure S1: Frequency vs. size distribution: (a) for Figure 3a; (b) for Figure 3b and (c) for Figure 5; Figure S2: C 1s spectral differences between (a) AuNPs/rGO and rGO; (b) AuNPs/G and G and (c) AuNPs/GO and GO; Figure S3: TEM image for (a)—G; (b)—rGO and (c)—GO; Figure S4: LSPR of AuNPs and G composite (SQ1 and SQ2) at t_0, t_{1w} and t_{2w}; Figure S5: LSPR of AuNPs and GO composite (SQ1 and SQ2) at t_0, t_{1w} and t_{2w}; Figure S6: LSPR of AuNPs and rGO composite (SQ1 and SQ2) at t_0, t_{1w} and t_{2w}.

Author Contributions: Conceptualization, E.C.B.A.A. and A.P.C.R.; investigation, A.P.G.C., A.F., A.M.F. and A.P.C.R.; writing—original draft preparation, A.P.G.C., E.C.B.A.A., A.F., A.M.F. and A.P.C.R.; writing—review and editing, A.P.G.C., E.C.B.A.A., A.F., A.M.F., A.M.B.d.R. and A.P.C.R.; supervision, E.C.B.A.A. and A.P.C.R. All authors have read and agreed to the published version of the manuscript.

Funding: This research was funded by EU funds through the FEDER European Regional Development Fund (project LISBOA-02-0145-FEDER-031311) and by Portuguese national funds by FCT—Fundação para a Ciência e a Tecnologia with projects PTDC/NAN-OPT/31311/2017, PTDC/QUI-QIN/29778/2017, project UIDB/00100/2020 of the Centro de Química Estrutural, LA/P/0056/2020 of Institute of Molecular Sciences, projects UIDB/04565/2020 and UIDP/04565/2020 of iBB and LA/P/0140/2020 of i4HB, project UID/EEA/00066/2020 from the Center of Technology and Systems, and from the Instituto Politécnico de Lisboa with IPL/2018/STREAM_ISEL and IPL/2020/AGE-SPReS_ISEL projects. APCR and AMF thank the Instituto Superior Técnico for the scientific employment contracts IST-ID/119/2018 and IST-ID/131/2018, respectively.

Institutional Review Board Statement: Not applicable.

Informed Consent Statement: Not applicable.

Conflicts of Interest: The authors declare no conflict of interest.

References

1. Wang, C.; Liu, M.; Wang, Z.; Li, S.; Deng, Y.; He, N. Point-of-care diagnostics for infectious diseases: From methods to devices. *Nano Today* **2021**, *37*, 101092. [CrossRef]
2. IUPAC. Biosensor Definition. In *Glossary for Chemists of Terms Used in Biotechnology*; Blackwell Scientific Publications: Oxford, UK, 1997. [CrossRef]
3. Yaseen, M.; Humayun, M.; Khan, A.; Usman, M.; Ullah, H.; Tahir, A.A.; Ullah, H. Preparation, Functionalization, Modification, and Applications of Nanostructured Gold: A Critical Review. *Energies* **2021**, *14*, 1278. [CrossRef]
4. Špačková, B.; Wrobel, P.; Bocková, M.; Homola, J. Optical Biosensors Based on Plasmonic Nanostructures: A Review. *Proc. IEEE* **2016**, *104*, 2380–2408. [CrossRef]
5. Ou, J.; Zhou, Z.; Chen, Z.; Tan, H. Optical Diagnostic Based on Functionalized Gold Nanoparticles. *Int. J. Mol. Sci.* **2019**, *20*, 4346. [CrossRef] [PubMed]
6. Khalil, I.; Julkapli, N.; Yehye, W.; Basirun, W.; Bhargava, S. Graphene–Gold Nanoparticles Hybrid—Synthesis, Functionalization, and Application in a Electrochemical and Surface-Enhanced Raman Scattering Biosensor. *Materials* **2016**, *9*, 406. [CrossRef] [PubMed]
7. Banerjee, A.N. Graphene and its derivatives as biomedical materials: Future prospects and challenges. *Interface Focus* **2018**, *8*, 20170056. [CrossRef]
8. Jung, I.; Dikin, D.A.; Piner, R.D.; Ruoff, R.S. Tunable Electrical Conductivity of Individual Graphene Oxide Sheets Reduced at "Low" Temperatures. *Nano Lett.* **2008**, *8*, 4283–4287. [CrossRef]
9. Guo, W.; Wu, L.; Fan, K.; Nie, D.; He, W.; Yang, J.; Zhao, Z.; Han, Z. Reduced Graphene Oxide-Gold Nanoparticle Nanoframework as a Highly Selective Separation Material for Aflatoxins. *Sci. Rep.* **2017**, *7*, 14484. [CrossRef]
10. Alegria, E.; Ribeiro, A.; Mendes, M.; Ferraria, A.; do Rego, A.; Pombeiro, A. Effect of Phenolic Compounds on the Synthesis of Gold Nanoparticles and its Catalytic Activity in the Reduction of Nitro Compounds. *Nanomaterials* **2018**, *8*, 320. [CrossRef]
11. Dortez, S.; González, M.C.; Crevillen, A.G.; Escarpa, A. Gold nanostructure-related non-plasmon resonance absorption band as a fingerprint of ortho-alkyl substituted phenolic compounds. *Microchem. J.* **2021**, *171*, 106788. [CrossRef]
12. Al-Radadi, N.S. Facile one-step green synthesis of gold nanoparticles (AuNp) using licorice root extract: Antimicrobial and anticancer study against HepG2 cell line. *Arab. J. Chem.* **2021**, *14*, 102956. [CrossRef]

13. Chelly, M.; Chelly, S.; Zribi, R.; Bouaziz-Ketata, H.; Gdoura, R.; Lavanya, N.; Veerapandi, G.; Sekar, C.; Neri, G. Synthesis of Silver and Gold Nanoparticles from Rumex roseus Plant Extract and Their Application in Electrochemical Sensors. *Nanomaterials* **2021**, *11*, 739. [CrossRef] [PubMed]

14. Verma, S.; Singh, A.; Shukla, A.; Kaswan, J.; Arora, K.; Ramirez-Vick, J.; Singh, P.; Singh, S.P. Anti-IL8/AuNPs-rGO/ITO as an Immunosensing Platform for Noninvasive Electrochemical Detection of Oral Cancer. *ACS Appl. Mater. Interfaces* **2017**, *9*, 27462–27474. [CrossRef]

15. Vashist, S.K.; Luppa, P.B.; Yeo, L.Y.; Ozcan, A.; Luong, J.H.T. Emerging Technologies for Next-Generation Point-of-Care Testing. *Trends Biotechnol.* **2015**, *33*, 692–705. [CrossRef] [PubMed]

16. Zeng, S.; Baillargeat, D.; Ho, H.-P.; Yong, K.-T. Nanomaterials enhanced surface plasmon resonance for biological and chemical sensing applications. *Chem. Soc. Rev.* **2014**, *43*, 3426–3452. [CrossRef] [PubMed]

17. Sharma, P.; Tuteja, S.K.; Bhalla, V.; Shekhawat, G.; Dravid, V.P.; Suri, C.R. Bio-functionalized graphene–graphene oxide nanocomposite based electrochemical immunosensing. *Biosens. Bioelectron.* **2013**, *39*, 99–105. [CrossRef] [PubMed]

18. Fantoni, A.; Vygranenko, Y.; Maçarico, A.F.; Serafinelli, C.; Fernandes, M.; Mansour, R.; Jesus, R.; Vieira, M. Arrayed graphene enhanced surface plasmon resonance for sensing applications. In *Optical Components and Materials XVIII*; Digonnet, M.J., Jiang, S., Eds.; SPIE: Bellingham, DC, USA, 2021.

19. Gao, X.-G.; Cheng, L.-X.; Jiang, W.-S.; Li, X.-K.; Xing, F. Graphene and its Derivatives-Based Optical Sensors. *Front. Chem.* **2021**, *9*, 615164. [CrossRef]

20. Galande, C.; Mohite, A.D.; Naumov, A.V.; Gao, W.; Ci, L.; Ajayan, A.; Gao, H.; Srivastava, A.; Weisman, R.B.; Ajayan, P.M. Quasi-Molecular Fluorescence from Graphene Oxide. *Sci. Rep.* **2011**, *1*, 85. [CrossRef]

21. Tsai, M.-F.; Chang, S.-H.G.; Cheng, F.-Y.; Shanmugam, V.; Cheng, Y.-S.; Su, C.-H.; Yeh, C.-S. Au Nanorod Design as Light-Absorber in the First and Second Biological Near-Infrared Windows for In Vivo Photothermal Therapy. *ACS Nano* **2013**, *7*, 5330–5342. [CrossRef]

22. Pramanik, A.; Fan, Z.; Chavva, S.R.; Sinha, S.S.; Ray, P.C. Highly Efficient and Excitation Tunable Two-Photon Luminescence Platform For Targeted Multi-Color MDRB Imaging Using Graphene Oxide. *Sci. Rep.* **2015**, *4*, 6090. [CrossRef]

23. Kalluru, P.; Vankayala, R.; Chiang, C.-S.; Hwang, K.C. Nano-graphene oxide-mediated In vivo fluorescence imaging and bimodal photodynamic and photothermal destruction of tumors. *Biomaterials* **2016**, *95*, 1–10. [CrossRef] [PubMed]

24. Song, X.; Li, S.; Guo, H.; You, W.; Shang, X.; Li, R.; Tu, D.; Zheng, W.; Chen, Z.; Yang, H.; et al. Graphene-Oxide-Modified Lanthanide Nanoprobes for Tumor-Targeted Visible/NIR-II Luminescence Imaging. *Angew. Chem. Int. Ed.* **2019**, *58*, 18981–18986. [CrossRef] [PubMed]

25. Lai, H.; Xu, F.; Zhang, Y.; Wang, L. Recent progress on graphene-based substrates for surface-enhanced Raman scattering applications. *J. Mater. Chem. B* **2018**, *6*, 4008–4028. [CrossRef]

26. Ouyang, L.; Hu, Y.; Zhu, L.; Cheng, G.J.; Irudayaraj, J. A reusable laser wrapped graphene-Ag array based SERS sensor for trace detection of genomic DNA methylation. *Biosens. Bioelectron.* **2017**, *92*, 755–762. [CrossRef] [PubMed]

27. Ilkhani, H.; Hughes, T.; Li, J.; Zhong, C.J.; Hepel, M. Nanostructured SERS-electrochemical biosensors for testing of anticancer drug interactions with DNA. *Biosens. Bioelectron.* **2016**, *80*, 257–264. [CrossRef]

28. Liu, Z.; Li, S.; Hu, C.; Zhang, W.; Zhong, H.; Guo, Z. pH-dependent surface-enhanced Raman scattering of aromatic molecules on graphene oxide. *J. Raman Spectrosc.* **2013**, *44*, 75–80. [CrossRef]

29. Yin, F.; Wu, S.; Wang, Y.; Wu, L.; Yuan, P.; Wang, X. Self-assembly of mildly reduced graphene oxide monolayer for enhanced Raman scattering. *J. Solid State Chem.* **2016**, *237*, 57–63. [CrossRef]

30. Shivananju, B.N.; Yu, W.; Liu, Y.; Zhang, Y.; Lin, B.; Li, S.; Bao, Q. The Roadmap of Graphene-Based Optical Biochemical Sensors. *Adv. Funct. Mater.* **2017**, *27*, 1603918. [CrossRef]

31. Yao, B.; Wu, Y.; Cheng, Y.; Zhang, A.; Gong, Y.; Rao, Y.-J.; Wang, Z.; Chen, Y. All-optical Mach–Zehnder interferometric NH3 gas sensor based on graphene/microfiber hybrid waveguide. *Sens. Actuators B Chem.* **2014**, *194*, 142–148. [CrossRef]

32. Gan, S.; Cheng, C.; Zhan, Y.; Huang, B.; Gan, X.; Li, S.; Lin, S.; Li, X.; Zhao, J.; Chen, H.; et al. A highly efficient thermo-optic microring modulator assisted by graphene. *Nanoscale* **2015**, *7*, 20249–20255. [CrossRef]

33. Some, S.; Xu, Y.; Kim, Y.; Yoon, Y.; Qin, H.; Kulkarni, A.; Kim, T.; Lee, H. Highly Sensitive and Selective Gas Sensor Using Hydrophilic and Hydrophobic Graphenes. *Sci. Rep.* **2013**, *3*, 1868. [CrossRef] [PubMed]

34. Xing, F.; Meng, G.-X.; Zhang, Q.; Pan, L.-T.; Wang, P.; Liu, Z.-B.; Jiang, W.-S.; Chen, Y.; Tian, J.-G. Ultrasensitive Flow Sensing of a Single Cell Using Graphene-Based Optical Sensors. *Nano Lett.* **2014**, *14*, 3563–3569. [CrossRef] [PubMed]

35. Jiang, W.-S.; Xin, W.; Xun, S.; Chen, S.-N.; Gao, X.-G.; Liu, Z.-B.; Tian, J.-G. Reduced graphene oxide-based optical sensor for detecting specific protein. *Sens. Actuators B Chem.* **2017**, *249*, 142–148. [CrossRef]

36. Gao, X.; Chen, G.; Li, D.; Li, X.; Liu, Z.; Tian, J. High-accuracy measurement of the crystalline orientation of anisotropic two-dimensional materials using photothermal detection. *J. Mater. Chem. C* **2018**, *6*, 5849–5856. [CrossRef]

37. Biris, A.R.; Pruneanu, S.; Pogacean, F.; Lazar, M.D.; Borodi, G.; Ardelean, S.; Dervishi, E.; Watanabe, F.; Biris, A.S. Few-layer graphene sheets with embedded gold nanoparticles for electrochemical analysis of adenine. *Int. J. Nanomed.* **2013**, *8*, 1429–1438. [CrossRef]

38. Aldewachi, H.; Chalati, T.; Woodroofe, M.N.; Bricklebank, N.; Sharrack, B.; Gardiner, P. Gold nanoparticle-based colorimetric biosensors. *Nanoscale* **2018**, *10*, 18–33. [CrossRef]

39. Giangregorio, M.M.; Jiao, W.; Bianco, G.V.; Capezzuto, P.; Brown, A.S.; Bruno, G.; Losurdo, M. Insights into the effects of metal nanostructuring and oxidation on the work function and charge transfer of metal/graphene hybrids. *Nanoscale* **2015**, *7*, 12868–12877. [CrossRef]

40. Aldewachi, H.; Woodroofe, N.; Turega, S.; Gardiner, P. Optimization of gold nanoparticle-based real-time colorimetric assay of dipeptidyl peptidase IV activity. *Talanta* **2017**, *169*, 13–19. [CrossRef]

41. Zoladek, S.; Rutkowska, I.A.; Blicharska, M.; Miecznikowski, K.; Ozimek, W.; Orlowska, J.; Negro, E.; Di Noto, V.; Kulesza, P.J. Evaluation of reduced-graphene-oxide-supported gold nanoparticles as catalytic system for electroreduction of oxygen in alkaline electrolyte. *Electrochim. Acta* **2017**, *233*, 113–122. [CrossRef]

42. Zhou, J.; Chen, M.; Xie, J.; Diao, G. Synergistically Enhanced Electrochemical Response of Host–Guest Recognition Based on Ternary Nanocomposites: Reduced Graphene Oxide-Amphiphilic Pillar[5]arene-Gold Nanoparticles. *ACS Appl. Mater. Interfaces* **2013**, *5*, 11218–11224. [CrossRef]

43. Chowdhury, A.D.; Nasrin, F.; Gangopadhyay, R.; Ganganboina, A.B.; Takemura, K.; Kozaki, I.; Honda, H.; Hara, T.; Abe, F.; Park, S.; et al. Controlling distance, size and concentration of nanoconjugates for optimized LSPR based biosensors. *Biosens. Bioelectron.* **2020**, *170*, 112657. [CrossRef] [PubMed]

44. Klekotko, M.; Matczyszyn, K.; Siednienko, J.; Olesiak-Banska, J.; Pawlik, K.; Samoc, M. Bio-mediated synthesis, characterization and cytotoxicity of gold nanoparticles. *Phys. Chem. Chem. Phys.* **2015**, *17*, 29014–29019. [CrossRef] [PubMed]

45. Parnianchi, F.; Nazari, M.; Maleki, J.; Mohebi, M. Combination of graphene and graphene oxide with metal and metal oxide nanoparticles in fabrication of electrochemical enzymatic biosensors. *Int. Nano Lett.* **2018**, *8*, 229–239. [CrossRef]

46. Liao, R.; Tang, Z.; Lei, Y.; Guo, B. Polyphenol-Reduced Graphene Oxide: Mechanism and Derivatization. *J. Phys. Chem. C* **2011**, *115*, 20740–20746. [CrossRef]

47. Clogston, J.D.; Patri, A.K. Zeta Potential Measurement. In *Characterization of Nanoparticles Intended for Drug Delivery*; Humana Press: Totowa, NJ, USA, 2011; pp. 63–70. [CrossRef]

48. Maier, S.A. *Plasmonics: Fundamentals and Applications*; Springer Science & Business Media: New York, NY, USA, 2007; ISBN 978-0-387-33150-8.

49. Mulvaney, P. Surface plasmon spectroscopy of nanosized metal particles. *Langmuir* **1996**, *12*, 788–800. [CrossRef]

50. Cittadini, M.; Bersani, M.; Perrozzi, F.; Ottaviano, L.; Wlodarski, W.; Martucci, A. Graphene oxide coupled with gold nanoparticles for localized surface plasmon resonance based gas sensor. *Carbon N. Y.* **2014**, *69*, 452–459. [CrossRef]

51. Beamson, G.; Briggs, D. *High Resolution XPS of Organic Polymers: The Scienta ESCA300 Database*, 1st ed.; Wiley: Chichester, UK; Wiley: New York, NY, USA, 1992; ISBN 0471935921.

52. Dias, A.; Bundaleska, N.; Felizardo, E.; Tsyganov, D.; Almeida, A.; Ferraria, A.M.; do Rego, A.M.B.; Abrashev, M.; Strunskus, T.; Santhosh, N.M.; et al. N-Graphene-Metal-Oxide(Sulfide) hybrid Nanostructures: Single-step plasma-enabled approach for energy storage applications. *Chem. Eng. J.* **2022**, *430*, 133153. [CrossRef]

53. Carapeto, A.P.; Ferraria, A.M.; Boufi, S.; Vilar, M.R.; do Rego, A.M.B. Ion reduction in metallic nanoparticles nucleation and growth on cellulose films: Does substrate play a role? *Cellulose* **2015**, *22*, 173–186. [CrossRef]

54. Boufi, S.; Ferraria, A.M.; do Rego, A.M.B.; Battaglini, N.; Herbst, F.; Vilar, M.R. Surface functionalisation of cellulose with noble metals nanoparticles through a selective nucleation. *Carbohydr. Polym.* **2011**, *86*, 1586–1594. [CrossRef]

55. Clifford, M.N. Diet-Derived Phenols in Plasma and Tissues and their Implications for Health. *Planta Med.* **2004**, *70*, 1103–1114. [CrossRef]

56. Tao, Y.; García, J.F.; Sun, D.-W. Advances in Wine Aging Technologies for Enhancing Wine Quality and Accelerating Wine Aging Process. *Crit. Rev. Food Sci. Nutr.* **2014**, *54*, 817–835. [CrossRef] [PubMed]

57. Bohren, C.F.; Huffman, D.R. *Absorption and Scattering of Light by Small Particles*; Wiley: Hoboken, NJ, USA, 1998; ISBN 9780471293408.

58. Fantoni, A.; Fernandes, M.; Vygranenko, Y.; Louro, P.; Vieira, M.; Alegria, E.C.B.A.; Ribeiro, A.; Texeira, D. A Simulation Study of Surface Plasmons in Metallic Nanoparticles: Dependence on the Properties of an Embedding a-Si:H Matrix. *Phys. Status Solidi* **2018**, *215*, 1700487. [CrossRef]

59. Fantoni, A.; Fernandes, M.; Vygranenko, Y.; Louro, P.; Vieira, M.; Silva, R.P.O.; Texeira, D.; Ribeiro, A.P.C.; Prazeres, M.; Alegria, E.C.B.A. Analysis of metallic nanoparticles embedded in thin film semiconductors for optoelectronic applications. *Opt. Quantum Electron.* **2018**, *50*, 246. [CrossRef]

60. Bhagwat, S.; Haytowitz, D.B.; Holden, J.M. *USDA Database for the Flavonoid Content of Selected Foods*; Release 3.1; U.S. Department of Agriculture: Beltsville, MD, USA, 2011; pp. 1–156.

61. ChemSpider. Available online: http://www.chemspider.com/Chemical-Structure.58575.html (accessed on 19 April 2021).

62. Jeon, H.B.; Tsalu, P.V.; Ha, J.W. Shape Effect on the Refractive Index Sensitivity at Localized Surface Plasmon Resonance Inflection Points of Single Gold Nanocubes with Vertices. *Sci. Rep.* **2019**, *9*, 13635. [CrossRef]

63. Zalyubovskiy, S.J.; Bogdanova, M.; Deinega, A.; Lozovik, Y.; Pris, A.D.; An, K.H.; Hall, W.P.; Potyrailo, R.A. Theoretical limit of localized surface plasmon resonance sensitivity to local refractive index change and its comparison to conventional surface plasmon resonance sensor. *J. Opt. Soc. Am. A* **2012**, *29*, 994. [CrossRef]

64. Bhuyan, M.S.A.; Uddin, M.N.; Islam, M.M.; Bipasha, F.A.; Hossain, S.S. Synthesis of graphene. *Int. Nano Lett.* **2016**, *6*, 65–83. [CrossRef]
65. Carapeto, A.P.; Ferraria, A.M.; do Rego, A.M.B. Unraveling the reaction mechanism of silver ions reduction by chitosan from so far neglected spectroscopic features. *Carbohydr. Polym.* **2017**, *174*, 601–609. [CrossRef]

biosensors

Communication

Real-Time Detection of LAMP Products of African Swine Fever Virus Using Fluorescence and Surface Plasmon Resonance Method

Hao Zhang [1], Yuan Yao [1], Zhi Chen [2], Wenbo Sun [2,*], Xiang Liu [3,*], Lei Chen [4], Jianhai Sun [3], Xianbo Qiu [1], Duli Yu [1] and Lulu Zhang [1,*]

[1] College of Information Science and Technology, Beijing University of Chemical Technology, Beijing 100029, China; zh1632414919@163.com (H.Z.); yy_xyyyyy@163.com (Y.Y.); xbqiu@mail.buct.edu.cn (X.Q.); dyu@mail.buct.edu.cn (D.Y.)

[2] Shandong Key Laboratory of Animal Disease Control and Breeding, Institute of Animal Science and Veterinary Medicine, Shandong Academy of Agricultural Sciences, Jinan 250100, China; charleschenzhi@163.com

[3] Aerospace Information Research Institute, Chinese Academy of Sciences, Beijing 100094, China; sunjh@aircas.ac.cn

[4] Shandong Provincial Key Laboratory of Animal Resistance Biology, College of Life Sciences, Shandong Normal University, Jinan 250014, China; chenlei@sdnu.edu.cn

* Correspondence: swbecho@126.com (W.S.); lx987@mail.ie.ac.cn (X.L.); llzhang@mail.buct.edu.cn (L.Z.); Tel.: +86-0531-66655096 (W.S.); +86-010-56535330 (X.L.); +86-010-64436719 (L.Z.)

Abstract: African swine fever (ASF) is a swine disease with a very high fatality rate caused by a complex double-stranded DNA virus. The fluorescence PCR detection method is widely used for virus nucleic acid detection. Surface plasmon resonance (SPR) is a label-free and real-time detection method, unlike the fluorescence PCR detection method. In this research, we detected the loop-mediated isothermal amplification (LAMP) products of the African swine fever virus by using the SPR and fluorescence methods separately and simultaneously. By comparing the positive and negative control results, we found that the SPR response unit is completely different before and after the LAMP process. In addition, the fluorescence results on a chip showed that with an increase in the concentration of the sample, the cycle threshold (CT) value decreased, which is consistent with commercial instruments. Both the decline rate of the SPR response unit and the CT value of the fluorescence realized were used to distinguish the positive control from the negative control and water, which indicates that the SPR method can be combined with fluorescence to detect LAMP products. This research provides a label-free and simple method for detecting LAMP products.

Keywords: African swine fever virus (ASFV); loop-mediated isothermal amplification (LAMP); surface plasmon resonance (SPR); fluorescence detection

Citation: Zhang, H.; Yao, Y.; Chen, Z.; Sun, W.; Liu, X.; Chen, L.; Sun, J.; Qiu, X.; Yu, D.; Zhang, L. Real-Time Detection of LAMP Products of African Swine Fever Virus Using Fluorescence and Surface Plasmon Resonance Method. *Biosensors* **2022**, *12*, 213. https://doi.org/10.3390/bios12040213

Received: 27 February 2022
Accepted: 31 March 2022
Published: 3 April 2022

Publisher's Note: MDPI stays neutral with regard to jurisdictional claims in published maps and institutional affiliations.

1. Introduction

African swine fever (ASF) has the features of acute and high contagiousness and a fatality rate of 100%. The subclinical infection of African wild boars, such as bush pigs and warthogs, can last for months or even years. In the last decade, the pig industry in about 50 countries around the world has been affected by this virus, making this industry precarious [1,2]. The African swine fever virus is a kind of complex double-stranded DNA virus with a size of 170–190 kbp [3]. Therefore, regular virus detection in pigs is a reasonable way to avoid risks.

The hemadsorption test, virus isolation, and real-time polymerase chain reaction (real-time PCR) are widely used methods for ASFV diagnosis [4]. Hemadsorption tests and virus isolation are reliable methods for virus detection, but their operations are complex and time consuming, which makes them unsuitable for rapid detection. Real-time PCR is recognized as the most sensitive and reliable method. However, it contains complex temperature-changing devices, which is not conducive to field rapid detection application.

Loop-mediated isothermal amplification (LAMP) is a technique for isothermal amplification at 60 °C–65 °C. Its strong points are its convenient operation, high efficiency, high specificity, and short reaction time [5]. It is suitable for detection in complex environments and can be well applied in the detection of African swine fever. At present, the real-time fluorescence detection method is widely used in LAMP. The following are some application examples: Jiang [6] used LAMP techniques to establish a rapid method for the detection of African swine fever. Xing [7] completed the LAMP and fluorescence detection of Zika viruses on a small microfluidic chip. A capillary-array microsystem with integrated DNA extraction, loop-mediated isothermal amplification, and fluorescence detection was used and, based on this system, successfully achieved the detection of Mycobacterium tuberculosis [8]. However, the fluorescence detection method needs to label the sample, and fluorescence bleaching and quenching especially lead to inaccurate detection results.

The use of surface plasmon resonance (SPR) technology is a label-free method used to measure refractive index changes on the surface of a sensor chip due to biological interactions caused by optical principles. SPR technology has the advantages of being simple to operate and label free as well as allowing real-time detection [9–12]. Bai [13] used a portable SPR biosensor to detect the H5N1 virus, but the detection concentration range has certain limitations. Wang [14] reported a label-free detection of single virus particles in solution after using surface plasmon resonance. Nguyen [15] reported a sandwich SPR biosensor detection method that can quickly and accurately detect the whole H5N1 avian influenza virus. Wang [16] developed an SPR biosensor platform based on fast and sensitive intensity modulation. It was used to detect H7N9, and its detection limit was about 20 times that of ELISA. Yoo [17] developed a reusable magnetic SPR sensor chip that can be used to repeatedly detect the H1N1 virus. However, the above SPR detection methods are based on the principle of immune interactions between biomolecules and are not combined with nucleic acid amplification and fluorescence detection.

The use of multiple testing methods together often leads to better results [18]. The combination of these technologies should improve detection sensitivity and specificity. We have realized the synchronous detection of microspheres and A549 cells by using the SPR and fluorescence methods with a homemade microscopic imaging system [19]. This system is complex in size and cannot realize the rapid amplification and detection of the virus on site. For this paper, a portable SPR imaging biosensor of the classic Kretschmann prism structure was designed to detect LAMP products [20]. By detecting the SPR response unit before and after LAMP, it was found that the decline rate of the SPR response unit for the positive sample can be used as an indication of the LAMP result. The independent fluorescence detection system was integrated with the heating module, and a microfluidic chip was used to complete the LAMP and fluorescence detection on the chip. The higher the concentration of the positive control was, the lower the CT response unit of fluorescence detection became. Finally, these two systems were combined to complete the LAMP on the chip, and the SPR and fluorescence detection methods were used to simultaneously detect the LAMP. This paper established a faster, more accurate, label-free, and highly sensitive detection method.

2. Materials and Methods

2.1. Materials and Reagent

The ASFV LAMP kit (HaiGene) and the ASFV LAMP kit (Quicking Biotech, Shanghai, China) were used in this paper. According to the requirements of the kits, 2 μL of different concentrations of positive control or negative control were added to 20 μL of reaction solution and enzyme mixture to configure different reaction mixtures. This took one minute per amplification cycle. The other reagents used in the experiments were purchased from HyperCyte Biomedical Co., Ltd., Beijing, China.

2.2. Systems of Detection

2.2.1. SPR Detection System

The schematic diagram of surface plasmon resonance (SPR) biosensors is shown in Figure 1a, which was designed by us [21]. A semiconductor red laser with a wavelength of 633 nm was used as a light source, and a CCD camera was used as a light-receiving device. The light source and CCD camera were symmetrical around the prism, and we could change the angle of the light source and prism by motor movement. The angle scan was used to find the SPR resonance angle at which the device could then be fixed for the experiment, and the CCD camera was used for the detection of the real-time SPR response unit. The chip and flow cell were located above the prism, and the liquid passed through the flow cell onto the surface of the chip.

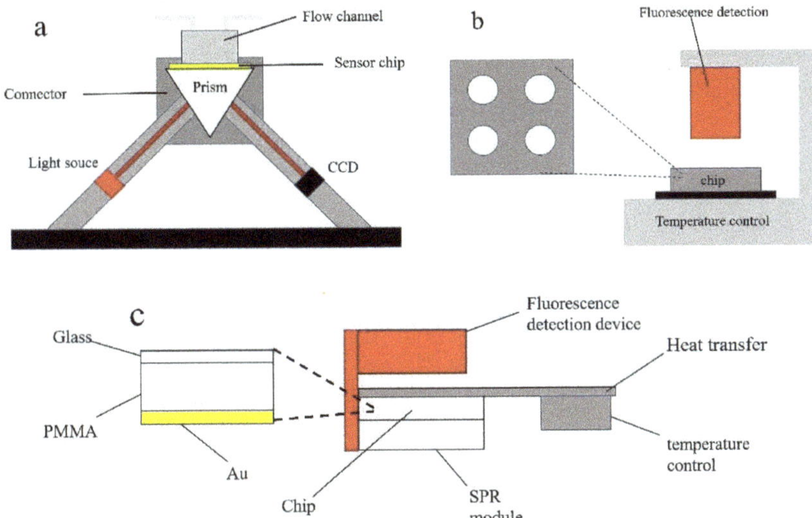

Figure 1. (**a**) Schematic of independent SPR system composed of light source, CCD camera, flow cell, prism and gold layer. (**b**) Schematic of the independent fluorescence detection system consisting of heating module and microfluidic chip. (**c**) Schematic of the integrated device of fluorescence detection, SPR module, temperature, and chip.

2.2.2. Fluorescence Detection System

The independent fluorescence detection system integrated with the heating module and microfluidic chip (abbreviated as the independent fluorescence system) was used for LAMP and fluorescence detection on the chip. The temperature control module was located at the bottom of the whole integrated device and was mainly used to provide the temperature required by the LAMP. The chip was fixed above the temperature control device by a metal fixing device so that the heat could be efficiently transferred to the chip. There were 4 reaction chambers on the chip designed for this experiment, and 4 groups of control experiments could be performed simultaneously. Before the experiment, the prepared reagents were added to different chambers on the chip, and then the chip was sealed with a transparent glass sheet. The fluorescence detection module was located directly above the chip and could collect the fluorescence response unit of the solution in the chip in real time.

2.3. Integrated Device of SPR Sensor Chip and Heating Module

To realize the combination of LAMP, SPR detection, and fluorescence detection, we designed an integrated device as shown in Figure 1c. The LAMP chip was fixed on the prism instead of the flow cell (in Figure 1a), and the aluminum block (which does not hinder fluorescence detection) was connected to the temperature control device and the chip, acting as a heat transfer device so that the chip could reach the temperature (64 °C) required by the LAMP. The independent fluorescence detection (in Figure 1b) was located directly above the chip and fixed with the SPR detection device (in Figure 1a). It could be controlled by software for temperature adjustment, SPR detection, and fluorescence detection. The LAMP-SPR chip was mainly composed of three layers, and the adjacent two layers were bonded by double-sided tape (purchased from 3M) to avoid liquid leakage. The bottom layer was glass sputtered with Au for SPR detection. The middle layer was the subject part made of PMMA material, which was used to hold liquid and carry out LAMP. On the top was a very thin, transparent layer of glass (0.13–0.17 mm) that could cover the entire chip without affecting the fluorescence collection.

2.4. Experiment Procedure

2.4.1. SPR Detection Procedure

To verify whether SPR can be used to detect LAMP products, the following experiments were carried out: According to the requirements of the ASFV LAMP kit (Quicking Biotech, Shanghai, China), reagents were configured. A PCR instrument was used to amplify the positive control sample for 15 cycles and 35 cycles, and the negative control sample for 35 cycles (abbreviated as pos-15, pos-35, neg-35). Biacore T100 (GE) was used to carry out the same SPR experiment. We also detected the LAMP products of different concentrations of positive control by using the SPR method. Positive nucleic acids of 0.5, 1, and 2 μL were prepared with an ASFV LAMP kit (Quicking Biotech, Shanghai, China), in which an appropriate amount of deionized water was added to form an equal volume of positive control for 2 μL. The SPR biosensor judges whether amplification occurs by detecting the change in the refractive index of the sample, so the mixed amplified sample was injected into the SPR biosensors before and after amplification to test the SPR response unit, and the difference in the response unit was calculated.

2.4.2. Fluorescence Detection Procedure

To demonstrate that LAMP on a chip gives similar results to LAMP in a PCR instrument, the following experiments were performed. According to the method in Section 2.4.1, the positive control solution with volumes of 0.2, 0.5, 1, and 2 μL and the negative control with a volume of 2 μL were prepared. The PCR instrument (Bio-rad) and independent fluorescence detection system (in Figure 1b) with a microfluidic chip were used to compare the fluorescence signals of LAMP. To compare the differences between different reagents, the ASFV LAMP kit (HaiGene, Haerbin, China) was used to configure the same concentration of reagents (in Figure 1b) and complete the LAMP and fluorescence detection with the independent fluorescence detection system.

2.4.3. SPR and Fluorescence Simultaneous Detection Procedure

The following experiments were used to prove the feasibility and stability of the integrated device for LAMP, SPR detection, and fluorescence detection: The mixture of the 2 μL positive control and 20 μL reaction solution and enzyme were prepared using the ASFV LAMP kit (Quicking Biotech, Shanghai, China), and then it was added to the chip and sealed. At the beginning of the experiment, we let the chip stand for 5 min to obtain the initial fluorescence and SPR response unit, and the temperature control device was used to heat the chip to 64 °C so that it started LAMP (this process took 10 min). The heating device was turned off after 50 min, and data were recorded until the device dropped to room temperature. This experiment was repeated three times.

3. Results and Discussion

3.1. SPR Detects LAMP Products

The LAMP process reduces the refractive index of the reaction solution [22]. The SPR method can theoretically be used to detect the LAMP products and judge whether LAMP has occurred by comparing the decline rate of the SPR response unit. In this paper, the response unit detected by the SPR method before amplification is defined as α, the response unit after amplification is defined as β, and the decline rate of the SPR response unit is defined as R.

$$R = (\alpha - \beta)/\alpha \times 100\% \tag{1}$$

To prove that SPR can detect LAMP products, the SPR response units before and after amplification were detected for comparison. As shown in Figure 2, the results of both of the instruments tested show a higher rate of decline in reaction units for pos-35 than pos-15, and both were much higher than neg-35. The volume of the positive control in this experiment was 2 µL. The detection results for the different concentrations of the positive control are shown in Figure 3, from which it can be seen that lower concentrations of positive control show a lower rate of decline. These results prove that the LAMP process reduces the refractive index of the reaction solution, and the SPR method can be used to judge whether LAMP has occurred.

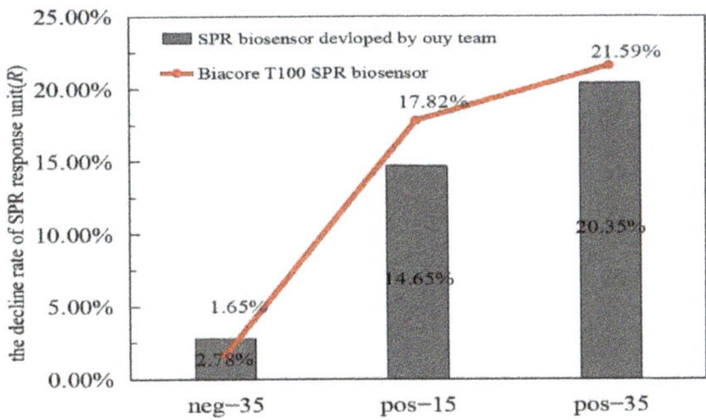

Figure 2. To compare the decline rates of SPR response units of the two instruments, the line graph represents the test results of Biacore T100 SPR biosensor, and the histogram represents the test results of the self-made SPR biosensor. This figure mainly compares the decline rate of SPR response unit (R) of positive control amplified for 15 cycles (pos−15), positive control amplified for 35 cycles (pos−35), and negative control amplified for 35 cycles (neg−35) of two instruments. The volume of the positive and negative controls in this experiment was 2 µL.

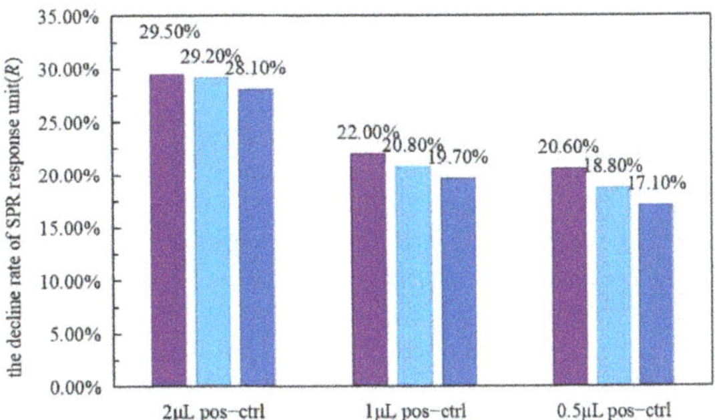

Figure 3. Tests were repeated three times for the decline rate of positive control (pos−ctrl) at different concentrations. It is clear that solutions with low concentrations of positive control have a small decline rate.

3.2. Fluorescence Detection of LAMP on a Microfluidic Chip

The independent fluorescence system (in Figure 1b) was used to perform LAMP on the chip, which is similar to LAMP on the PCR instrument (Bio-rad). The reaction solution of the negative control and the different concentrations of the positive control were tested in the two instruments. Figure 4a,b show the detection process and the results of the PCR instrument and our home-made fluorescence detection system, respectively. The fluorescence detection CT values of PCR and the independent fluorescence system are shown in Figure 4c. It can be seen that the amplification response unit, after 20 min in the PCR machine, did not have a linear relationship with the concentration. Therefore, the CT value was used as the basis for judging the nucleic acid concentration. The CT value changes according to the law, that is, the higher the concentration, the lower the CT value. Although their error bars will intersect slightly, this does not affect the overall trend. The result of the independent fluorescence system on-chip is the same as the trend of the PCR instrument, which also illustrates the success of the LAMP on-chip. We performed three experiments with the same reagents (positive control at a volume of 2 μL) in the standard PCR instrument, and the results are shown in Figure 4d. It can be seen that the CT values of the three experiments are similar, but the responses after 20 min are very different, which proves that the final fluorescence response cannot determine the results of LAMP. The CT values of the two kits are similar, indicating that the two kits have little effect on the experiment.

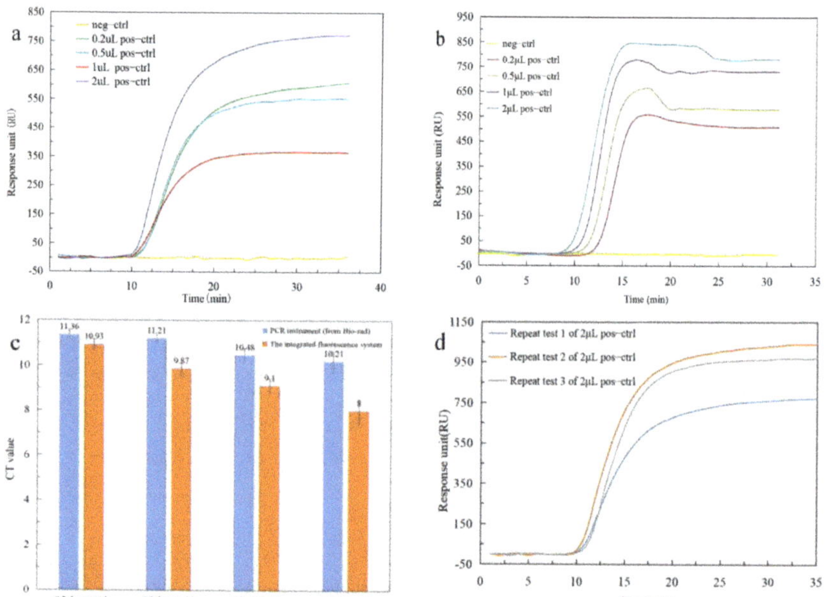

Figure 4. Comparing the LAMP and fluorescence detection results of the independent fluorescence system and the PCR instrument. (**a**) The fluorescence detection process of different concentrations of positive control (pos−ctrl) with PCR instrument (Bio−rad). (**b**) The independent fluorescence detection system on a chip for the same experiments as in Figure 4a. (**c**) CT values of different pos−ctrl with two instruments. (**d**) Comparison of the results of three fluorescence detections using the same concentration of reagent (positive control at a volume of 2 μL). The CT values were similar, but the final fluorescence responses were very different.

3.3. Simultaneous Detection of LAMP Products by SPR and Fluorescence on a Microfluidic Chip

LAMP reagent (Quicking Biotech, Shanghai, China) was added to the integrated device to verify whether it could be detected by SPR and fluorescence at the same time. The results are shown in Figure 5a. The volume of the positive control in this experiment was 2 µL. It can be seen that the fluorescence response unit of water did not change much. However, the fluorescence response unit of the positive control is similar to the previous experiment (data in Section 3.2). This illustrates the success of LAMP on the chip in the integrated device. Observing the SPR response unit, we found that when the temperature changes (region II in Figure 5a), both the water and the positive reference SPR response unit change sharply. After this (region III in Figure 5a), the water response unit gradually stabilizes while the positive control response unit slowly rises. This proves that the amplification reaction is taking place. After the amplification (region IV in Figure 5a), the SPR response unit begins to change sharply due to temperature changes and finally stabilizes. After stabilization, the SPR response unit of water almost returns to the initial value (region I in Figure 5a), but the value of the positive control drops greatly. When the temperature changes, the SPR response unit is very sensitive, resulting in a slight deviation in the detection of the LAMP process by the SPR. It is better to judge this using the decline rate of the SPR response unit. The decline rate of the SPR response unit and the CT values of fluorescence for the positive control and water samples are shown in Figure 5b,c. By comparing the SPR response units before and after amplification, it can be seen that the rate of response unit decline of the positive control is 18.93%, which is similar to the previous results (data in Section 3.1), whereas the response unit of water does not change much. The results of the SPR response unit also proved that the amplification was successful and that the amplification products were simultaneously detected by SPR and fluorescence. The same experiment was performed three times, which shows that the experiment is repeatable and also proves the feasibility of detecting LAMP products by SPR and fluorescence.

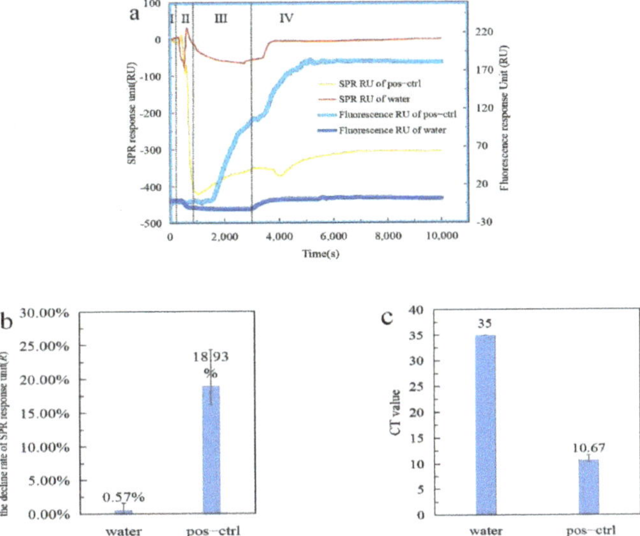

Figure 5. The results of simultaneous fluorescence and SPR detection in the integrated device are shown in this figure, (**b**,**c**) show the results of three replicates. The volume of positive control in this experiment was 2 µL. (**a**) The results of SPR and fluorescence response unit of water and positive control during the LAMP process; the left ordinate in this figure is the SPR response unit, and the right is the fluorescence response unit. Region I of the figure shows the process of detecting the baseline, region II is the pre−warming part, region III is the amplification part, and region IV is the process of cooling to room temperature after the amplification is completed. (**b**) The decline rate of SPR response unit of water and positive control. (**c**) CT values of fluorescence detection (because water has no fluorescence response unit, it is assumed that its CT value is 35).

3.4. Discussion

Traditional loop-mediated isothermal amplification methods for detecting fluorescent signals in nucleic acids require fluorescent labeling, and fluorescent molecules have problems such as easy quenching and photobleaching, which affect the accuracy of the detection. In this paper, we demonstrate through the experiments of SPR signal detection before and after LAMP that the SPR method can detect the refractive index change of amplification products and that the SPR signal can be detected in real time during the amplification process. In addition, we demonstrate that the difference in the SPR refractive index signal can be measured at different amplification cycles and different concentrations of the positive control to be tested and that the amplification reagents do not need to be labeled, which can avoid the problem of fluorescence quenching that occurs with fluorescence detection techniques. However, the SPR detection method is very sensitive to temperature, and fluctuations in temperature can lead to inaccurate signals, so an accurate temperature control system is required, and this is where further development and improvement are needed. SPR and fluorescence signals are currently combined to analyze the reaction process of nucleic acid amplification, both of which can be monitored in real time, but because the fluorescence method takes a signal every minute to avoid the attenuation of fluorescence intensity, whereas the SPR signal can be monitored every second in real time (or even at a higher frequency), the combination of the two methods ensures that the amplification process is efficient and allows for a more detailed and precise analysis of the kinetic change process in amplification. In particular, the fluorescence signal rises exponentially during the stage at which the signal starts to rise, and the precision sampling of the SPR is a more accurate response to the rapid change in the amplification peak. Thus, the work in this paper on the simultaneous detection of LAMP products by two methods provides a platform and means for multidimensional information analysis of nucleic acid detection, which can be extended to the nucleic acid detection of other infectious diseases in addition to its application in the detection of African swine fever, such as detecting Corona Virus Disease 2019.

4. Conclusions

In this paper, LAMP products were measured separately and simultaneously by two methods. The results show that the higher the level of positive control, the lower the CT value. SPR reaction units before and after the amplification of the LAMP product can be used to determine whether LAMP has occurred. The feasibility and stability of an integrated instrument for LAMP, SPR, and fluorescence detection are demonstrated. This instrument allows for a rapid, label-free, real-time multi-detection method that can be used in complex environments. Although some issues require improvement, they provide a reference for future research.

Author Contributions: Conceptualization, L.Z.; methodology, W.S.; software, J.S.; validation, Y.Y. and Z.C.; formal analysis, J.S.; investigation, H.Z.; resources, W.S.; data curation, L.Z.; writing—original draft preparation, H.Z.; writing—review and editing, L.Z.; visualization, L.C.; supervision, X.L.; project administration, X.L.; funding acquisition, X.Q., D.Y. and L.Z. All authors have read and agreed to the published version of the manuscript.

Funding: The authors greatly acknowledge the financial support from the National Natural Science Foundation of China (61971026), the Natural Science Foundation of Beijing Municipality (4202053), the Agricultural Animal Breeding Project of Shandong Province (2021LZGC001), the Joint Project of BRC-BC (Biomedical Translational Engineering Research Center of BUCT-CJFH) (XK2020-15), the National Natural Science Foundation of China (81871505) and the Agricultural Scientific and Technological Innovation Project of Shandong Academy of Agricultural Sciences (CXGC2021B03).

Institutional Review Board Statement: Not applicable.

Informed Consent Statement: Not applicable.

Data Availability Statement: Not applicable.

Conflicts of Interest: The authors declare no conflict of interest.

References

1. Basto, A.P.; Portugal, R.S.; Nix, R.J.; Cartaxeiro, C.; Boinas, F.; Dixon, L.K. Development of a nested PCR and its internal control for the detection of African swine fever virus (ASV) in *ornithodoros erraticus*. *Arch. Virol.* **2006**, *151*, 819–826. [CrossRef] [PubMed]
2. Hu, L.; Lin, X.Y.; Yang, Z.X.; Yao, X.P.; Li, G.L.; Peng, S.Z. A multiplex PCR for simultaneous detection of classical swine fever virus, African swine fever virus, highly pathogenic porcine reproductive and respiratory syndrome virus, porcine reproductive and respiratory syndrome virus and pseudorabies in swines. *Pol. J. Vet. Sci.* **2015**, *18*, 715–723. [CrossRef] [PubMed]
3. Luther, N.J.; Udeama, P.G.; Majiyagbe, K.A.; Shamaki, D.; Owolodun, O.A. Polymerase chain reaction (PCR) detection of the genome of African swine fever virus (ASFV) from natural infection in a nigerian baby warthog (phacochoereus aethiopicus). *Niger. Vet. J.* **2007**, *28*, 63–67. [CrossRef]
4. Schoder, M.E.; Tignon, M.; Linden, A.; Vervaeke, M.; Cay, A.B. Evaluation of seven commercial African swine fever virus detection kits and three Taq polymerases on 300 well-characterized field samples. *J. Virol. Methods* **2020**, *280*, 113874. [CrossRef] [PubMed]
5. Tomita, N.; Mori, Y.; Kanda, H.; Notomi, T. Loop-mediated isothermal amplification (lamp) of gene sequences and simple visual detection of products. *Nat. Protoc.* **2008**, *3*, 877. [CrossRef]
6. Jiang, Z.Y.; Zhu, H.F. Establishment of a Loop-mediated Isothermal Amplification Assay for the Detection of African Swine Fever Virus. *China Anim. Husb. Vet. Med.* **2009**, *36*, 72–74.
7. Chen, X.; Zhang, L.L.; Cui, D.F.; Yin, H. Miniaturized System with a Facile Isothermal Amplification Microfluidic Chip for Rapid Detection of Zika Viruses. In Proceedings of the 2nd International Conference on Biomedical and Biological Engineering 2017, Guilin, China, 26–28 May 2017.
8. Liu, D.Y.; Liang, G.; Zhang, Q.; Chen, B. Detection of mycobacterium tuberculosis using a capillary-array microsystem with integrated DNA extraction, loop-mediated isothermal amplification, and fluorescence detection. *Anal. Chem.* **2013**, *85*, 4698–4704. [CrossRef]
9. Guner, H.; Ozgur, E.; Kokturk, G.; Celik, M.; Esen, E.; Topal, A.E. A smartphone-based surface plasmon resonance imaging (SPRi) platform for on-site biodetection. *Sens. Actuators B Chem.* **2017**, *239*, 571–577. [CrossRef]
10. Sipova, H.; Homola, J. Surface plasmon resonance sensing of nucleic acids: A review. *Anal. Chim. Acta* **2013**, *773*, 9–23. [CrossRef]
11. Qu, J.H.; Dillen, A.; Saeys, W.; Lammertyn, J.; Spasic, D. Advancements in SPR biosensing technology: An overview of recent trends in smart layers design, multiplexing concepts, continuous monitoring and in vivo sensing. *Anal. Chim. Acta* **2020**, *1104*, 10–27. [CrossRef]
12. Lin, L.M.; Xue, J.C.; Xu, H.F.; Zhao, Q.; Zhang, W.B.; Zheng, Y.Q.; Wu, L.; Zhou, Z.K. Integrating lattice and gap plasmonic modes to construct dual-mode metasurfaces for enhancing light–matter interaction. *Sci. China Mater.* **2021**, *64*, 3007–3016. [CrossRef]
13. Bai, H.; Wang, R.; Hargis, B.; Lu, H.; Li, Y. A spr aptasensor for detection of avian influenza virus H5N1. *Sensors* **2012**, *12*, 12506–12518. [CrossRef] [PubMed]
14. Wang, S.P.; Shan, X.N.; Patel, U.; Huang, X.; Lu, J.; Li, J.; Tao, N. Label-free imaging, detection, and mass measurement of single viruses by surface plasmon resonance. *Proc. Natl. Acad. Sci. USA* **2010**, *107*, 16028–16032. [CrossRef] [PubMed]
15. Nguyen, V.T.; Seo, H.B.; Kim, B.C.; Kim, S.K.; Song, C.S.; Gu, M.B. Highly sensitive sandwich-type SPR based detection of whole h5nx viruses using a pair of aptamers. *Biosens. Bioelectron.* **2016**, *86*, 293–300. [CrossRef] [PubMed]
16. Wang, S.F.; Wang, W.H.; Chang, Y.F.; Yuan, R.Y.; Hong, Y.W.; Su, L.C. Detection of emerging avian influenza a H7N9 virus based on a rapid and sensitive intensity-modulated spr biosensor. *J. Virus Erad.* **2018**, *4*, 4. [CrossRef]
17. Yoo, H.; Shin, J.; Sim, J.; Cho, H.; Hong, S. Reusable surface plasmon resonance biosensor chip for the detection of h1n1 influenza virus. *Biosens. Bioelectron.* **2020**, *168*, 112561. [CrossRef]
18. Zhang, L.; Fu, Q.; Tan, Y.; Li, X.; Deng, Y.H.; Zhou, Z.K.; Zhou, B.; Xia, H.Q.; Chen, H.J.; Qiu, C.W.; et al. Metaoptronic multiplexed interface for probing bioentity behaviors. *Nano Lett.* **2021**, *21*, 2681–2689. [CrossRef]
19. Zhang, L.L.; Miao, G.J.; Liu, L.Y.; Gong, S.; Chen, X. Development of a surface plasmon resonance and fluorescence imaging system for biochemical sensing. *Micromachines* **2019**, *10*, 442. [CrossRef]
20. Giergiel, J.; Reed, C.E.; Hemminger, J.C.; Ushioda, S. Surface plasmon polariton enhancement of Raman scattering in a Kretschmann geometry. *J. Phys. Chem.* **1988**, *92*, 5357–5365. [CrossRef]
21. Huang, Y.F.; Zhang, L.L.; Zhang, H.; Li, Y.C.; Liu, L.Y.; Chen, Y.Y.; Qiu, X.B.; Yu, D.L. Development of a Portable SPR Sensor for Nucleic Acid Detection. *Micromachines* **2020**, *11*, 526. [CrossRef]
22. Tan, T.N.; Trinh, K.; Yoon, W.J.; Lee, N.Y.; Ju, H. Integration of a microfluidic polymerase chain reaction device and surface plasmon resonance fiber sensor into an inline all-in-one platform for pathogenic bacteria detection. *Sens. Actuators B Chem.* **2017**, *242*, 1–8. [CrossRef]

MDPI

St. Alban-Anlage 66

4052 Basel

Switzerland

Tel. +41 61 683 77 34

Fax +41 61 302 89 18

www.mdpi.com

Biosensors Editorial Office

E-mail: biosensors@mdpi.com

www.mdpi.com/journal/biosensors